THE AMERICAN CHESTNUT

The American Chestnut

AN ENVIRONMENTAL HISTORY

Donald Edward Davis

THE UNIVERSITY OF GEORGIA PRESS *athens*

Paperback edition, 2025

Athens, Georgia 30602
www.ugapress.org

Designed by Erin Kirk
Set in Adobe Caslon Pro

Most University of Georgia Press titles are
available from popular e-book vendors.

Printed digitally

Library of Congress Control Number: 2021939542
ISBN: 9780820360454 (hardcover)
ISBN: 9780820374529 (paperback)
ISBN: 9780820360461 (epub)
ISBN: 9780820369501 (PDF)

for *Castanea dentata*

Now we read of this,
of the hundred-foot tree that once, in airy-
white blossom or heavy
with pods of food,
distilled the American earth for us.
At Mount Vernon,
the Washingtons' roast wild turkey
was sewn stuffed with chestnuts.
Cattle and deer fattened
or held to life on the winter meat of chestnuts.
Now one book of trees says
"only a few scattered
sucker growths remain."

Nothing to replace it, maybe
never again:
that fruit
a brown dimension
like nothing else in nature;
when newly fallen,
of such a swirled,
teardrop shine
that even our children's eyes
seemed made
with less skill.

WILLIAM HEYEN,
from *The Chestnut Rain* (1986)

contents

illustrations

FIGURES

PLATES

(following page 80)

François André Michaux, *The North American Sylva*, vol. 3 (Paris: C. d'Hautel, 1819). Image courtesy of the Biodiversity Heritage Library.

preface

On December 21, 2013, when best-selling author Bernd Heinrich announced in a *New York Times* op-ed that he would be eating American chestnuts for Christmas, the response was predictable. Readers assumed he had his facts wrong, believing the nut-bearing trees on his property were foreign or hybrid varieties. Heinrich admitted the chestnuts might not be native, but was hopeful they indeed were, as he had purchased them from a Michigan grower in 1981. Not only did the trees thrive after being planted on his Maine farm, one of the original seedlings was now thirty-five feet tall. More impressive was the fact that the trees had produced one hundred and fifty progeny, seedlings scattered across some two hundred acres of forest. The young trees, discovered Heinrich, had been planted by jays after the birds failed to unearth their cache of buried nuts. As a result, the trees had become a significant component of the forest ecosystem and even afforded him the luxury of "roasting chestnuts by the fire."

Although the response to Heinrich's op-ed was largely positive and replete with nostalgic comments about the trees and their possible return, some doubted the veracity of the report or took objection to his comments about genetically engineered (GE) chestnuts. Among the objectors was William Powell of the SUNY College of Environmental Science and Forestry. Heinrich was aware that Powell had planted GE chestnuts at the New York Botanical Garden in 2012 and made it known that he did not want the genome of the native trees altered. "Yes, I would love to see the American chestnut restored to our forests," wrote Heinrich, "but do we need to alter the chestnut's genome—the code of life that has evolved over millenniums? I don't think so, and I worry that the feel-good campaign in the Bronx could be a Trojan horse that may seduce the public into accepting other genetically engineered trees."[1]

To appease those doubting the origins of his chestnuts, thinking they were not native trees, Heinrich located their true source. In his 2014 book *The Homing Instinct: Meaning and Mystery in Animal Migration*, Heinrich recounted their journey to his Maine farm, stating the nuts had been purchased from the American Chestnut Council in Cadillac, Michigan. For decades, the American Chestnut Council had access to groves of native blight-resistant chestnuts and thus was able to ship seed stocks across the United States.

Heinrich was elated that the trees were the native variety, but was uncertain about their resistance to the blight fungus. "Could some have survived because they were resistant to it," he asked, "or were they saved simply because of their isolation?"[2] Heinrich's query is not an uncommon one and represents the many questions that continue to be asked about the species, even by well-informed naturalists. This book was written in part to answer such queries as well as shape public policy regarding the future reintroduction of the tree. Presently, the return of the American chestnut shows as much possibility of failure as it does promise, despite media reports claiming breakthroughs have been made toward that end.

Correcting certain misconceptions about the trees, including their future chances for survival, was an important aim of this book. To claim the trees were "King of the Eastern Forest," as Heinrich and others have done, is also not entirely factual. The trees were certainly a dominant if not foundational species across the Appalachians, but elsewhere the trees seldom comprised more than one-tenth of any wooded area. And when trees did reach higher densities outside the so-called "chestnut belt," it was often due to human activities, such as forest clearance, charcoal production, or anthropogenic fire.

Another common misconception about the American chestnut is that its natural range is relatively fixed and will remain so in the near and distant future. Many have assumed, for example, that the map produced by dendrologist Elbert E. Little for the U.S. Department of Agriculture in 1977 represents the true historic range of the American chestnut. However, Little's map—which is recognized by government agencies as well as the American Chestnut Foundation—does not include areas where the trees were found prior to the introduction of *Phytophthora* disease, a subject discussed at length in chapter 5. Before that time, the trees were found along the Mid-Atlantic coastline and much of the southern Piedmont. Nor does Little's map include the small but important chestnut groves that resulted from the naturalization of the tree outside its historic range, such as those found in Michigan and Wisconsin.

Indeed, the most current attempts to redraw the range of the America chestnut—using botanical collections, field surveys, and historical references—fail to capture the full range of *Castanea dentata* prior to the nineteenth century. Although the map published in *The North American Plant Atlas* in 2015 shares many similarities with the range map found in this volume (see figure 3.1), a number of the outlying trees identified in the *Atlas* are likely Allegheny or Ozark chinquapins or even American chestnut/chinquapin hybrids. Nevertheless, historical evidence suggests the natural distribution of the American chestnut extended across more than four hundred thousand square miles of territory, an area stretching from eastern Maine to southeast Louisiana.

Determining the historic range and prevalence of the tree is important for a number of reasons. First of all, such knowledge tells us that all chestnut habitats are not created equal. Some parts of eastern North America were less suitable for the species and did not support large numbers—even before the arrival of both *Phytophthora* and the *Cryphonectria parasitica* fungus. The awareness that extreme temperature and moisture levels also influence chestnut mortality is another important lesson for those seeking to resurrect the native tree. Obviously, more emphasis should be placed on growing chestnuts in areas with the most optimal soil types, rainfall amounts, and temperatures, as well as the fewest recorded instances of pests and diseases.

For this reason, the southern Appalachians—which were once home to the largest and densest chestnut forests in the eastern United States—may not be the best "regeneration niche" for completing this task. Maine and Michigan could actually be more suitable for reviving the American chestnut, since environmental conditions in those states, at least in the twenty-first century, appear more conducive to the tree's growth and vigor, yet, at the same time, unfavorable to both *Phytophthora* and the *Cryphonectria* fungus. The findings of a 2019 study support such claims, as the authors anticipate a northerly shift in the tree's range. Their findings, based on climate change and species-distribution modeling, predict not only a reduction in suitable habitat for the species, but also the emergence of new habitat niches. After 2050, portions of southern Ontario and Nova Scotia, as well as the island of Newfoundland, may possess the most suitable growing conditions for *Castanea dentata*.[3]

Equally important to this discussion is an informed understanding of how and when the trees spread across the North American landscape. Pollen recovered from the bottom of perennial ponds and sand dunes provides evidence the trees inhabited a very small geographic area during the last ice age. Only after the hemisphere warmed and precipitation levels increased did the American chestnut shift its range northward. It did so, on average, 110 yards annually due to hundreds of thousands of nut movers, including mice, squirrels, jays, crows, and passenger pigeons. Passenger pigeons and crows were most likely responsible for introducing the trees into the Appalachian uplands, as both birds were known to visit high-elevation mountaintops during feeding and roosting forays. Even with the assistance of birds and mammals, it took five thousand years for the trees to colonize what is today the state of Georgia and another ten millennia for the trees to reach northern New England and Canada.

Such lessons in ecological history are particularly relevant for those who believe American chestnut restoration can be accomplished in a single century. Even with the assistance of armies of tree planters and the most ideal growing conditions, the trees will need at least a millennium to become reestablished in the eastern hardwood forest. Moreover, as I argue in the final chapter, not all restoration efforts are created equal. In fact, some attempts to restore the tree to its former range might even threaten its long-term survival. There are always pitfalls in restoring landscapes to past desiderata, as ecosystems change considerably over time, with or without human intervention. It is important to know how the trees evolved across both time and space, particularly for decision makers wanting to return the species to its native habitat—or determine if such efforts are even warranted.

Documenting the rise and fall of the American chestnut over the *longue durée* and across different human and geographic landscapes is no easy task. In North America humans lived among chestnuts for more than ten millennia and each influenced the other in unique and important ways. Needless to say, an environmental history of the American chestnut is timely, if not long overdue, especially for those wishing to ensure the future of this iconic tree species.

THE AMERICAN CHESTNUT

INTRODUCTION

giving character to the landscape

Humans love trees. They shade our homes in summer, provide convenient locations for play or rest, and give us warmth in winter. They symbolize fertility and regeneration, especially in springtime, when bursting buds and blossoms ensure us a new season will soon prevail. Trees connect us to specific places, providing us with a sense of direction, and in some locales, where a single species dominates the landscape, community identity. Wooded places evoke, as the geographer Yi-Fu Tuan once observed, joy, fear, mystery, grief, tradition, and childhood memories.[1] Urban dwellers also have an affinity for trees and often go to great pains to make sure city parks and streetscapes are safe and permanent homes for their sylvan companions.

Ironically, the tree that most piqued the emotions of nineteenth- and early twentieth-century Americans—the American chestnut—has virtually disappeared from the eastern United States. As is discussed in subsequent chapters, the American chestnut became functionally extinct during the first four decades of the twentieth century after the introduction of an exotic fungus on Japanese chestnut nursery stock.[2] Before that time, the tree played a central role in the ecology, economy, and material culture of the eastern United States. From Maine to Mississippi, the American chestnut evoked memories of street vendors, community gatherings, picnics, holiday feasts, small- and big-game hunting, fence building, shingle splitting, livestock husbandry, and even moonshining. For residents of Appalachia, where the trees defined the pre–World War II landscape, the loss of the American chestnut even served as a metaphor for the passing of a self-sufficient and forest-dependent way of life.[3]

Thousands of communities in the United States and Canada remain home to places bearing the chestnut name, including streets, cemeteries, schools, churches, and post offices. Numerous mountains, ridges, hills, knolls, valleys, streams, and ponds are also prefaced by the chestnut adjective.[4] Although some locales were home to only a single grove of trees, many areas possessed large and impressive stands. George Ramseur, who lived in southeast Tennessee during the 1930s, recalled that chestnut trees atop the Cumberland Plateau were "as common as the moon rising and sun setting."[5] The mountainous portions of Kentucky, West Virginia, North Carolina, and Pennsylvania also contained large numbers of trees, but they could also be found in lower elevations. Chestnut Neck, New Jersey, for example, a colonial village that was the site of an important Revolutionary War battle, got its name from the many chestnuts that once grew in the township, even though the terrain there is less than ten feet above sea level.[6]

During the second half of the nineteenth century in America's largest cities, chestnut trees were planted along major thoroughfares, where they shaded urban pedestrians in warmer months and fed them during colder ones. In the summer of 1859, a *New York Times* writer editorialized that in parts of Manhattan, where the wealthy still "own land by the block," individuals were planting "avenues of chestnuts and elms."[7] In America's first planned suburb, Baltimore's Roland Park, native chestnut trees were touted as a drawing card for future residents. During the mid-1890s, a frequently posted advertisement in the *Baltimore Morning Herald* announced, "The more you see of Roland Park, the more it grows on you. It is an improved piece of property with . . . beech and ash, sycamore and chestnut, which shade while they shelter—all are here, for they have been here for years and years, and are no less delightful on that account."[8] The trees remained a visible part of the Roland Park landscape for at least another decade, as evidenced by a 1902 article from the *Baltimore Sun* society pages. "Mr. and Mrs. William M. Ellicott gave a tea and chestnut hunting party yesterday afternoon to the instructors and pupils of the Arundel School at her residence," states the notice. "An experienced raccoon hunter was engaged to climb and thresh the chestnut trees, and the nuts were eagerly gathered from the ground by the children, regardless of leaves and burrs falling upon their heads."[9]

Smaller townships also encouraged the planting of native chestnuts, as local residents perceived the trees as natural capital that might pay future dividends. In 1893, Maine tavern keeper Samuel Farmer pleaded for his neighbors to invest in a chestnut orchard, citing the success of Temple, a town in the western part of the state that planted American chestnut trees from Massachusetts in the late 1840s. "Forty-five years later," proclaimed Farmer, "the trees are . . . over two feet in diameter, and in height and general size have outgrown all other trees in their vicinity. It is a valuable tree for timber, and is used for telephone poles, railroad poles, railroad ties, fence posts, sawed timber and plank."[10] In fact, the American chestnut's many uses made it one of the country's favorite species, causing several writers to christen it "the perfect tree."[11]

Indeed, in 1915, when New England forester Philip L. Buttrick discussed the tree's importance to the U.S. economy, he announced it possessed more uses than all other American hardwoods, exclaiming that "it touches almost every phase of our existence."[12] To fully bolster his argument, he added that the tree "serves as a shade and ornamental tree on [*sic*] our parks and estates. Its wood is used in the building and decoration of our houses and the manufacture of our furniture. We sit down in chairs made of chestnut and transact our business at desks . . . of chestnut veneered with oak, we receive messages from the distance over wires strung on chestnut poles. We sit in a railroad train and read newspapers into whose composition chestnut pulp has gone, while our train travels over rails supported on chestnut ties and over trestles built of chestnut piles, along a track whose right-of-way is fenced by wire supported on chestnut posts. On the same train travel goods shipped in boxes and barrels made of chestnut boards and staves. Even the leather for our shoes is tanned in an extract made from chestnut wood. . . . At last when the tree can serve us no longer in any other way it forms the basic wood . . . to make our coffins."[13]

Perhaps the most pleasurable memories associated with the American chestnut involved the annual consumption of nuts, which started in late September and continued through the colder winter months. In urban areas along the eastern seaboard—from Washington, D.C., to Boston—the motley-dressed sidewalk vendor who roasted and sold chestnuts was seen as the harbinger of the holiday season.[14] By mid-November, the aroma from chestnut vendors' pushcarts was nearly inescapable, making it difficult for city dwellers to keep their spare change. In 1898, a reporter for the *Boston Evening Transcript* commented on the sensual allure of roasting chestnuts, wryly stating it was "the incense" of the vendor's trade that was his best advertisement. "There are few who can permanently resist the sweet savor sent up by the chestnut roaster," observed the writer. "They may wish the vender had a little cleaner hands and a little more wholesome attire, but one sense contends against the other, and at last they are likely to shut their eyes and the sense of smell triumphs."[15]

Chestnuts were also important in rural areas during the winter season, feeding both the eyes and stomachs of those fortunate enough to live near a stand of nut-bearing trees. However, in parts of southern Appalachia, the nuts were less likely to be eaten than to be bartered for needed provisions at the crossroads store or fed to livestock for winter fattening.[16] But even in the most remote areas of the tree's native range, it would be difficult to find a single individual who did not taste a handful of chestnuts before year's end. Northeast Alabama resident Marie Smith Washburn recalled that as late as the 1920s her family tried to maintain their annual store of chestnuts until Christmas Day: "Well, as far back as I can remember we always had chestnuts to eat. And when I got big enough to go up to the field up there where they were, Daddy would have us picking them up. He'd sack them up and . . . try to hide them from us but we'd find them. And he'd say 'now don't eat them all up, we've got to have some for Christmas.'"[17]

In late spring or early summer, depending upon the geographic location of the stand, the American chestnut again assaulted the senses of those living near, or passing by, the largest trees. After chestnut trees fully leaf out, they produce hundreds of long delicate catkins that turn from green to white in a matter of a few weeks. Accompanying the catkins is a strong pungent odor, which some commentators, including Henry David Thoreau, found disagreeable or even offensive.[18] Others were more kind, however, using adjectives ranging from "delicate" to "heavy" when describing the odor of the blossoms.[19] Visually, the trees were a sight to behold, turning entire mountainsides a creamy yellow and then, as the catkins began to release their pollen, a sugary white that from a distance resembled snow. Nineteenth-century travel writers Wilbur Zeigler and Ben Grosscup, who spent considerable time exploring the mountains of western North Carolina, said it was "the glory" of the chestnut blossom that was responsible for them "giving character to the landscape."[20] As a result, the annual chestnut blossom was the impetus for the naming of dozens of mountains and ridges in the Appalachians, including Yellowtop Mountain, North Carolina; Whitetop Knobs, Tennessee; Yellow Mountain, Georgia; and Little Yellow Mountain, Virginia.[21]

As significant as the American chestnut was to the American populace, the tree was not without detractors, especially those who believed its annual harvest could be improved. In some parts of the tree's range, it was an unpredictable nut producer, rarely making a bumper crop for more than two or three years in a row.[22] And even though newly arriving immigrants found native chestnuts sweeter in taste, their smaller size made them cumbersome to gather and process for the kitchen table. Moreover, eighteenth- and nineteenth-century America was settled largely by Europeans, all of whom had personal memories of much larger chestnuts in Italy, France, Spain, and the Balkans. In those countries, the European chestnut (*Castanea sativa*) was the dominant producer of nuts, which also explains their early importation to America.[23] As early as 1773, Thomas Jefferson grafted five "French Chestnuts" onto the stock of two American trees with hopes of producing a more marketable nut crop.[24] Three decades later, Éleuthère Irénée du Pont, at his Eleutherean Mills estate near Wilmington, Delaware, planted what was likely the first European chestnut orchard in the United States. The du Pont trees flourished for more than a century, as well as established the Wilmington-Philadelphia corridor as a center of commercial chestnut growing.[25]

While most Americans referred to European chestnuts by their country of origin—"Spanish chestnuts," for example—few realized the trees had their origins in the Caucasus Mountains of Georgia and Turkey. In fact, the name of the scientific genus for all chestnut species—*Castanea*—derives from Kestane, a village in northwest Turkey where the trees became established after the late Bronze Age. Although paleobotanists believe humans were responsible for spreading European chestnuts westward into Greece and the southern Balkans before Roman times, there is evidence the trees existed in northern Italy and southern Germany before the eleventh century BCE, surviving there as a relict species during the last glaciation.[26] Although Romans may have taken chestnuts as far north as England, where they thrive in naturalized stands, it is more likely the trees were introduced into Britain during the eleventh century, when they first appear in the historical record.[27]

The America chestnut also received competition from the Japanese chestnut (*Castanea crenata*), a species brought to America in 1876 by the Parsons Commercial Nursery of Flushing, New York. In 1882, horticulturalist William Parry imported one thousand trees from Japan, grafted specimens which he planted at his New Jersey orchard along the Delaware River.[28] The smaller size of the Japanese tree made it ideal for home orchards, and its extremely large nuts made it a favorite among commercial nut growers. By the end of the century, hundreds of acres of Japanese chestnuts were growing at a dozen nurseries devoted to chestnut production, including the Elm City Nursery in New Haven, Connecticut, the Mammoth Chestnut Company of Riverton, New Jersey, and the Storrs and Harrison Company of Painesville, Ohio.[29] The Japanese chestnut did have its detractors, including New England forester Ernest A. Sterling, who complained the nuts were covered with "a bitter skin which had to be removed before eating."[30] Although Sterling found Japanese chestnuts more palatable after proper preparation and cooking, and thought the tree a valuable acquisition to American horticulture, he recommended the "better flavored" European varieties for propagation.[31]

Imported chestnuts had an enormous impact on the native species, influencing their rapid decline and, ultimately, their renewed resurgence.[32] In fact, it was the fin de siècle popularity of Japanese chestnuts that led to their unmonitored importation into the United States and it was those trees that were infected with chestnut blight, setting into motion, sometime before 1900, what one biologist has called the "greatest ecological catastrophe since the last ice age."[33] The end result? The loss of an estimated five billion chestnut trees and two billion tons of biomass across more than 320 million acres of forests.[34] In the twenty-first century, it is perhaps ironic that the Chinese chestnut (*Castanea mollissima*), provides one of the best hopes of returning the American chestnut to the eastern deciduous forest, as breeding programs that cross the two species have produced trees with significant levels of blight resistance.[35]

As an environmental history of the American chestnut, this book surveys the ecological history of the tree over the past twenty thousand years and provides an informed discussion about human-chestnut relationships for half that period. For those unfamiliar with environmental history, the discipline emerged in the 1970s when communities in the United States were undergoing numerous environmental problems.[36] Although the first environmental historians were trained in the study of history, individuals from other academic disciplines also joined their ranks. In the main, environmental history is simply the study of human-nature relationships over time, in a particular geographic setting. It documents the impact of humans on the natural world, as well as the "role and place of nature in human life."[37] Initially, environmental historians focused on such topics as resource use, wilderness protection, air and water pollution, pesticides, and urban sprawl. Some also chronicled the impact of plants and animals on human history, including Alfred Crosby, who wrote about the "Columbian Exchange" as early as 1972.[38]

To date, the most notable environmental history of an individual tree species is *The Tanoak Tree*, a book about the acorn-producing hardwood native to California and Oregon. The tanoak has declined in recent years due to sudden oak death, an introduced pathogen of unknown origin.[39] In *The Georgia Peach*, William Thomas Okie carefully documents the history of peach production in the American South, which influenced everything from labor relations to popular culture.[40] Jared Farmer's *Trees in Paradise* focuses on not one but three tree species: the California redwood, the eucalyptus, and the orange tree. According to Farmer, California history is so intertwined with these trees that it would be a very different place if they were absent from the landscape.[41] And finally, in *Mesquite*, Gary Paul Nabhan celebrates the cultural and ecological importance of one of the most common tree species in the American southwest. As Nabhan notes, the mesquite tree possesses edible seed pods, provides the material for boutique furniture, and even gives unique flavor to smoked meats.[42]

Although environmental history has, to date, been largely the work of academicians, the discipline also has real-world implications for the natural environment. Proponents not only amplify the importance of nature in our collective past, but suggest ways in which humans, in the present and future, might *enrich* the natural world.[43] The successful reintroduction of a blight-resistant American chestnut, for example, could dramatically alter the composition of the eastern deciduous forest as well as increase populations of wild turkey, ruffed grouse, and black bear—game species that could in turn alter human consumption and recreation patterns.[44] Local communities would also benefit from the restoration of the tree, as numerous businesses might be inspired by its return, including commercial nut growing and furniture manufacturing. With the increasing planting of blight-resistant chestnut trees on public and private lands, it is possible that North Americans will once again be roasting chestnuts over open fires and purchasing chestnut timber at local lumberyards.

It is also possible that current attempts to restore the American chestnut will prove unsuccessful. Not only does the problem of chestnut blight need to be solved, but the landscape itself has to be more favorable to the reintroduction of the tree. The species does not thrive in all soils and terrains and new pests and pathogens will undoubtedly play havoc with restoration efforts. Although the trees colonized large geographic areas in relatively short periods of geologic time, they also quickly vanished from forested landscapes—even in a single growing season. Over the millennia, the American chestnut has indeed been a species on the move, changing not only its natural range, but rising and falling in numbers.

PART ONE

Chestnuts on the Move

CHAPTER 1

the evolutionary history of the species

The story of the American chestnut begins some ninety million years ago, when a family of trees known as *Fagaceae*, the ancestors of all oak and beech species, became prevalent in temperate forests. At that time, such trees could be found across much of Laurasia, the supercontinent linking North America to both Europe and Asia.[1] Although fossil evidence tells us the first plants of the *Castanea* genus appeared around eighty-seven million years ago, a tree resembling the modern species did not inhabit the planet until the end of the Cretaceous period, some twenty-two million years later.[2] Evolutionary biologists agree that chestnuts likely evolved from a closely related oak species occupying the eastern border of Laurasia, near what are today the islands of Japan.[3]

Around sixty million years ago, after the K-T meteor event led to the extinction of dinosaurs, chestnuts spread westward toward Europe as well as eastward, beyond the Bering Strait into Alaska.[4] Five million years later, as a result of fluctuating temperatures and geographic isolation, chestnuts in what is today Asia developed a new lineage or clade, resulting in a tree closely akin to the Chinese chestnut. Both species coexisted over the next twelve million years, extending their ranges across the Laurasian continent.[5] Forty-three million years ago, chestnuts near the western edge of Laurasia developed even more novel traits, becoming the progenitors of the European chestnut, a tree sometimes misidentified in the United States as the American species.[6] Finally, around forty million years before the present period (BP), chestnut trees inhabiting the landmass connecting Europe to North America again changed their genetic structure, becoming established just before the separation of the supercontinent. It was those trees that spread both westward and southward across much of North America.[7]

Among the North American lineage of trees was the American chestnut (*Castanea dentata*), one of nine *Castanea* species now populating the planet.[8] Diversification of the *Castanea* genus occurred again at the end of the Oligocene epoch, some twenty-four million years BP, when chinquapins appeared alongside chestnuts in North American forests. The Ozark chinquapin (*Castanea ozarkensis*), a smaller version of the American chestnut, traces its genetic lineage to this clade of trees. Finally, about twenty million years ago, the Allegheny chinquapin (*Castanea pumila*), a small tree producing a single nut per burr, became established in North American woodlands.[9]

Chestnuts were common in the deciduous forests of the Northern Hemisphere, reaching their peak in number about five million years ago.[10] At that time, a tree closely resembling the American chestnut inhabited the Rocky Mountains of northern Idaho, as verified by leaf imprints recovered at the Clarkia fossil site (see fig. 1.1).[11] As the climate cooled before the first glaciations, chestnuts vanished from the northernmost latitudes of North America, Europe, and Asia, with some species becoming extinct about 2.6 million years BP.[12] With the onset of what geologists refer to as the Quaternary period, enormous glaciers descended over the North American continent, initiating ice ages that lasted for as long as fifty thousand years. Interglacial warming followed the cooling periods, which influenced the range and prevalence of North American flora, including the American chestnut. Climatologists tell us these warming-cooling cycles have repeated themselves no fewer than eight times over the last eight hundred thousand years, with the last major glaciation ending some twenty-one thousand years ago.[13]

Although it seems impossible today, at the peak of the last ice age there were virtually no American chestnuts living within the tree's native range. At that time, the Laurentide ice sheet—a glacier three miles high at its peak—covered portions of present-day New England, New York, Pennsylvania, and northern Ohio. The enormous glacier helped in lowering the Atlantic Ocean by as much as four hundred feet, making the Florida peninsula twice its current width.[14] Below the glacier was a narrow band of tundra, a treeless plain that abruptly gave way to a boreal forest of birch, spruce, fir, and jack pine. Snowfields, krummholz, and alpine meadows covered the Appalachian peaks as far south as north Georgia, high-altitude areas entirely void of trees. Below the boreal forest was an expanse of conifers and northern hardwoods, an area stretching from the southern Appalachians to the Mississippi River. An oak-hickory and evergreen forest occupied the area extending from the thirty-third parallel to northern Florida, where it was gradually replaced by savannahs, salt marshes, and submerged wetlands.[15] A diversity of tree species inhabited these southernmost forests,

FIG. I.I. Fossilized chestnut leaf imprint from northern Idaho, Tertiary Period. Image courtesy of William C. Rember of the Tertiary Research Center, University of Idaho, Moscow.

particularly along the Florida panhandle and Georgia barrier islands. It is precisely this location where one finds the American chestnut some twenty-five thousand years ago.[16]

We know the America chestnut survived in the Deep South during the last ice age thanks to palynologists, individuals who study the distribution of fossilized plant pollen in freshwater lakes and ponds, or, in some instances, the soil beneath undisturbed moss beds or sand dunes.[17] Each year, trees release trillions of pollen grains into the atmosphere and this yellow rain not only coats the surfaces of automobiles, sidewalks, and picnic tables, it also falls to the bottom of lakes and ponds, where it becomes buried under layers of silt. To recover and study this pollen, palynologists sink metal tubes into the bottoms of lakes and ponds, effectively coring the sediment layers. After the mud or peat cores are extracted, the grains are examined, measured, and preserved. Organic matter taken from the cores is then carbon-dated, allowing researchers to ascertain when the pollen was deposited. After the palynologist has, with the aid of a microscope, counted and identified

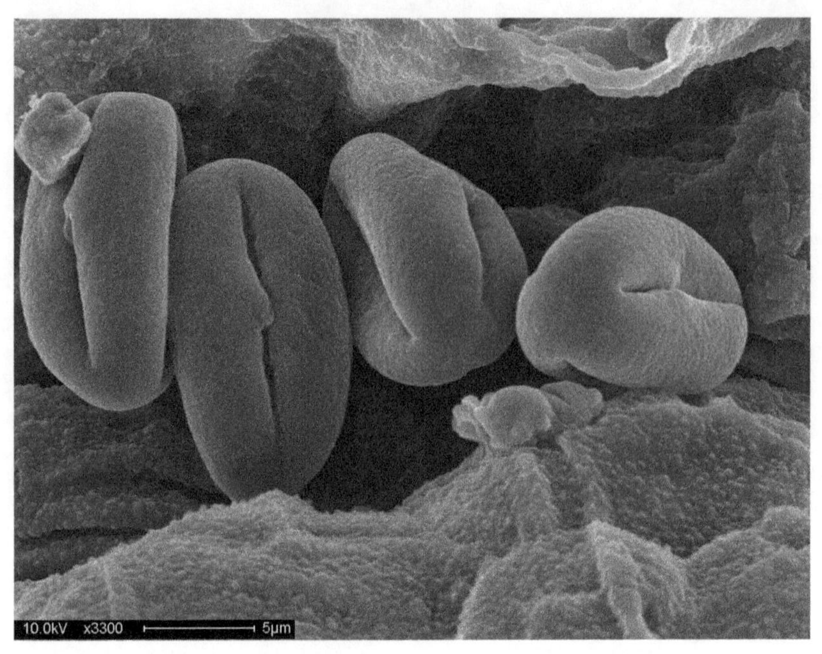

FIG. 1.2. American chestnut pollen grains. Image courtesy of Danilo D. Fernando, SUNY College of Environmental Science and Forestry, Syracuse, New York.

the pollen grains, he or she is able to reconstruct, in the near and distant past, the ecological composition of the surrounding forest (fig 1.2).[18]

Although their findings are not widely known, palynological studies challenge many commonly held notions about the impact of climate change on native flora and fauna, as well as the ability of trees to repopulate large expanses of territory in relatively short periods of geologic time. Harvard Forest director David R. Foster has stated that such migrations are among the greatest biological stories of the North American landscape, demonstrating both the individualistic and adaptive nature of plant species.[19] Although the American chestnut has a long history of environmental adaptation, fossil pollen places the tree at only a handful of locations at the peak of the last ice age.

One of these sites is Goshen Springs in southern Alabama, a perennial lake referred to as the "Forty-Acre Pond" before it was drained in 1951.[20] In the mid-1970s, the lakebed was cored by paleoecologists Paul and Hazel Delcourt, as the pair believed the surrounding area was a place of refuge for temperate hardwoods. The Delcourts postulated that the colder temperatures and drier climate of the ice age drove many deciduous trees into the bluffs and river bottoms of the Gulf coastal plain, where they remained until the Laurentide ice sheet retreated some twenty thousand years ago. The pollen record at Goshen Springs corroborates their hypothesis, revealing that twenty-nine thousand years BP the coastal forest was prime habitat for a variety of hardwoods, including the American chestnut.[21] In a summary of their findings published in the journal *Ecology* in 1980, Paul Delcourt estimated that as much as 28 percent of the sandy woodlands surrounding Goshen Springs was comprised of chestnut and—in descending order of frequency—pine, oak, sweetgum, hickory, blackgum, beech, and elm.[22]

Another place of refuge for the species was Camel Lake, a perennial pond located forty miles west of Tallahassee. The lake was cored for pollen in 1986 by William A. Watts, a paleoecologist who also served as the provost of Trinity College in Dublin and later the president of the Royal Irish Academy.[23] At Camel Lake, Watts found the greatest amount of chestnut pollen around thirty thousand years BP, when the trees comprised 20 percent of the surrounding area. Two thousand years later, chestnuts began disappearing from the site, before vanishing completely from the pollen record around twenty-six thousand years ago.[24] Although it seems unlikely a dominant species would be exterminated in only four or five thousand years, a similar phenomenon occurred at Goshen Springs, where chestnut decline started around twenty-nine thousand years BP and ended several millennia later, when the trees vanished entirely from the area.[25]

To understand why Goshen Springs and Camel Lake were unsuitable for chestnuts, a review of the meteorological conditions is in order. Twenty-eight thousand years ago, rainfall amounts were considerably lower across the southeastern United States. There was also more persistent cloud cover, limiting the amount of sunshine for photosynthesis. Carbon dioxide levels had declined by as much as 35 percent, further disadvantaging broad-leafed trees like the American chestnut. Evergreen trees were not as stressed by the drier climate as they could process CO_2 year-round and thus maximize their water-use efficiency.[26] By the time the Laurentide ice sheet reached its southernmost limit, moisture-loving trees had migrated well below the fall line or retreated into the braided shoals and ravines bordering the streams and rivers of the Gulf and Atlantic coasts. In fact, these were the only locations in eastern North America where a drought-intolerant tree species could find safe haven, as annual rainfall amounts had dropped below thirty inches as far south as the Alabama and Georgia fall lines.[27]

Coupled with the drier conditions was also a dramatic decline in temperatures. In northern Florida, mean January temperatures were as low as 10°F and daytime temperatures did not rise above freezing for several weeks. The extreme temperatures limited the growth of the trees as well as their ability to sprout leaf buds after winter freezes. In fact, recent studies have found the species less cold-hardy than both red oaks and sugar maples, suffering permanent stem and bud damage at −18°F.[28] Needless to say, the drier and colder climate of the last glacial maximum, coupled with the sandy, nutrient-poor soils of the Gulf coastal plain, made the survival of the American chestnut tenuous at best.[29]

Where conditions were most favorable were the areas closer to the present coastline and along the submerged shelf of the Gulf and Atlantic coasts. As noted earlier, sea levels dropped considerably during the peak of the last ice age, causing the Gulf and Atlantic shorelines to extend sixty miles beyond their present location. According to University of South Florida professor emeritus Richard A. Davis, the sea-level drop exposed more than sixty-two thousand square miles of terrain along the Gulf of Mexico, including areas that would have been suitable to the species.[30] While much of the southeast interior was in a deep freeze during the coldest winter months, conditions near and beyond the present shoreline were both milder and wetter.

The waters of the Atlantic Ocean and Gulf of Mexico served as a buffer to the effects of the Laurentide ice sheet, as they were only two degrees cooler during the glacial peak.[31] Consequently, when southerly winds swept across the warm sea surface, they created a thermal enclave that was beneficial to warm-temperate tree species occupying the coastline.[32] Isochronic weather maps of the period reveal a narrow band extending from the Savannah, Georgia, to Mobile, Alabama, areas where surface temperatures and rainfall amounts fall within the range of viability for all *Castanea* species. In fact, the atmospheric conditions there were not unlike those found at the northern end of the tree's present range in New England, where mean January temperatures fluctuate between 18°F and 25°F and rainfall amounts between thirty-five and forty inches annually.[33]

If chestnuts did occupy the Gulf and Atlantic shorelines and the now-submerged ocean floor at the end of the last ice age, then why is there so little pollen evidence supporting that claim? One reason is that lake-bottom sediments in coastal areas are often disturbed by erosion, droughts, hurricanes, severe flooding, and the large roots of trees and woody plants. Along the coastline, pollen grains and the corresponding materials for carbon dating accumulate so sporadically that sediment layers and their contents cannot always be linked to specific periods of geologic history.[34] Moreover, the more open woodlands of the Late Pleistocene allowed for the wider dispersal and diffusion of chestnut pollen, so trees located greater distances from a palynological study area would have left little trace of their presence.[35]

Evidence amassed in recent decades, however, suggests the American chestnut did, in fact, occupy areas near and beyond the present shoreline during the peak of the last ice age. Fredrick J. Rich, professor of geology at Georgia Southern University, for example, recovered chestnut pollen at several barrier islands and other offshore locations.[36] In the St. Mary's River drainage basin, near the ancient shoreline, *Castanea* pollen was recorded at rates as high as 6.4 percent.[37] Pollen grains dating from the glacial maximum were also removed from the ocean floor at Gray's Reef National Marine Sanctuary, with one location registering more than 2 percent of the total pollen sum.[38] At St. Catherines Island, located just offshore in the Ogeechee River basin, chestnuts were the third most common tree species, falling behind oaks and pines in frequency.[39] Taken together, these findings provide conclusive evidence that, at the peak of the last ice age, fluvial river systems located near and along the continental shelf were the primary refuge areas for the species.[40]

This also means that the chestnuts inhabiting the Camel Lake area twenty-eight thousand years ago migrated southward along the Chattahoochee and Apalachicola Rivers before putting down roots somewhere near or beyond the present shoreline. The chestnuts at Goshen Springs also moved southward, following both the Conecuh and Chattahoochee Rivers before stopping, as topography and atmospheric conditions allowed, below present-day Panama City. In fact, all of the terminal riverine floodplains of the coastal southeast—from Mobile Bay to the Savannah River—would have provided suitable habitat for the trees, as these were the only areas possessing the climate and topography required for their long-term survival.[41] While these locations may seem like unlikely places of refuge for a species associated with the Appalachian highlands, some of these areas remain, even today, important sanctuaries for mixed mesophytic hardwoods, including white oak, hickory, persimmon, basswood, tuliptree, ironwood, beech, ash, and holly.[42]

For eight thousand years, the American chestnut survived in the coastal forests of the southeast, expanding or shrinking its range as meteorological and soil conditions allowed. Chestnuts never reached dominance anywhere in the forest canopy, however, as competition from other species, including conifers, limited their growth and spread.[43] The variety and number of species sharing the landscape with chestnuts is remarkable, evidence that Late Pleistocene forest communities were truly unique, with no ecological parallel in present-day North America.[44] On St. Catherines Island, chestnuts lived alongside northern and southern conifers, ash, birch, beech, oaks, hickories, maples, sweetgum, alder, elm, and numerous other trees.[45] Some of the *Castanea* pollen recovered at these sites was also left by chinquapins, as pollen grains from the two species are virtually indistinguishable. However, because the pollen percentages consistently remain above 2 percent in the palynological record, the American chestnut is the most likely source. Allegheny chinquapins seldom produce large amounts of pollen as the trees are smaller in size, seldom grow in large stands, and nearly always occupy the forest understory, thereby limiting pollen dispersal. In fact, in the modern era, palynological investigations of areas dominated by chinquapins seldom record pollen rates higher than 0.5 percent of the total sum.[46]

Around twenty thousand years ago, the Laurentide ice sheet slowly began retreating, reversing direction in erratic starts and stops. As the frozen ice pack melted and moved incrementally northward, southern biomes shifted their borders, with some decreasing, some expanding, and others disappearing entirely. It is doubtful the American chestnut immediately followed the shrinking glacier northward, as the climate did not appreciably warm for several more millennia. Indeed, it would take two thousand years of climate change to significantly alter the forest composition of the southeastern United States, although certain tree species were certainly expanding or reducing their range as early as nineteen thousand years BP.[47] At that time, the total area inhabited by chestnuts was less than fifty thousand square miles, which increased the likelihood that refugia populations exchanged pollen, and thus hybridized, with other *Castanea* species. In fact, studies by phylogeographers—individuals who use molecular DNA to document the geographic and evolutionary histories of plants and animals—have shown that some chestnut trees do share a portion of their genome with both Allegheny and Ozark chinquapins.[48] Such findings certainly complicate "the evolutionary history of the species," as Fenny Dane, professor emeritus at Auburn University states it, but they also say a great deal about the genetic diversity needed before the tree can be successfully reintroduced into the eastern deciduous forest.[49]

agents of tree seed dispersal

Eighteen thousand years ago, the American chestnut was again on the move, pushing northward as environmental conditions allowed. The tree's path away from the shrinking coastline was influenced by a number of variables, including rainfall, temperature, soil type, and summer winds, which aided pollination. Although chestnuts are monoecious, possessing both male and female flowers, pollen is needed from another tree before successful fertilization can occur. Bottomland estuaries, which had previously been safe havens for chestnuts, were less welcoming as glacial runoff and heavier rainfall amounts caused periodic flooding and the expansion of wetlands. Major river systems also served as barriers to east-west and west-east movement, providing natural corridors for the trees as they moved away from their ice age homelands.[50] This also explains why phylogeographic studies of *Castanea* species reveal a pattern in their distribution, with Ozark chinquapins dominating the western edge of the American chestnut's range, and Allegheny chinquapins, the eastern flank. Not surprisingly, where the American chestnut had the greatest opportunity to coexist with chinquapins is where one finds the most hybridization between the species. In areas unbroken by mountains or large watersheds, there are trees that morphologically resemble the American chestnut, but with DNA haplotypes identifying them as Allegheny or Ozark chinquapins or American chestnut/chinquapin hybrids.[51]

The advancement of trees northward also required the presence of birds and mammals, as nut-bearing trees have no other means of moving themselves across the landscape. Indeed, in order for the species to arrive at journey's end eighteen millennia later, the species had to extend its range no less than 110 yards annually.[52] While squirrels, chipmunks, and woodland mice accounted for a portion of that shift in territory, blue jays, American crows, and passenger pigeons transported the greatest number of trees. Blue jays were the most consistent movers of chestnuts, as they not only have the ability to store numerous chestnuts in their esophagus, but also cache them in the ground, thus increasing the likelihood of germination.[53] In fact, recent studies have found that a single flock of jays "scatter-hoard" tens of thousands of acorns or beechnuts in a single season, over distances ranging from 110 yards to 2.5 miles. More importantly, blue jays often store nuts in young, successional patches, ideal locations for chestnut regeneration.[54] In the late nineteenth century, when the American naturalist John Burroughs observed jays hiding chestnuts outside his cabin window, he noted they "brought them from a near tree and covered them up in the grass, putting but one in a place." When later reflecting on this behavior, Burroughs mused that "the jays were really planting chestnuts instead of hoarding them."[55]

Another bird responsible for the northerly migration of trees was the American crow, which has received little attention as a mover of chestnuts, but undoubtedly transported them over considerable distances. In a study of the American chestnut in southern Maryland published in 1904, forester Raphael Zon noted that "chestnut seedlings are most frequently found at the tops of hills. . . . The nuts are brought there by crows and squirrels."[56] Crows were certainly regular consumers of chestnuts, as D. D. Stone of Oswego, New York, noted in 1911. "Crows are great lovers of chestnuts and do lots of mischief in chestnut groves, stealing the nuts from the burrs that are open," wrote Stone. "A few miles south of here are many chestnut groves, and in fall the crows are more numerous there than here."[57] Crows, like their blue jay cousins, also cached nuts in the ground, and there are published studies recording how many they buried for later use. A study of the caching behavior of crows in California, for instance, found the birds buried as many as two thousand walnuts in a single season. According to William & Mary biologist Daniel Cristol, who directly observed this behavior, "most (73%) crows that found a walnut left the foraging site with it, and most (77%) of these crows cached their nuts."[58] More importantly, added Cristol, 5 percent of the crows carried walnuts long distances—more than 1.3 miles—before burying them in unforested habitat. For this reason, believes Cristol, American crows were extremely important "agents of tree seed dispersal."[59]

Another bird responsible for moving chestnuts was the now-extinct passenger pigeon. While some scholars have questioned their role in nut-tree migrations, the trees were undoubtedly moved northward by the gregarious birds.[60] It should be noted that passenger pigeons probably only numbered in the millions during the Late Pleistocene (as opposed to billions, their estimated numbers in the nineteenth century), as their favorite foods—beechnuts, acorns, chestnuts—were in more limited supply.

Although passenger pigeons might swallow as many as a dozen chestnuts in a single feeding, the nuts moved rapidly from crop to gizzard and were thus rendered infertile in a matter of hours.[61] The only way, then, for the birds to serve as nut-dispersal agents was for them to regurgitate the nuts after locating another food source or as the result of sudden death due to predators. Although both possibilities seem unlikely to modern observers, predation was not an uncommon occurrence. As John James Audubon observed in the nineteenth century, pigeon roosts were regularly visited by foxes, bears, raccoons, opossums, and "eagles and hawks of different species."[62]

When Drew University biologist Sara Webb studied the role of passenger pigeons in nut-tree migrations, she concluded that even though predation rarely occurred "on a per-bird basis," it happened with such frequency that it directly impacted North American "plant populations and plant biogeography."[63] Her thinking is that if only two hundred passenger pigeons were killed by hawks or owls each autumn, several thousand chestnuts would have been scattered over the landscape as a result. And since passenger pigeons were constantly on the move, "foraging 40 km [25 miles] or more from roosting areas," chestnuts dispersed in this way would have been transported over many miles.[64] Passenger pigeons may not have moved as many chestnuts as blue jays or squirrels, but they did so over far greater distances when such relocations did occur.

Whatever the means by which the American chestnut traveled northward, the trees had shifted their range some two hundred miles by fifteen thousand years BP. The trees seldom retraced their southward journey, however, although pollen records reveal the trees did pass through formerly occupied areas. At Camel Lake, Florida, for example, chestnuts returned to the area around seventeen thousand years ago before abruptly disappearing two millennia later.[65] At Fort Jackson, near Columbia, South Carolina, *Castanea* pollen was recovered from sediment cores dating to fourteen thousand years BP, and at Cahaba Pond, in north-central Alabama, around thirteen thousand years BP. Although some chestnuts, and possibly Ozark chinquapins, appear to have reached portions of western Tennessee during that period, nowhere was the tree found in large numbers in the upper South until about twelve thousand years ago, at the end of the Pleistocene epoch.[66] At that time, the largest stands of American chestnut were probably in north Georgia, in an area extending above what are presently the Tennessee and North Carolina borders.[67]

Around 11,500 years ago, at the beginning of what is commonly referred to as the Holocene epoch, the American chestnut was established across most of the southeastern United States. The trees were present as far north as the Cumberland Plateau in Middle Tennessee and southeast Kentucky and the Blue Ridge Mountains of North Carolina near the Virginia border. In western North Carolina, chestnuts represented as much as 5 percent of all tree species and even occupied some high altitude terrains. In fact, the southern Blue Ridge Mountains would serve as a seedbed for the species over the next several millennia, allowing trees to continue their northward migration as the climate continued to moderate.[68] The early presence of chestnuts in the Appalachian highlands suggests the trees were brought there by passenger pigeons or American crows, since both birds visited higher-elevation mountaintops during their seasonal wanderings.

With the onset of the Holocene and the return of more temperate weather conditions, humans were poised to take advantage of the growing number of chestnut trees inhabiting the American Southeast. Although there is little archeological evidence supporting human-chestnut encounters during the Late Pleistocene, Paleo-Indians were undoubtedly aware of the trees and certainly brought chestnuts into their encampments during nomadic forays.[69] The first documented cases of humans consuming chestnuts occur during the Archaic period, however, when native groups began relying less on megafauna species like mammoths and mastodons and more on deer, fish, shellfish, and nutmeats.[70] Archaeological evidence amassed in recent years provides ample evidence that peoples of the Archaic and Woodland periods regularly consumed chestnuts and in some places even located their settlements near individual groves of trees.

PART TWO

Chestnut Encounters

CHAPTER 2

the seasonal bounty of nuts and acorns

After the dawn of the Holocene, some eleven thousand years ago, the American chestnut was having a direct impact on the lives of humans. Excavations of Archaic settlements in the Tennessee Valley reveal native peoples were seasonally consuming chestnuts, supplementing a diet of deer, small game, fish, and shellfish. At Icehouse Bottom, a well-known archaeological site in Monroe County, Tennessee, chestnut nutshells were recovered at a floodplain encampment dating to 10,500 BCE. The finding is not surprising, since a pollen study done at the site revealed the trees comprised 42 percent of the surrounding woodlands.[1] Nutshells from the same period were also uncovered inside several large rock shelters on the Cumberland Plateau in Kentucky.[2] Archaeological excavations at both locations reveal chestnut limbs and logs were also burned for warmth and cooking fires, although never in considerable quantities.[3]

The first human-chestnut encounters were likely infrequent, as daily life during the Early Archaic period was dictated by the weather, movement of game, and the peculiarities of local terrain. Like their Paleo-Indian ancestors, Archaic groups moved often—at least several times per year—in order to find food and make appropriate shelter. Eventually, however, they developed patterns of movement that more predictably placed them in their wooded environs.[4] According to University of Florida archaeologist Kenneth Sassaman, after the landscape was more densely populated and the warming climate favored mast-producing trees (oaks, chestnuts, hickories) over boreal species, inhabitants of the eastern woodlands became reliant on "the seasonal bounty of nuts and acorns."[5] As nut gathering became more central to their lives, new technologies were developed to process larger quantities of mast, including chestnuts. Generally referred to as groundstone tools, these innovations included millstones and stone pestles used to grind nutmeats into meal or flour. Around 8,000 BCE, such tools were commonplace across most of eastern North America.[6]

As humans adapted to changing weather conditions, so did the American chestnut. With moderating temperatures and increasing rainfall amounts, the trees continued their northerly migration, reaching the upper North Carolina Piedmont, near the central Virginia border, some 10,500 years ago. A millennium later, the trees had moved well into West Virginia, where they became an integral part of the forest canopy. According to William A. Watts, the trees comprised as much as 15 percent of what is today the Cranberry Glades Botanical Area in Pocahontas County. That percentage is significant, as it means the area also served as a staging ground for the spread of the trees northward.[7] It is doubtful the West Virginia chestnuts were the only source of their northerly movement, however, as the trees were also extremely prevalent in the Blue Ridge Mountains of North Carolina.[8]

The northerly migration of oak, chestnut, beech, and hickory trees certainly made life easier for those struggling to survive the long, harsh winters of the early Holocene. Indeed, the colder climate suppressed population levels in the northern latitudes, and did so for several more millennia. However, by the mid-Holocene, the geologic period beginning around 8,400 BCE, both humans and chestnuts occupied areas as far north as central Pennsylvania.[9] At the Sheep Rock Shelter archaeological site, which lies buried beneath Raystown Lake in Huntington County, chestnuts and chestnut hulls were uncovered in human bedding dating to the Middle Archaic, the archaeological period beginning about 8,000 BCE.[10] These findings are not surprising given the prevalence of chestnuts in the area. A pollen study done at Panther Run Pond, a location sixty miles northeast of Raystown Lake, found that, seventy-five hundred years ago, 6 percent of the area was comprised of the American chestnut.[11]

Accompanying the mid-Holocene were much warmer temperatures and even lower rainfall amounts. Climatologists refer to the period as the Hypsithermal interval, which was characterized by a two-degree temperature rise across the Northern Hemisphere and weather patterns not unlike those witnessed in the twenty-first century. The changing climate also allowed oak and hickory species—which prefer warmer and drier conditions—to become even more dominant across the eastern United States.[12] Some deciduous trees like the American chestnut escaped the hotter and drier weather by expanding their range northward, as well as moving higher into the Appalachian mountains. As a result, seven thousand years ago chestnuts could be found as far north and east as what is today Longswamp Township in eastern Pennsylvania, where chestnuts comprised 20 percent of the forest.[13] At Panthertown Valley, North Carolina (3,400 feet above sea level), they occupied as much as 35 percent of the mountain area, a curious fact since a portion of the valley floor was also a wetland bog.[14]

At the height of the Hypsithermal interval, some six thousand years ago, oak and hickory species were established as far north as southern Canada. Several southern pine species also extended their range, moving northward along the Atlantic coastal plain into southern New Jersey.[15] Chestnuts were not yet present in New England, although a few trees possibly entered the Catskill Mountains of southeastern New York and the Berkshires of Connecticut before 5,000 BCE.[16] The increase in oaks, hickories, and chestnuts was significant as the trees provided Archaic peoples with a predictable source of protein and carbohydrates during the coldest winter months. The numerous and larger stands of mast-bearing trees also meant humans were able to occupy a location for much longer periods.[17] As a result, native groups began to make extended settlements in places where nut mast was abundant, including higher-altitude locations in the Appalachians. This strategy was doubly advantageous as deer, bears, turkeys, squirrels, and passenger pigeons were also found in areas possessing large numbers of mast-bearing trees.[18]

Among those benefiting from the proliferation of nuts and acorns were the Lamoka, a Middle Archaic people occupying what is today the Susquehanna River valley in southern New York. Excavations at the Lamoka Lake archaeological site, a national historic landmark, revealed that five thousand years ago, chestnuts were contributing directly to the Lamoka diet.[19] Unique to the site are groundstone tools used to process nuts as well as hundreds of storage pits presumably used to store chestnuts and acorns.[20] Interestingly, chestnuts appeared rather suddenly at the site, traveling more than two hundred miles in less than a millennium. Pollen studies also confirm their presence at the site, although not elsewhere in the Susquehanna River valley. It is very possible the trees were brought to the site by passenger pigeons, as oak and beech trees were already present at the location.[21] As noted in the previous chapter, passenger pigeons could travel long distances in a relatively short amount of time, so if they perished with crops full of chestnuts, the nuts would have fallen directly onto the forest floor.[22] Passenger pigeons were certainly important to the Lamoka diet. Archaeological excavations found their remains more frequent than all other bird species combined.[23]

There is also the possibility the trees were brought to New York by the Lamoka peoples themselves. According to Appalachian State University anthropologist Thomas R. Whyte, during the Middle Archaic period inhabitants of the Susquehanna River watershed had direct ties with native groups living in the southern Appalachians. Evidence for this includes shared mortuary practices, projectile points, and linguistic similarities between the Iroquois and Cherokee languages. According to Whyte, the Lamoka were proto-Iroquois who were among the groups moving up and down the Appalachians as mast resources allowed.[24] Mitochondrial DNA studies support Whyte's theory and provide additional evidence that an "ancestral Iroquoian population lived in southeastern North America."[25] This also means the Lamoka Lake people could have brought nuts and seedlings directly to the area or simply obtained them from cultural intermediaries living farther southward.

factors ranging from 5 to 200

The Lamoka were not the only native community consuming chestnuts during the Middle Archaic period. Dozens of groups gathered, processed, and stored the nuts, devoting several weeks or more for that purpose (see plate 2). However, the archaeological record offers conflicting evidence regarding the level of chestnut consumption during the Archaic and Woodland periods. Some archaeologists believe they played a vital role in subsistence activities, while others are more reluctant to make such claims. Many argue that acorns and hickory nuts were the most important source of plant nutrition, an assumption supported by the physical evidence.[26]

Richard A. Yarnell and M. Jean Black, for example, in one of the earliest and most comprehensive studies of plant remains in the archaeological record, found less than 1 percent of all excavated nutshell fragments in the Southeast came from the American chestnut. The most common nutshells were hickory and pecan, comprising, on average, 86 percent of all excavated fragments. Nutshells from walnuts and butternuts were the third and fourth most common plant food remains, with acorns comprising only about 6 percent of the identified nutshell. Curiously, in the same study, Yarnell and Black—for reasons discussed below—multiplied the weight of all excavated acorn nutshell fragments by a factor of fifty. As a result, they concluded acorns comprised "75% of the nut food overall during Archaic and Woodland times with hickory nut contributing roughly 20%."[27]

Another study of Native American nut consumption suggests that prior to the development of agriculture, hickory nuts exceeded both acorns and chestnuts in dietary importance. Paul Gardner, a regional director of the Archaeological Conservancy, argues that the flavor and nutritional value of hickory nuts gave them a distinct advantage over acorns, as the latter had to be boiled or continuously soaked in water as a result of their high tannin content, thus giving them a bitter taste. Not only are hickory nuts "superior to both maize and acorns as a source of nine of the ten amino acids," notes Gardner, only twelve ounces of hickory nutmeat were needed to meet the USDA's recommended daily intake of 2,200 calories. Acorns require fifteen ounces of nutmeat to reach the same caloric threshold, he adds, whereas twenty-two ounces of corn had to be eaten to reach that level.[28] Although Native Americans were not calorie counters or lay dietitians, they likely deduced which foodstuffs provided the most nutritional returns for the least amount of exerted energy, as their lives depended upon such decisions. For those and other obvious reasons, concludes Gardner, the "abundant, nutritious, and easily stored hickory nut" was the nutmeat of choice for Archaic peoples.[29]

To appreciate why chestnuts were just as important to Native American diets, as well as underrepresented in the archaeological record, a more in-depth discussion of the above two studies is in order. Yarnell and Black admit that all archaeological data regarding plant remains suffer from "inadequacy of recovery, analysis, synthesis, and interpretation," as well as "distortions imposed by differential preservation."[30] In layman terms, this means archaeologists not only misidentify nutshell, but uncover nutshells preserved in specific ways or possessing a particular size or shape, as those characteristics most influence their capture during soil screenings. Many nutshell fragments taken from archaeological digs are also carbonized, which means they were charred after exposure to hearth or campfires.[31] If the nutshell is larger, thicker, or composed of denser material, the likelihood it will be preserved at a dig site increases exponentially. Walnut, hickory, pecan, and butternut nutshells all fit that description, which explains why they are overrepresented in the archaeological record. Acorn and chestnut nutshells, on the other hand, because of their thin pericarps and smaller size, often disintegrate into unrecoverable ash when placed in fire pits.[32]

Conceding this fact, Yarnell and Black defer to an earlier study by archaeobotanist Neal Lopinot, who claimed acorn nutshell is underrepresented relative to hickory nutshell "by factors ranging from 5 to 200."[33] As noted above, Yarnell and Black decided to multiply the weight of recovered acorn shell by a factor of fifty, which they believed is the most "reasonably appropriate way of comparing the relative food quantities."[34] What is curious about their study is that chestnuts did not get the same treatment as acorns in the analysis and are thus relegated to the status of "rarely important foods."[35] However, the pericarps of American chestnuts are thinner than those of most acorns and comprise only a small fraction of their total weight.

This explains why chestnuts are scarcer than both acorns and hickory nuts in the archaeological record. It also means they deserve far higher-weighted parity when assessing their ubiquity. By the early 1990s, some archaeologists were in fact using "food-to-shell ratios" in their reports, calculations designed to correct the bias toward thicker-shelled hickory nuts and walnuts. Using this method, the focus shifted from counting the number of nutshell fragments to calculating the quantity of eatable nutmeat each nutshell represented. In her summary of an excavation done at the Cold Oak rock shelter in eastern Kentucky, for example, Ohio State University anthropologist Kristen Gremillion determined that at the end of the Late Archaic, some thirty-two hundred years ago, 70 percent of all consumed nutmeat at the shelter came from chestnuts, 25 percent from hickory nuts, and the remaining 5 percent from acorns.[36]

Gardner also sees bias in the archaeological record regarding the preservation and recovery of nutshell. He concedes hickory nuts are overrepresented, but finds them more utilized than acorns, and presumably chestnuts, during the Archaic period. Gardner bases his conclusion on a number of factors, including the nutritional value of hickory nuts, their large yields and availability, and the relative ease with which the nuts are gathered and stored.[37] However, chestnuts are important contenders and surpass all rivals—including hickory nuts—on many of those same scores. Regarding nutrition, chestnuts possess all nine essential amino acids, including methionine, an amino acid not found in most hickory nuts and acorns. While not as calorically superior as hickory nuts, the energy value of chestnuts equals or surpasses that of acorns, requiring about sixteen ounces of nutmeat to reach the USDA's recommended daily intake of 2,200 calories.[38] Protein, crude fiber, and carbohydrate levels are also high in chestnuts, as are important minerals, including potassium, phosphorus, manganese, and copper. Unlike acorns, chestnuts can also be eaten raw, with minimum processing, and if preserved by parching or drying can be stored for many months, if not years.[39]

The American chestnut also surpassed its nut-bearing rivals in annual yields. In a carefully executed study simulating mast production in a southern Appalachian forest, the tree was the most stable producer of mast over a ten-year period, yielding more nuts than all other species combined. After base production figures were calculated and applied to conditions in the preblight forest—where chestnuts comprised 58 percent of the study area and oaks and hickories the remainder—the American chestnut was the dominant food producer, accounting for "64% of the total hard mast output."[40] Whereas the tree contributed between 37 and 96 percent of the annual mast crop, oaks or hickories never produced more than 20 percent of the total mast in a single year. The projected annual yields for the species were also remarkable, reaching 320 pounds per acre in two separate years.[41] Historical narratives and published oral histories provide additional evidence chestnuts produced more mast than either oaks or hickories, with one anecdotal account claiming very large mature trees bearing as many as nine bushels.[42]

Although one must be careful when linking mast yields to conditions existing on the ground four thousand years ago, chestnuts undoubtedly offered vital sustenance to Native Americans. They also encouraged, if not made entirely possible, permanent human habitation. In fact, at a Late Archaic settlement near Townsend, Tennessee, adjacent to the Great Smoky Mountains National Park, there is evidence for such claims. Archaeologists excavated as many as eighty-two storage silos at the site, all used for chestnuts, hickory nuts, and acorns.[43] According to University of Tennessee archaeologist Kandace Hollenbach, many of the silos were so large that, when filled to capacity, they would have fed "1000 people at one feasting event, or a family of five for several months."[44] Although it is impossible to know how many silos were annually filled with nut mast (the site was occupied for more than a millennium), if only eight or nine pits were annually used by the group, occupation could have lasted well into summer. Indeed, little barley (*Hordeum pusillum*) seeds were found embedded in cooking utensils at the location, suggesting an occupation there as late as June, when the seeds were most likely harvested.[45]

Given such evidence, chestnuts were more than just seasonal fare: they were agents of social stability, providing the food surpluses needed to develop new subsistence strategies, and eventually more complex systems of social organization. A similar view is advanced by archaeologists Stephen Carmody and Kandace Hollenbach, who have studied the consumption of hickory nuts in Late Archaic settlements. In their book, *Barely Surviving or More than Enough? The Environmental Archaeology of Subsistence, Specialization and Surplus Food Production*, the pair argue that the intensification of nut use was extremely significant, resulting in both population gains and greater human densities. "An increase in the availability and reliability of high-quality foodstuffs," claim Carmody and Hollenbach, "leads to a decrease in childhood mortality and quite simply can support larger numbers of people. Because plants foods such as nuts can be easily stored, for months if not years, their availability can be enhanced and extended in ways that most animal resources cannot."[46]

Carmody and Hollenbach do not specifically mention chestnuts, due to their relative scarcity at the various study sites. However, two of the locations were outside the tree's historic range and several sites were in low-lying floodplains where the trees were less common. Moreover, as noted above, the absence of chestnut nutshell at a particular archaeological site does not mean they were consumed in lesser quantities. Nor does the presence of hickory nutshell prove the nutmeat was eaten in corresponding amounts. Several hickory species, including the mockernut and pignut hickory, have incredibly thick shells and required considerable processing.[47] Hickory trees were also less prevalent, seldom exceeding 10 percent of any forest canopy. In fact, the shellbark hickory, the most productive species in terms of mast yields, never grows in large stands, preferring a solitary existence among other deciduous trees.[48] That chestnuts would have been an undesired food source, when they were more abundant and easier to process, is highly unlikely. Chestnuts were undoubtedly consumed as much if not more than all other nutmeats and in many locations exceeded them in dietary importance.

very liable to destruction by fire

Although evidence for chestnut use during the Archaic period is partly circumstantial, it is less so for the archaeological phase known as the Woodland period. It should be noted that the transition from the Late Archaic to the Woodland period did not happen all at once, as some groups made the transformation more rapidly than others.[49] However, by 3,000 BCE many native groups practiced rudimentary forest management and were directly engaged in horticulture, including the growing of sunflowers, goosefoot, and squash in small garden plots (maize had not yet been introduced into the Southeast).[50] The adoption of horticulture also lessened the problem of food scarcity, resulting in unprecedented levels of social stability. In fact, most Woodland peoples lived in permanent, year-round settlements, which allowed them to develop new kinds of ceramic pottery, stone and bone tools, and textile fabrics. Consequently, Woodland peoples were able to intensify food-gathering activities, engage in economic exchanges with others, and improve their overall ability to store food.[51] This led to higher birth rates, larger settlements, and eventually the emergence of tribal forms of social organization.

Woodland Indians also lived in closer proximity to chestnuts, which allowed them to improve the yields of individual trees and groves. Initially this was done indirectly, as the result of burning leaves underneath the trees to aid in nut gathering and to kill the chestnut weevils that penetrated and consumed the nutmeat.[52] Over time, this practice restricted the growth of competing trees as well as the plants and shrubs in the surrounding understory.[53] These fires also parched and dried the chestnuts, which better preserved them for future use.

Using fire to improve chestnut harvests is corroborated in the archaeological record, although from an unlikely source.[54] Remarkably, palynologists are able to recover microscopic particles of charcoal in the very same sediments containing fossil pollen. By counting the number of charcoal particles in the sediments, they are able to determine the historical frequency of local and regional fires: larger particles signify local fires, and smaller ones, more distant fires.[55] In the 1990s, at Horse Cove in Macon County, North Carolina, Hazel and Paul Delcourt found local fires increasing at the end of the Late Archaic, when chestnuts comprised only 7 percent of the area. Over the next several millennia, local fires and chestnuts steadily increased, peaking around 900 AD. At that time, the species comprised 35 percent of the forest, evidence the fires had encouraged their growth and spread.[56] The Delcourts were confident such fires were linked to mast consumption, particularly on upper slopes and ridgetops where, they noted, "people hunted and gathered hickory nuts, acorns, and chestnuts seasonally" (see fig. 2.1).[57]

Similar associations between fire use and chestnut spread have been documented elsewhere, including Greenbrier County in southern West Virginia. In the Buckeye Creek watershed, human-induced fires started increasing around 2,500 BCE, before peaking around 900 AD. Graphs depicting the prevalence of fire and chestnuts at Buckeye Creek closely align, making the case that fire use and chestnut consumption were related phenomena.[58] Less certain is how the woods burning promoted the long-term growth of chestnut, as historical evidence suggests frequent fires

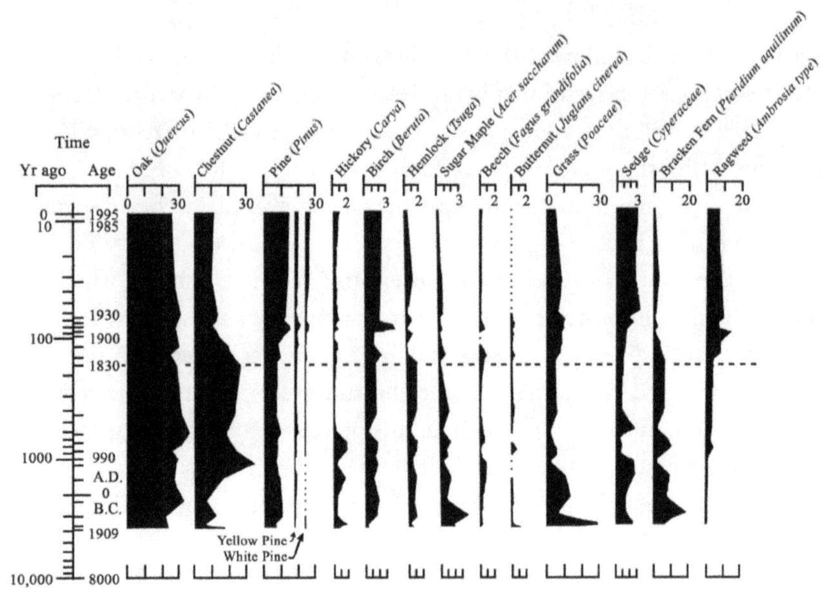

FIG. 2.1. Pollen diagram showing the historical prevalence of American chestnut at Horse Cove, Macon County, North Carolina. Hazel R. Delcourt and Paul A. Delcourt, "Pre-Columbian Native American Use of Fire on Southern Appalachian Landscapes," *Conservation Biology* 11, no. 4 (1997): 1012. Reproduced with permission from the Society for Conservation Biology.

killed seedlings and young saplings. Chestnut leaves are highly flammable due to their shape and moisture content, and would have promoted hotter and more intense burns.[59] Oak regeneration studies find that acorns often become infertile after prescribed fires, even when executed during the dormant season. Acorns buried deep in the leaf duff or cached by birds or mammals often escape harm, although they, too, are damaged if exposed to temperatures above 400°F.[60]

If chestnuts are impacted in the same way as acorns, the trees would have become *less abundant* as a result of frequent fires. If, on the other hand, Woodland Indians avoided burning the same location each year, the saplings could have increased in number, as the remaining ash, which contains essential minerals and nutrients, would have promoted their growth. In order for the trees to reach maturity, however, the fires would have to be absent for three or four consecutive years due to the thin bark of the young saplings, which lowers their rate of survivability.[61] At the same time, other fire-intolerant tree species (American holly, sugar maple, eastern hemlock, eastern white pine, black cherry, blackgum) would have been suppressed by the fires, so if the burning did suddenly stop, chestnuts would have had an advantage due to their faster growth and tolerance of sunlight. Larger trees injured by fire would have also regrown from coppice sprouts. All of these things together explain why the trees increased in number after exposure to human-induced fires.

Those findings do not justify the use of prescribed burns in twenty-first-century chestnut restoration efforts, although many modern-day foresters —who see fire as an important management tool—will undoubtedly use such evidence to promote their use.[62] Most of these fires were set in autumn during the driest time of the year, and likely burned the surrounding understory, and even the forest canopy, if unfavorable conditions prevailed. This also explains why late nineteenth- and early twentieth-century foresters were highly critical of such practices.[63] In 1894, after witnessing the effects of human-set fires near the Cherokee reservation in North Carolina, William W. Ashe noted their detrimental impact on the trees. "Burning the woods has been practiced in [Graham County] and in Cherokee [C]ounty ever since they were settled, and before that time the Indians practiced it," observed Ashe. "The trees in many places, especially the chestnuts, have been scorched on one side and then hollowed out from the effects of the fires. Much other timber and young growth is injured. Many of the mountains in Graham and Swain counties were burned over by the Indians during the past year."[64]

Perhaps the broadest indictment of intentionally set fires is found in a timber survey compiled by Ashe and Horace B. Ayers in 1901. Published by the U.S. Geological Survey in 1905, *The Southern Appalachian Forests* contains no fewer than fifty references to the impact of human-set fires on chestnuts. In Smyth County, Virginia, for example, Ashe and Ayers wrote, "fires have been frequent, and along the spurs and ridges of the divides have greatly injured the forests," adding that the "seedlings of oak, chestnut, cherry, etc., start freely and are very liable to destruction by fire."[65] In Macon County, North Carolina, the pair noted that nearly all chestnuts sprouting from harvested stumps had been "fire killed," but those protected from flames were abundant and healthy.[66] In a later published volume, Ashe even explains the cause of the mortality, noting that "chestnut suffers severely from fire because of its thin bark. Sprout trees not only have thinner bark, but are likely injured through the burning of old stumps."[67]

If chestnuts were indeed harmed by frequent fires, Woodland Indians likely avoided igniting, for several years, the same groves. This seems probable, as the amount of leaves accumulating on the forest floor would not have covered all fallen nuts in a single season. If they did annually set fire to the same stands, wind and ground conditions likely pushed the flames in different directions, so the shape and size of the burns varied from year to year. Undoubtedly, locations nearest village settlements or encampments saw the greatest fire exposure, while remote areas only saw fire once or twice a decade, especially if mast levels were low or negligible. Over the long term, the American chestnut appears to have benefitted from the incendiary practices, becoming more dominant in the forest canopy. However, it is only in the Late Woodland period (800 AD) that we find charcoal particles at such frequencies to suggest that large expanses of woodlands were burned for the sole purpose of gathering chestnuts.[68] It is also possible that some chestnut increases had nothing to do with intentionally set fires, as the species also increased in number in areas with little or no human habitation. Three thousand years ago at Flat Laurel Gap, North Carolina, for example, a location nearly five thousand feet in elevation, chestnuts comprised 10 percent of the forest canopy. A thousand years later they occupied 23 percent of the wooded area—apparently with little or no direct assistance from Native Americans.[69]

Of course, naturally occurring fires likely caused a portion of the chestnut increase at Flat Laurel Gap, as they did in other U.S. regions. When the species first arrived in southern New England about 2,500 years ago, its distribution remained uneven, if not negligible, for nearly a millennium. Pollen studies reveal the trees moved relatively slowly across Connecticut and Massachusetts and then only sporadically into Vermont, New Hampshire, and Maine. According to former Harvard Forest director David Foster, the trees were prevalent across New England only after 500 AD, when there is an increase in fires associated with drier conditions.[70] As with human-set fires, naturally occurring fires would have eliminated competing species in the forest understory, as well as provided the early successional habitat required for maximum chestnut growth and proliferation.

Elsewhere in North America—with or without the aid of fire—chestnuts continued to reclaim forest clearings, colonizing landscapes that for millennia had been void of such trees. By 1000 AD, the species was well established in northern Ohio, central Indiana, southern Illinois, and the woodlands of southern Ontario. In the Atlantic region, from Chesapeake, Virginia, to northern New Jersey, chestnuts were found along the terminal coastline, flourishing in terrain only a few feet above sea level. The species also had moved into the higher-elevation coves of the Appalachians, where moisture levels and soil conditions favored their growth and regeneration.[71] In some parts of the mountain region, the proliferation of chestnuts appears to have even delayed the adoption of large-scale agriculture. In upland areas, circular pits for storing mast were common well into the Late Woodland period, as were the stone tools used for processing chestnuts, acorns, and hickory nuts.[72] Eventually, however, the number and size of Native American villages increased in these uplands areas, causing a migration toward river floodplains where the trees were less common.[73]

When the Woodland period came to an end across eastern North America, native peoples could claim a long and intimate relationship with the American chestnut. In the Early Archaic, the encounters had been seasonal and infrequent, as the trees were less ubiquitous and even absent in the northern latitudes. By the Late Archaic, chestnuts were commonly eaten as well as stored and processed into meal. During the Early Woodland period, Native Americans intentionally inhabited areas where the trees were abundant and cached large quantities of chestnuts in underground silos. By the Late Woodland period, Native Americans routinely harvested chestnuts with the aid of fire, which better preserved them and extended their use. They also made unleavened bread using chestnut flour, a practice that continued well into the twentieth century.[74] With each passing millennium, human-chestnut encounters became more frequent, nuanced, and intimate.

After the Woodland period, Native Americans were more dependent on the cultivation of corn, beans, and squash, agricultural staples that required preparing and planting large fields.[75] While such changes made chestnuts no less important, their role as a dominant food source lessened as maize and other annual crops became more commonplace. Although a lesser source of nutritional sustenance, the trees continued to provide critical habitat for game animals as well as the raw material for homes, meeting houses, and canoes. Well before European contact, the tree had entered the realm of mythology and folklore, and, among groups like the Natchez, the inspiration for seasonal festivals.[76] And although one can never be fully certain about the details of daily life in North American prehistory, the preponderance of evidence suggests that Native Americans had developed a unique and indelible relationship with the American chestnut. Later eyewitness accounts confirm this fact, shedding additional light on the importance of the tree during the colonial and frontier periods of American history.

CHAPTER 3

wherever there are mountains

In 1491, the year before Christopher Columbus came ashore in the West Indies, the natural range of the American chestnut included four hundred thousand square miles, an area that today falls within the borders of twenty-six states, the District of Columbia, and the province of Ontario. This was, in essence, the modern range of the tree, a vast territory extending from southern Maine to southeast Louisiana (see figure 3.1).[1] The trees also grew sporadically below the Atlantic fall line, and could be found from the southern terminus of the Chesapeake Bay to northwest Florida. Although the Mississippi River served as a natural barrier for the movement of the trees westward, a few stands flourished beyond its banks, possibly as the result of birds, humans, or favorable water currents. By the end of the fifteenth century, the American chestnut was one of the more common species in the eastern deciduous forest, commonly reaching diameters of seven feet or more in the southern Appalachians. The tree was not ubiquitous across its range, however, and even infrequent in parts of New England, western Tennessee, and eastern North Carolina. Its lack of ubiquity made it no less important to Native Americans, who continued to utilize the trees for sustenance, warmth, and shelter.[2]

The first Europeans to observe the American chestnut were likely Spaniards, as they explored and laid claim to areas where the trees were commonly found. The slave raiders Francisco Cordillo and Pedro de Quejo may have seen chestnuts after coming ashore north of Charleston in the summer of 1521, although it is doubtful that they traveled far enough inland to observe large mature trees.[3] Returning to the same area in 1525, Pedro de Quejo sailed northward along the North Carolina coast before entering the Chesapeake and Delaware Bays, locations where chestnuts would have been visible beyond the shoreline.[4] Estêvão Gomes, the Portuguese pilot who sailed for the Kingdom of Spain, navigated the New England coastline that same year, a voyage that influenced mapmakers for decades. According to geographer Thomas Suárez, the map drawn by cartographer Juan Bellero in 1554 not only documents Gomes's

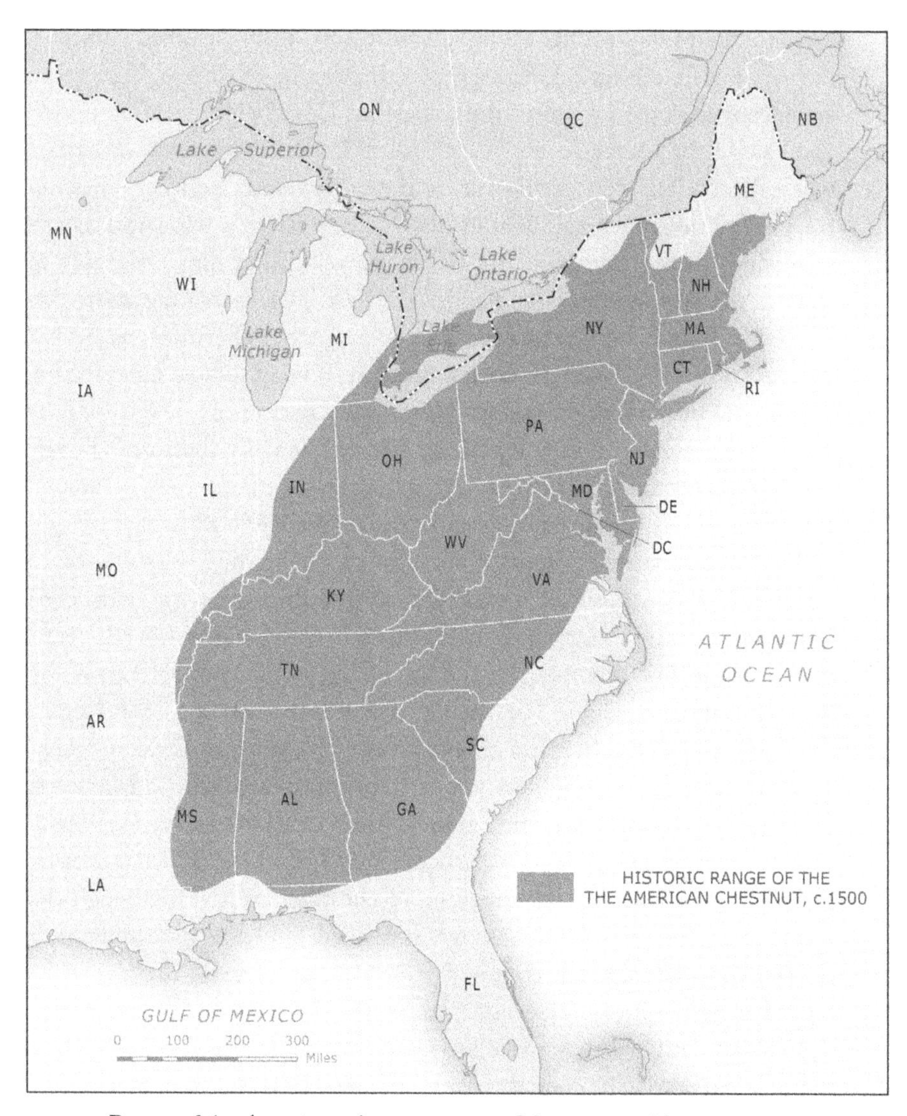

FIG. 3.1. Range of the American chestnut, c. 1500. Map prepared by Will Fontanez, Southeast Cartography, St. Marys, Georgia.

journey, but includes the location of a chestnut grove along the eastern coast of Maine.[5]

The Italian Giovanni da Verrazzano also very likely observed chestnuts during his visit to North America in 1524. In March of that year, Verrazzano set anchor near the mouth of the Cape Fear River before heading northward along the Outer Banks.[6] On April 27 he entered New York harbor, where he undoubtedly witnessed, near the shoreline, budding chestnut trees. After taking a small boat up the Hudson River, a "violent unfavorable wind" forced him to return to his caravel, giving him little time to explore the surrounding woodlands.[7] A week later, at Narragansett Bay in Rhode Island, Verrazzano certainly encountered chestnuts, as he not only spent fifteen days in the harbor area but traveled sixteen miles inland. Verrazzano does mention seeing "Nutte trees" during his inland foray as well as the presence of large dugout canoes.[8] Incidentally, the canoes he saw were very likely made from chestnut logs, as the craft were common among the Narragansett Indians during the early colonial period.[9]

The earliest written record of the American chestnut is attributed to Spanish historian Gonzalo Fernández de Oviedo (1478–1557), who lived and worked in the West Indies during the sixteenth century. Oviedo refers to the tree in his recounting of the life of Lucas Vázquez de Ayllón, the Spanish licentiate, who in 1526 tried to establish the colony of San Miguel de Gualdape on the South Carolina coast. According to Oviedo, whose narrative is based on eyewitness accounts, the forests beyond San Miguel de Gualdape possessed "many pine and oak trees [and] chestnuts, with small fruit."[10] In another passage, Oviedo noted the Indians of the interior were great archers who made "sturdy bows of chestnut wood."[11] Although the Vázquez de Ayllón account was not published until the mid-nineteenth century, the original manuscript was written during Oviedo's lifetime, making it among the first references to the North American tree.[12]

The Spaniard perhaps most associated with the species, however, is Hernando de Soto, the conquistador who explored the southeastern United States from 1539 to 1542. Although there were four chroniclers of the de Soto expedition, only two accounts are by actual eyewitnesses.[13] Perhaps the most reliable is attributed to Rodrigo Ranjel, de Soto's private secretary during the expedition. Based on his own diary, the story was also retold by Oviedo in the *General and Natural History of the West Indies*.[14] Although Oviedo is alleged to have corresponded with Ranjel to get a complete accounting of the expedition, the narrative is probably embellished, containing the thoughts of both individuals. Regarding Ranjel's descriptions of North American flora, much of the information is considered trustworthy, as it is based on his firsthand observations.[15]

Ranjel's first mention of chestnuts occurs early in the expedition—August 25, 1539—after de Soto and a smaller contingent of his men entered the Robinson Sinks area of northern Florida. In a "fair-sized village" the late anthropologist Charles Hudson believed to be the Alachuan town of Cholupaha, Ranjel states the men saw "many small chestnuts, dried and delicious."[16] He noted the plants bearing them were only "two palms high," which means they were Allegheny chinquapins, the *Castanea* species that in northern Florida seldom obtains heights of more than three feet.[17] Ranjel added that there were "other chestnuts in the land, which the Spaniards saw and ate, which are like those of Spain, and grow on as tall chestnut trees."[18] He then offers a description of the American chestnut, perhaps the first penned to paper: "The trees themselves are big and with the same leaf and burrs or pods [as Spanish chestnuts], and the nuts are rich and of very good flavor."[19]

The second eyewitness account of the de Soto expedition, and the first chronicle available to the general public, is attributed to the "Gentleman of Elvas," a Portuguese cavalier who survived the ordeal and retold his story to the Spanish publisher André de Burgos. When his *A True Account of the Travails Experienced by Governor Hernando de Soto* was published in Evora, Portugal, in 1557, it was thought to be based on actual notes Elvas kept during the expedition.[20] Although a few scholars have questioned the true source of the narrative, it does provide more detail than the Ranjel account and even mentions vegetation, crops, and food supplies found in Native American villages.[21] However, there are only two references to the tree in the Elvas account and one is a brief comment about roasted pumpkins that "taste of chestnuts."[22] The other observation is found in the chapter entitled "The Diversities and Peculiarities of Florida," the name the Spanish commonly used to refer to the present-day southeastern United States.[23] In the book's final chapter, on one of its very last pages, Elvas proclaims, "Wherever there are mountains, there are chestnuts" (see fig. 3.2).[24]

In the first English version of the volume, published in 1609, historian Richard Hakluyt translates the entire sentence as, "Where There Be Mountaines, there be chestnuts: they are somewhat smaller than the chestnuts of Spaine."[25] According to Michael J. Ferreira, associate professor of romance philology and linguistics at Georgetown University, the passage should actually read, "Where there are mountains, there are chestnuts, although they are somewhat smaller than the colarinha chestnuts of Hispania."[26]

Ferreira finds it perplexing that the Portuguese term for mountains (*montanhas*) is not used, especially if Elvas only observed the trees in high-altitude areas. However, there is also the possibility a typo was made, as the printer, says Ferreira, could have wrongly used the letter *s* (forming the word *serras*) instead of the letter *t* (which would have given the word *terras*), since the two fonts are virtually identical in sixteenth-century typescript.[27] Given that chestnut trees could be found in both lowlands and uplands at the time of the de Soto *entrada*, the Portuguese term for lands—*terras*—may have been the correct choice of words.[28] His use of *serras* (mountain ranges), on the other hand, implies that chestnuts were in their highest concentrations in the uplands, an observation that in the sixteenth-century Southeast would not have been entirely contrary to other accounts or the archaeological evidence.[29] The mention of colarinha chestnuts is less surprising, as the cultivar was commonly grown in the town of Elvas.[30] Regarding "Hispania," Elvas is referring not to Spain, but the entire Iberian Peninsula, including the Kingdom of Portugal.[31]

Other explorers confirm the presence of chestnuts in the Southeast during the sixteenth century, documenting both their ubiquity and usefulness. In August 1566, for example, the governor of Florida, Menéndez de Avilés, ordered Captain Juan Pardo to lead an expedition deep into the backcountry interior. Four months later, on December 11, 1566, Pardo and 125 men left Santa Elena (Parris Island, South Carolina) for the mountains of western North Carolina.[32] The entourage stopped at several Native American villages along the way, including Cofitachequi, near present-day Camden, South Carolina, and Joara, a native township fifty miles east of Asheville. At Joara, Pardo and his men built Fort San Juan, the first European settlement in the interior United States.[33] Excavations at the site reveal the soldiers' barracks were constructed with chestnut poles placed intermittently within the exterior walls. A large chestnut plank was also found inside the structure, a badly decayed board measuring fifty inches long, ten inches wide, and one-and-a-half inches thick. Covered with river-cane matting, the plain-sawn board was perhaps used as an interior bench and represents the first documented use of chestnut planking in North America.[34]

Pardo returned to Santa Elena after arriving in Joara but left behind thirty men to maintain and protect Fort San Juan. Months later, on

¶ Relaçam verdadeira dos trabalhos q̃ ho gouernador dõ Fernãdo d' ſouto ⁊ certos fidalgos portugueſes paſſarom no dſcobrimẽto da prouincia da Frolida. Agora nouamẽte feita per hũ fidalgo Deluas.

¶ Foy viſta por ho ſeñor inquiſidor.

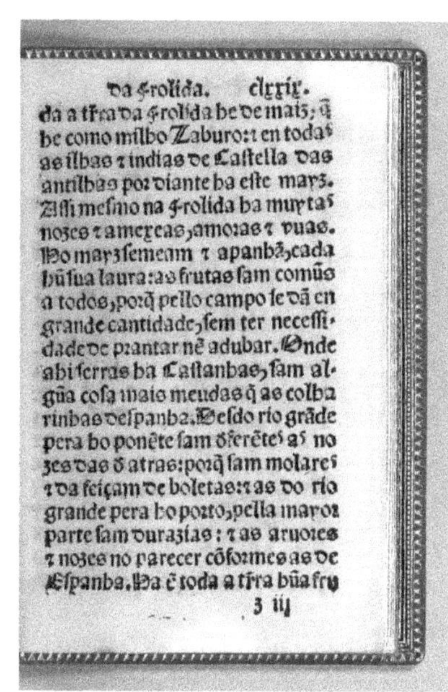

da Frolida. clxxix.

da a tr̃ra da Frolida he de mais; q̃ he como milho Zaburo: ⁊ en todas as ilhas ⁊ indias de Caſtella das antilhas por diante ha eſte mayz. Aſſi meſmo na Frolida ha muytas nozes ⁊ amexeas, amoras ⁊ vuas. Ho mayz ſemeam ⁊ apanhã, cada hũ ſua laura: as frutas ſam comũs a todos, porq̃ pello campo ſe dã en grande cantidade, ſem ter neceſſidade de prantar nẽ adubar. Onde ahi ſerras ha Caſtanhas, ſam algũa coſa mais meudas q̃ as colharinhas deſpanha. Deſdo rio grãde pera ho ponẽte ſam dferẽtes as nozes das d' atras: porq̃ ſam molares ⁊ da feiçam de boletas: ⁊ as do rio grande pera ho porto, pella mayor parte ſam durazias: ⁊ as aruores ⁊ nozes no parecer cõformes as de Eſpanha. Ha ẽ toda a tr̃ra hũa fru

3 iij

FIG. 3.2. Title page, *The Gentleman of Elvas: A True Account of the Travails Experienced by Governor Hernando de Soto* (Evora, Portugal, 1557), left. The right-hand image contains the quote, "Where There Are Mountains, There Are Chestnuts, although They Are Somewhat Smaller than the Colarinha Chestnuts of Hispania." Courtesy of the John Carter Brown Library, Providence, Rhode Island.

September 11, 1567, Pardo began a second expedition to Joara, a journey that included overnight stays in Native American villages.[35] According to eyewitnesses, when Pardo and his entourage arrived at each township, the Indians offered them numerous provisions, as their storehouses were filled with "maize, beans, pumpkins, and chestnuts for two or three years ahead."[36] Among the eyewitnesses was Teresa Martín, a native Catawban who later married Juan Martín de Badajoz, one of Pardo's soldiers. In testimony provided to Florida governor Gonzalo Mendez de Canço in 1600, Martín reported the chestnuts given to Pardo were *castaña apilado*, meaning they were smoke-dried over a smoldering fire.[37] This method of chestnut preservation not only extended the shelf life of the nuts, but also of the flour or meal made from them.[38] The term *castaña apilado* is used several times by members of the Pardo expedition, evidence that chestnuts were indeed eaten as dried foodstuffs well beyond the fall and winter months.[39]

When Pardo finally arrived at Fort San Juan, he found the fort virtually abandoned and his principal sergeant, Hernando Moyano de Morales, under attack at a distant location. With hopes of rescuing Moyano, Pardo led a foray of men over the Blue Ridge Mountains into what is today the Tennessee Valley of East Tennessee. Pardo's notary, Juan de la Bandera, noted the presence of "very good land" during their journey as well as "many chestnuts, walnuts, and quantities of other fruits."[40] After rescuing Moyano near the confluence of the French Broad and Nolichucky Rivers, Pardo returned along the flank of Chilhowee Mountain, perhaps to avoid detection. In mid-October, the men camped near the junction of Walden and Cove Creeks, where they were visited by three principal chiefs, among them Otape Orata. According to anthropologist Charles Hudson, *orata* translates as chief or headman, whereas *otape* is "the Muskogean word referring to chestnut trees or a place where they may be found."[41] Thus, it is very possible the chief's name was "Chestnut Tree." Hudson believes his village was located near the headwaters of the Little Pigeon River near the Great Smoky Mountains National Park, an area known for possessing large and numerous chestnut trees.[42]

Upon returning to Joara, Pardo garrisoned another thirty or so men at Fort San Juan before abruptly leaving for Santa Elena. Six months later, Fort San Juan was destroyed by Indians, allegedly after Spanish soldiers took native women as concubines.[43] As a result, Spain lost its strategic foothold in the interior and, soon afterward, along much of the Atlantic coast. This series of events allowed the English to settle at Roanoke Island, a location explored by English navigators Philip Amadas and Arthur Barlowe in 1584.[44] On August 17, 1585, Sir Ralph Lane and more than a hundred colonists were left inside a small fort on the island and told to develop stronger ties with the Algonquin Indians. Among the settlers was Thomas Harriot (or "Hariot"), the historian and ethnographer assigned to catalog the natural resources to be claimed by the Virginia colony.[45] Harriot departed the island in the summer of 1586, however, after Sir Francis Drake offered him and the colonists safe passage back to England.[46] A year later, Harriot was living in Ireland, writing *A Briefe and True Report of the New Found Land of Virginia.* Published in 1588, the document provides ample evidence that chestnuts were eaten by the Algonquins as well as the very first American colonists.

in diverse places great store

Harriot's reference to chestnuts is found in the second part of *A Briefe and True Report*, which possesses the lengthy title, "Of suche commodities as Virginia is knowne to yeelde for victuall and sustenance of mans life, usually fed upon by the naturall inhabitants: as also by us during the time of our aboad."[47] Under the heading "Of Fruites," the first item mentioned is chestnuts, which Harriot explains, were found "in divers places great store: some they use to eat rawe, some they stampe and boile to make spoonemeate, and with some being sodden they make such a manner of dowe bread as they use of their beanes."[48] This suggests the Algonquins had two ways of preparing chestnuts beyond eating them raw: (1) they made a polenta-like porridge by adding ground chestnuts to boiling water, or (2) they added water to chestnut meal to make a dough that was boiled in water or baked over hot coals.[49] Although there are no eyewitness accounts explaining the precise method of preparing chestnut bread, the governor of Jamestown, Captain John Smith, believed the Algonquins did so by boiling the nuts for several hours. This apparently softened them for the dough-making process, which ultimately yielded, according to Smith, "bread for their chiefe men, or at their greatest feasts."[50]

Chestnut bread was also a staple in the Southeast interior, which is verified in the historical record. On October 28, 1540, de Soto and his entourage were offered large amounts of chestnut bread at a palisaded village along the Alabama River. The bread was delivered by messengers from the chiefdom of Mabila, another populous township located near present-day Camden, Alabama.[51] According to Rodrigo Ranjel, there were "many and good chestnuts" in the area, a fact later corroborated by the naturalist William Bartram. Bartram, who traveled across southern Alabama in 1777, observed an abundance of chestnut in the vast open forests that once comprised Alabama's Black Belt region.[52] Chestnut bread was also regularly consumed by the Cherokees, remaining a dietary staple well into the nineteenth century. By the twentieth century, however, Cherokees often added corn meal to the nuts when making chestnut bread. According to one eyewitness, the two ingredients were formed into "hand-sized pones," wrapped in corn leaves, and then placed in boiling water "for an hour or so."[53]

Boiled or dried chinquapins were also added to chestnut bread, although their smaller size made them difficult to process in large quantities.

John Smith was the first Englishman to mention their use among Native Americans, and, in doing so, reveal the Algonquin origins of the word. Smith describes the nuts as "a small fruit growing on little trees, husked like a Chestnut, but the fruit most like a very small acorne. This they call *Chechinquamins*, which they esteeme a great daintie."[54] Smith was equally impressed with the flavor of native chestnuts, equating them with European varieties. "In some parts," he wrote, "were found some Chestnuts whose wild fruit equalize the best in France, Spaine, Germany, or Italy," especially to those individuals who "had tasted them all."[55] The Jamestown captain noted that chestnuts and chinquapins were "dryed to keep," claiming the preservation method allowed them to "live a great part of the yeare."[56]

Although there is little evidence chestnut bread or meal was eaten by the early American colonists, the nuts were certainly a welcomed resource. The nuts were a familiar form of sustenance and their exceptional sweetness perhaps negated any criticism harbored against them due to their smaller size. Although known for hyperbole, Edward Johnson of Woburn, Massachusetts, went as far as to claim the "boil'd chesnuts" of the Narragansett were "as sweet as if they were mixt with sugar."[57] John Josselyn, the English naturalist who lived for a brief period in southern Maine, similarly found the nuts "very sweet in taste."[58] Not surprisingly, some even presented chestnuts to neighbors as gifts. In September 1638, the wife of Roger Williams sent a handful to the wife of Massachusetts Bay Colony governor John Winthrop, promising her a larger basket of nuts should "Mrs. Winthrop love them."[59]

In some places, the demand for chestnuts was so great it sometimes resulted in the felling of entire trees. In his well-known description of the New Netherlands, Adriaen van der Donck lamented that the Dutch and Indians of the Hudson River valley sometimes cut down the trees in chestnut season to more easily gather the nuts.[60] The practice of felling trees laden with chestnuts says more about their widespread availability than avarice on the part of Native Americans or colonists, however. Without knowing the location or size of the felled trees (smaller trees suggests intentional thinning; larger trees implies forest clearance for croplands, pasture, or fencing), it is impossible to draw firm conclusions about what appears, in the twenty-first century, to be wasteful and profligate behavior.

By the 1660s, chestnuts were in such demand that Massachusetts colonists purchased them from neighboring Algonquins for twelve pence a bushel.[61] In fact, they were so valuable that writer Thomas Morton included the tree in his "catalogue of commodities," an annotated list of natural resources he believed would make the Massachusetts colony prosper. For Morton, "chestnutt" was not only a timber in "very greate plenty," it was also a "very good commodity, especially in respect of the fruit, both for man and beast."[62]

Although Morton's observations referred to southern New England, the trees were certainly plentiful elsewhere in the colonies. As noted in the introduction, the American chestnut inhabited a diversity of locales and habitats, including lakeshore and riverine environments. During the colonial period, chestnut-laden woodlands could be found across northern New Jersey, southern Vermont, eastern Maryland, southern New Hampshire, and southeastern New York.[63] On the island of Manhattan the trees reportedly comprised as much as "half of the wood of the forest."[64] However, pollen studies conducted north of the island near Piermont suggest they comprised only 20 percent of the wooded canopy in the early seventeenth century.[65]

In Canada, the trees were well established along the northern shore of Lake Erie, particularly the interior woodlands between what are today Port Burwell and Port Dover. Samuel de Champlain also witnessed chestnut groves during his excursion into the St. Lawrence River Valley, including trees near the lake he later named for himself. In a journal entry dated July 14, 1609, the French explorer observed "many chestnut trees" near Lake Champlain's eastern shoreline.[66] Although he admitted he had not yet seen any trees "except for above the edge of this lake," their size and number warranted specific mention.[67] Historians believe the grove was located near Burlington Bay just south of the Winooski River, as the species was observed there in the mid-1800s.[68] Further evidence the trees were in the Burlington Bay area comes from nineteenth-century physician Clinton H. Merriam, who witnessed red squirrels swimming across Lake Champlain toward that very location "in years when the yield of chestnuts is large."[69]

Champlain's most endearing reference to the tree is found in *The Voyages of Samuel de Champlain*, a book published in Paris in 1613. The volume contains a folded map of New France with a decorative cartouche that features several North American plants, including the American chestnut (see fig. 3.3).[70] According to museum curator Victoria Dickenson, Champlain most likely created the cartouche to illustrate the "food or medicinal plants of the Aboriginal inhabitants." She believes it also represents those things

FIG. 3.3. American chestnut burr and leaf detail in Samuel Champlain, "Map of New France . . . 1612." The map, a folded insert, was published in *Les Voyages du Sieur de Champlain* [The voyages of Samuel de Champlain] (Paris: Jean Berjon, 1613).

he "dried and sketched, or even attempted to bring back to France to be planted in Old World gardens."[71] There is little evidence Champlain or others took the trees back to Europe, however, as many thought them identical to the European variety. The nuts were also smaller than those in Europe, providing little incentive to transport trees back across the Atlantic. Europeans had been cultivating chestnuts in orchard settings for centuries and possessed long-held cultural biases regarding their proper growth and management. When Frenchman René Goulaine de Laudonnière described "the trees of Florida" in 1565, he noted that the American chestnut was "more wild than those of France."[72]

Regardless of their ultimate opinion about the trees, colonial settlers greatly benefited from them during the seventeenth century. Not only an important food source, the wood from the trees was also durable, easily worked, and relatively lightweight. Indeed, Thomas Morton claimed chestnut was "excellent for building," an opinion he also shared with the Algonquins, who used it to construct their homes and dwellings.[73] In some respects, felled trees were more useful than living ones, which explains why colonists sometimes chopped down nut-bearing specimens. The trees were transformed into building timbers, fencing, shingles, and even seaworthy vessels. Native Americans used the largest logs for dugout canoes and the pole timbers for framing dwellings and council houses. As important as the nuts were to North American inhabitants, the wood and bark from the trees had perhaps even greater value.

covering for their houses

One of the most important byproducts of the American chestnut was its thick bark, which Native Americans employed in the construction of various kinds of buildings.[74] Seventeenth-century accounts provide fairly detailed descriptions of these structures, which could be large and substantial. One such dwelling was observed by the Dutch travelers Jasper Danckaerts and Peter Sluyter in the vicinity of what is today the Fort Hamilton neighborhood of Brooklyn, New York. On October 10, 1679, the pair observed an Algonquin wigwam sixty feet long and fifteen feet wide, constructed from reeds and "the bark of chestnut trees."[75] The doors of the wigwam were also made of reeds and "flat bark," presumably of chestnut. In 1683, William Penn observed similar dwellings among the Lenape Indians of Delaware, structures that were made, he noted, of the bark of mostly chestnut trees, "set on poles."[76]

According to anthropologist Peter Nabokov, Native Americans gathered chestnut bark in late spring, after the sap had risen. The bark was removed in long sheets, then flattened under heavy rocks and moistened to prevent warping and cracking. After it was fully cured, the slabs could be bent, trimmed, and interwoven into the walls, ceilings, and doors of pole-framed structures. On rooftops, the bark was arranged as overlapping shingles, providing protection from the rain and wind and insulation from the heat and cold.[77] It was also an integral component in Cherokee dwellings, as William Bartram observed during the eighteenth century. According to Bartram, when Cherokees built log cabins, they first plastered the interior and exterior walls with wet clay tempered with dried grass. Afterward, the entire structure was covered and roofed "with the bark of the chestnut tree or broad shingles."[78]

Similar structures were described by Luisa Mendez, the female cacique who in the late sixteenth century lived near the south fork of the Holston River near Saltville, Virginia. In testimony provided to Florida governor Gonzalo Mendez de Canço in 1600, Luisa Mendez recalled that the houses of her township—which she considered a large and populous *pueblo*—were made of timbers covered with "slabs of chestnut bark and cedar boards."[79] Although the use of chestnut for such purposes often destroyed living trees, most of the damage occurred near the immediate village environs. In the New Netherlands, however, the deforestation was visible enough to cause van der Donck to bluntly remark, "chestnuts would be plentier if it were not for the Indians, who destroy the trees by stripping off the bark for covering for their houses."[80]

Colonial settlers also used the trees to construct their homes and barns, although the evidence for this is still primarily anecdotal for the seventeenth century. Hugh Morrison, the noted historian of American colonial architecture, believed chestnut was among the woods used for shingles and framing timbers but could only base his claim on homes built after 1700.[81] While there is consensus that hewn beams and roofing shakes made of chestnut were used in many seventeenth-century dwellings, not enough structures have been found yet to verify that assumption. In 1993, investigators at the Towne Neck archaeological site near Annapolis, Maryland, found "chestnut wood charcoal" in the remains of an "earthfast dwelling" built c. 1660, which is suggestive but not conclusive.[82] The earliest home actually known to feature chestnut in its construction, with documented provenance, is the Sarum estate of Charles County, Maryland. Although the colonial dwelling was previously thought to have been built in the late seventeenth century, a tree-ring analysis revealed the structure was built no earlier than 1717. According to architectural historian Mark R. Edwards, the studs, rafters, and clapboards of the home are made of riven chestnut, as are the "round-end roof shingles."[83] In truth, the use of chestnut timber for home construction was probably not commonplace until the second decade of the eighteenth century, when colonial settlement spread inland, away from the Atlantic coastline. The trees were more common in the coastal interior, so chestnut would have been much more readily available.[84]

Despite its relatively minor role in home and barn construction during the 1600s, the American chestnut did suffer losses from human use during the early colonial period. Statutes requiring the fencing of croplands and gardens were passed by colonial lawmakers as early as 1623 and chestnut was considered one of the best woods for making post-and-rail fences.[85] As a result, the trees likely declined in considerable numbers after 1650, as many township woodlots were cleared for that purpose. In 1659, the Quaker merchant John Bowne—one of the first residents of Flushing, New York—sold "117 good substantial 5-hole chestnut posts" to his neighbor Nicholas Parcells, who purchased, in the same exchange, an additional two hundred chestnut rails.[86] As with colonial architecture, evidence for the use of chestnut fencing does not always come from eyewitnesses, but from later published sources. In 1812, Lewis Mills Norton, a resident of Goshen,

Connecticut, recorded that "for the last few years there has been an increased attention to the building of stone fences; till which time chestnut rails were mostly used and the timber was fast decreasing."[87]

Paling fences were also made from chestnut, especially where lawns and gardens needed protection from livestock, rabbits, or rodents. In 1982, when four chestnut palings were found in an archway attached to a Fredericksburg, Virginia, plantation built c. 1690, authorities announced they were "the oldest surviving wood fence pickets in the country."[88] Some chestnut timber was also removed for firewood, which was in growing demand by the middle of the seventeenth century. On the island of Manhattan, for example, firewood was so scarce that the common council noted already in 1683 that "moste of the firewood braught to the citte is cut in other parts of the province."[89] Although it would take a full century before the woodlands of Manhattan were, as George Washington later claimed, "totally stripped of Trees, & wood of every kind," considerable amounts of chestnut were no doubt removed by firewood-starved colonists long before that time.[90]

This was especially true in New England, where cutting trees for firewood was a year-round task. As environmental historian William Cronon has noted, a single colonial household consumed "thirty or forty cords of firewood per year," or roughly an acre of mature woodland.[91] By the end of the seventeenth century, twenty million cords of firewood were annually harvested in eastern North America, representing some half a million acres of impacted forests.[92] Only a small percentage of the harvested timber was chestnut, however, although it did rank as a leading fuel wood in the "East Gulf" region, while still falling behind ash, gum, hickory, oak, and pine in popularity.[93] Because chestnut has one of the lowest heat generation values among hardwoods, it does not burn as hot as oak, hickory, maple, or even beech. Uncured chestnut logs and limbs also frequently sparked in open-hearth fireplaces, making the wood a fire hazard if left unattended.[94] This is why the Iroquois Confederacy forbade its use during official assemblies. "When the Lords are assembled," states its five-hundred-year-old constitution, "the Council Fire shall be kindled, but not with chestnut wood."[95]

When chestnut was cut for firewood, it quickly resprouted from the felled stumps, giving it an advantage over its forest competitors. The avoidance of chestnut for firewood, along with the clearing of oak, maple, hickory, and pine for that same purpose, certainly benefited the species, negating losses it incurred from other uses.[96] In fact, the American chestnut appears to have expanded its range during the second half of the seventeenth century, increasing in numbers in New England and the Mid-Atlantic. This was due in part to the natural spread of the trees into disturbed areas caused by summer and winter storms, as well as the clearing of woodlands by European colonists. The decline of beaver in New England and the Mid-Atlantic as a result of the fur trade also had a positive impact on the species, as the animals were responsible for diminishing some stands. Heavier annual rainfall also increased the number and size of chestnuts, particularly along the Atlantic seaboard, which had undergone a century of severe and periodic droughts.[97] Abandoned cropland was another reason for the chestnut increase, as the introduction of smallpox, measles, and influenza into North America decimated entire Native American communities, eliminating as much as 90 percent of their population in some areas. In the decades that followed, thousands of acres of croplands returned to copse and woodland, a reforestation process that lasted in some places a half-century.[98] Not only did trees appear where there were none before, forest composition also changed, as Native American Old Fields favored the regeneration of faster-growing and more light-tolerant trees like the American chestnut.[99]

In the colonial period, the American chestnut remained an important forest species well into the eighteenth century. Evidence for the tree's ubiquity is found in witness-tree data gathered from land-surveys, as well as the material-culture artifacts of early American life. From Jamestown, Virginia, to Falmouth, Maine, chestnut wood was daily transformed into fences, shingles, and building timbers as well as numerous household items, including chairs, benches, and shelving.[100] The nuts continued to be eaten and bartered by settlers and Native Americans even after maize became more important to local diets. European livestock, especially hogs, also consumed chestnuts, turning the mast crop into another important product of the forest commons: pork.[101]

In northern Massachusetts and southern Maine, the trees spread into areas cleared for other purposes, such as the cutting of community woodlots or commercial timbering operations. Of course, the impact of human settlement on the wooded landscape varied from region to region, occurring first in New England and the Mid-Atlantic and later in the southern United States. As forest historian Michael Williams has expertly noted, forest clearance first happened along the Atlantic coast before moving westward and northward toward the fall line.[102] Even in important agricultural areas like the Chesapeake Tidewater, timber removal did not happen at once: prior to 1740, nearly 80 percent of the interior woodlands remained intact.[103] While that percentage changed substantially after the mid-1700s, before that period the American chestnut was not greatly impacted by timber removal.

In the backcountry, upland forests saw even fewer declines, providing opportunities for explorers and frontier settlers to observe old-growth trees in their native splendor. In the southern Appalachians, chestnuts occupied large swaths of territory, including high-altitude areas generally void of human settlement. These oak-chestnut forests remained essential habitat for big and small game, including elk, deer, bear, grouse, passenger pigeons, and several species of squirrel. As human settlements moved westward, these animals provided nutritional sustenance for tens of thousands of individuals, as well as incomes for settlers and fur traders alike. Chestnuts and chestnut timber were important fixtures throughout the frontier period, and influenced, albeit indirectly, the entire backcountry economy.

CHAPTER 4

the most celebrated hunting grounds

At the beginning of the eighteenth century, good arable land was becoming scarce in the original colonial settlements. By 1710, newly arriving settlers were seeking their fortunes above the Atlantic fall line and, in New England, as far inland as Brattleboro, Vermont. After 1720, demand for new ground became competitive even in the interior, causing immigrants to move deeper into Indian territory. Surveyors also traveled into the backcountry during this period, not only to settle boundary disputes between colonial governments, but to claim lands for themselves and the British crown. Among those providing descriptions of the well-wooded landscape was William Byrd II, who in 1728 was given the task of surveying the boundary line between North Carolina and Virginia.

Byrd's writings make frequent reference to the American chestnut and offer ample evidence the trees provided important food and habitat to large game animals, especially black bears. He also informs us the tree was not strictly an upland species and even inhabited, in some locales, bottomland hardwood forests. Much of Byrd's commentary about chestnuts is found in *History of the Dividing Line*, a first-person account of his expedition published in 1841.[1] A less polished version, *The Secret History of the Dividing Line*, contains additional information not found in the longer and more carefully worded tract.[2] A handwritten journal provides additional details about his backcountry travels, including daily encounters with native flora and fauna.[3]

Although the original survey began on March 16, it was suspended after a month due to, in Byrd's own words, "the danger of Rattle snakes in this advanced season" and "the great fatigue already undergone."[4] By April 15, the survey crew had traveled only seventy-three miles, from the Currituck Inlet on the Atlantic coast to the bank of the Meherrin River near present-day Franklin, Virginia. There is no specific mention of chestnut trees during the journey, as their trajectory took them through flooded swamps, submerged wetlands, wooded pocosins, and other low-lying terrain.[5]

The survey continued five months later, on October 1, with Byrd and his men heading westward toward the Roanoke River. The number of wild game animals observed by the men is quite remarkable and indicative of the large number of oak and chestnut trees present in the wooded backcountry. Indeed, nearly every game animal mentioned by Byrd was a mast-dependent species, meaning their populations are adversely affected when there are few chestnuts and acorns in the forest.[6] Byrd's first reference to the tree occurs on October 23, when he notes in *The Secret History* that "we judg'd by the great Number of Chestnut Trees that we approach't the Mountains."[7]

The surveyors were near present-day Danville, Virginia, where they had camped just beyond the Dan River at a location not more than seven hundred feet in elevation. Although Byrd noted the evening before that they had seen mountains in the distance, it would take them another week of travel before they encountered the Appalachians proper, where elevations exceed a thousand feet. What the men most likely saw were White Oak, Turkey Cock, and Smith Mountains, narrow ridges rising twenty miles to the north and west.[8] Evidence that chestnuts were prevalent in the lower-elevation forests of Pittsylvania County is found in colonial land patents, such as the one issued to Isaac Cloud in 1748. The parcel surveyor recorded its corner boundary as lying "at a Hollow Chestnut Tree in which said Cloud and [Gideon] Smith used to camp, on the grounds between a branch of Banister and Turkey Cock Creek."[9] The chestnut tree was so large, wrote historian Maud Carter Clement, "as to admit sleeping quarters for the two men."[10]

Byrd also mentions black bears on that date, exclaiming in *The Secret History* that they were "great Lovers of Chestnuts."[11] A few days earlier, Byrd had seen chestnut trees with outward-broken branches, and when he inquired about the phenomenon was told bears did not "trust their unwieldy bodies on the smaller limbs of the tree, that would not bear their weight; but after venturing as far as is safe, which they can judge to an inch, they bite off the end of the branch." After the limbs and burrs had fallen from the tree, the bears were "content to finish their repast upon the ground."[12] That bears directly benefitted from chestnuts is evident from the incredible number of bears taken from the Dan River location forward. From the first crossing of the river to the end of the survey line near present-day Stuart, Virginia, the men killed twenty-two bears. Byrd remarked that one of the bears was "so prodigiously fat, that there was no way to kill Him but by fireing in at his Ear."[13] Over the same two-week period, the twenty-one man crew shot another nineteen turkeys, ten deer, and one raccoon. This represents a remarkable number of mast-eating game animals, as the total distance traveled by the men was less than fifty miles, encompassing a land area of no more than seventy thousand acres.[14]

On October 30, Byrd makes mention of another consumer of chestnuts—the passenger pigeon. After fording Matrimony Creek, not far from the present town of Eden, North Carolina, the group observed a "prodigious flight of wild pigeons" which, Byrd added, "flew high over our heads to the southward."[15] The birds were coming from the direction of Chestnut Knob, a narrow ridge rising above the horizon about five miles from where the men were camped. Byrd appears knowledgeable about the bird's habits and his description of them is similar to other eighteenth- and nineteenth-century accounts. "The flocks of these Birds of Passage are so amazingly great, Sometimes," he wrote, "that they darken the Sky; nor is it uncommon for them to light in such Numbers on the Larger Limbs of Mulberry Trees and Oaks as to break them down." He added that the largest flocks caused "vast Havock among the Acorns and Berries of all Sorts, that they waste whole Forrests in a short time, and leave a Famine behind them for most other Creatures" (see plate 3).[16]

The passenger pigeon was likely more friend than foe to the American chestnut, although the ultimate determination requires more information than we currently possess about the two keystone species. The largest flocks consumed thousands of bushels of mast in a single day, making it unlikely that any regeneration of trees occurred the following spring.[17] However, when the birds did depart from oak-chestnut forests, they left behind huge quantities of phosphorous-rich manure. Byrd himself witnessed the phenomenon, noting that "under Some Trees where they light, it is no Strange thing to find the ground cover'd three Inches thick with their Dung."[18]

Over time, the acidic and mineral-rich deposits were beneficial to chestnuts, encouraging their blossoming and further reproduction. The spiked burrs covering the nuts also made it less likely that birds would roost in the upper canopy, thus minimizing destruction to large limbs and branches. The destruction of other tree species by flocks of pigeons was also advantageous to the American chestnut, especially where beech and oak trees shared the same landscape. And, as noted in chapter 1, the gregarious birds were capable of transporting chestnuts to distant locations when they fell prey to predators en route to evening roosts. Thus, as long as passenger pigeons did not return to the same location each year, the birds had only a minor impact on the total number of chestnut trees in the eastern deciduous forest.

The noted permaculturalist Peter Bane even suggests that the nineteenth-century demise of the passenger pigeon had a direct influence on the health of the American chestnut, making the tree more susceptible to diseases such as chestnut blight. According to Bane, nutrient flows from pigeon droppings were as high as "10 billion kilograms [22 billion pounds] per year."[19] Pigeon excrement also contained high amounts of phosphorous, including the mineral zinc, a substance that allowed isolated trees, Bane holds, to remain blight-free. While Bane's hypothesis is not empirically testable, it is intriguing nonetheless, and underscores not only the complexity of North American ecosystems but the challenges facing restoration ecologists. In some cases it may not be possible to reintroduce rare plants or animals into modern landscapes, as conditions no longer support their successful growth and reproduction.

Byrd makes no further mention of chestnuts in *The History of the Dividing Line*, but the trees do appear in *A Journey to the Land of Eden*, a shorter tract chronicling his tour of a twenty-thousand-acre parcel Byrd purchased from the North Carolina commissioners, lands he hoped to sell to Swedish immigrants. On October 11, 1733, Byrd and a small entourage bivouacked at the end of a peninsula at the confluence of the Dan and Smith Rivers, a location that is today within the city limits of Eden, North Carolina. After getting a freestyle swimming lesson from his Indian guide (Byrd knew only the breast stroke), he observed several large trees full of chestnuts near the riverside encampment.[20] His subsequent remarks are reminiscent of those Adriaen van der Donck made a century earlier, who complained about the method in which Indians and colonial settlers gathered chestnuts. "Our men were too lazy to climb the trees for the sake of the fruit, but, like the Indians," wrote Byrd, "chose rather to cut them down, regardless of those that were to come after. Nor did they esteem such kind of work any breach of the Sabbath, so long as it helped to fill their bellies."[21]

Although Byrd's plans for developing the Eden estate never materialized, the failed venture did not stop the flow of individuals from entering the backcountry. By 1740, both fur traders and longhunters were exploring this hunter's paradise, seeking deer, elk, and buffalo hides, as well as beaver, wolf, fox, panther, and raccoon pelts.[22] All were astounded by the number of game animals in the upland interior, including Thomas Walker, a Virginia physician who owned an estate near Monticello and later served as mentor to Thomas Jefferson. Walker was also a close friend of Thomas Jefferson's father, Peter Jefferson, the surveyor and planter who extended

Byrd's survey line ninety miles to the west.[23] In 1750, while surveying for the Loyal Land Company, Walker crossed over the Appalachian divide at Cumberland Gap, becoming the first Englishman to do so. The journal describing his journey not only mentions numerous large trees, but also incredible numbers of large game animals. The final journal entry, penned July 23, 1750, reads, "We killed in the Journey 13 Buffaloes, 8 Elks, 53 Bears, 20 Deer, 4 Wild Geese, about 150 Turkeys, besides small Game." Perhaps even more astonishing is Walker's statement that "we might have killed three times as much meat, if we had wanted it."[24]

There is little doubt that this bounty of game was sustained by mature oak-chestnut forests, as no other landscape could have supported such a density of large mammals. Evidence for such claims is found in the mountains of southwestern Pennsylvania, an area also known for its chestnut-dominant woodlands. In fact, for much of the eighteenth and nineteenth centuries, the mountain chain stretching from Morgantown, West Virginia, to Nanty-Glo, Pennsylvania, was known locally as Chestnut Ridge.[25] According to historian Samuel P. Hildreth, who was born in 1783, the ridgetop was named for "the immense forests of chestnut trees that clothe its sides and summit." In 1814, he stated that the sandy and rocky soils of the mountains were "so exactly adapted to the growth of this tree, that no part of the world produces it more abundantly." In recounting the life of John Rouse, who traveled across Pennsylvania in 1788, Hildreth remarked that "in fruitful years," residents drove their hogs to the top of Chestnut Ridge to "fatten on its fruit." Black bears, wild turkeys, elk, and deer were also frequent visitors to "this nut-producing region," making it a prime location for big- and small-game hunting. "The congregation of wild animals, on this favored tract," noted Hildreth, "made it one of the most celebrated hunting grounds, not only for the Indians, but also for the white man, who succeeded him in the possession of these mountain regions."[26]

Similar observations were made by George W. Featherstonhaugh, who passed through Virginia's chestnut-dominant forests in the 1830s. "When mast is plentiful," observed Featherstonhaugh, deer retreat to the mountains to consume "chestnuts and the acorns of the white oak." "The panther" also withdrew to such areas, he added, "not because he eats chestnuts, but because he knows that he shall find deer there."[27] Bears were also regularly drawn to the mast-laden woodlands; so much so, that backcountry residents affectionately called them "bear gardens."[28] Even the American bison was a beneficiary of the mast crop, as the animals also browsed on chestnuts and acorns during fall and winter months.[29]

It is little wonder such areas became prime destinations for Daniel Boone and John Findley, individuals who would make their life fortunes on the ecological bounty of the backcountry wilderness. Bear, deer, and bison not only supplied meat for the pioneer table, but also skins and furs—commodities that drove the entire frontier economy. In fact, for the period comprising the first half of the eighteenth century, no other export product exceeded furs and skins in economic value (in Europe, deer hides were transformed into hats, gloves, shirt-sleeve linings, book bindings, purses, coat linings, and even men's breeches).[30] By 1765, when the deerskin trade peaked in the southern colonies, as many as a half million leather hides, valued at some four million dollars, were annually leaving Atlantic ports.[31] The claim that the fur and skin trade played a major role in the settlement and evolution of early America, if not determined the "course of empire," as one author stated it, is hardly hyperbole.[32] The trade in furs and skins was instrumental in the founding of such institutions as the College of William & Mary and directly supported the growth and development of many frontier trading posts, including Augusta, Georgia, and Columbia, South Carolina.

The American chestnut, as a key component of the oak-chestnut forest, influenced frontier history in a unique and important way. After the long-hunters and Native Americans intensified their participation in the deerskin and fur trade, they increasingly relied on the interior uplands, where oaks and chestnuts comprised 50 percent or more of the forested landscape, as their source of animals.[33] As a result, the "oak-chestnut forest region"—as forest ecologist E. Lucy Braun famously called it—was a major catalyst driving frontier expansion.[34] Chestnut-laden forests provided sustenance and refuge for millions of fur-bearing animals and, in doing so, supplied the natural capital upon which the frontier economy was based. Yes, the trees and nuts were extractable and merchantable commodities, but they were also essential components of a dynamic, living landscape.

Native Americans definitely shared this view, as the trees were vital to their existence, providing not only nutritional sustenance but mast for wild game and the raw materials for homes and dwellings. However, for some indigenous groups, the American chestnut was a part of their very cultural identity. The trees and nut harvest had become the subject of tribal folklore and even marked the passage of time. And in one Native American creation story, chestnuts were *the first living thing* brought forth by the Creator.

in the natural state and also in the prepared

According to interviews conducted by nineteenth-century ethnologist Albert Gatschet, the Choctaws believed that when the earth was first made, "the hills were formed by the agitation of the waters and winds on the soft mud." The earth then brought forth "the chestnut, hickory nut and acorn; it is likely that maize was discovered, but long afterward, by a crow."[35] For other Native American communities, the chestnut harvest was so important it corresponded with their lunar calendars. The Creeks, for example, celebrated the event during the Big Chestnut Moon in October, whereas the Little Chestnut Moon (September) was dedicated to harvesting the smaller and earlier-bearing chinquapin.[36] The Chickasaw and Natchez observed the Chestnut Moon during the month of February (the twelfth of the year, by their reckoning), when they feasted on the nuts and offered the "delicacies to their honored Chief, the Great Sun."[37] Although both groups considered the twelfth moon to be "of chestnuts" (the New Year began in March), the timing of the festival implies the nuts had been dried and stored for later use.[38] As Antoine-Simon Le Page du Pratz observed in his *History of Louisiana*, first published in 1758, "the fruit had been collected a long time ago, but nevertheless this month bears the [chestnut] name."[39]

Although Native American folklore involving chestnuts was recorded after European contact, it is assumed—due to the conservative nature of culture change—the stories originated much earlier. The fact they were part of oral tradition is evidence the trees did, in fact, enjoy special status among Native Americans.[40] This is especially true for the Cherokees, who include the tree in several folktales and legends.[41] In a story entitled "The Bear-Man," for example, a hunter pursues a black bear he is unable to kill, despite having injured it with numerous arrows. The bear, who apparently possesses magical powers, eventually turns to the hunter and says, "It is of no use for you to shoot at me, for you can not kill me. Come to my house and let us live together."[42]

The hunter agrees and follows him into a cave where a bear council is formed to address the problem of food scarcity. Two bears announce to the group that they had found a place "where there were so many chestnuts and acorns that mast was knee deep." To celebrate the news, they hold a ceremonial dance. Afterward, the hunter enters the den of his new comrade and asks him for something to eat. Standing on his hind legs, the bear rubs his belly and gives the hunter "both paws full of chestnuts." After the bear produces more nuts, the hunter decides to live with the bear all winter. In the early spring, however, the bear tells the man that hunters "will come to this cave and kill me." When the hunters arrive, they kill the bear but find the hunter, whose hair has grown long and shaggy, alive inside the cave. Returning to the village, the man refuses food and drink for seven days. Although his wife attempts to revive him, the man dies before the end of the week, as he "still had a bear's nature and could not live like a man."[43]

Another noteworthy folktale is the Iroquois fable "The Lad and the Chestnuts." The story begins with a small boy who is puzzled by his older brother's desire to eat only after the boy has fallen asleep. He later learns the older brother has access to a magic kettle that, when rubbed, produces an infinite number of chestnuts. When the younger brother finds the kettle and learns how supper is prepared, the magic spell is broken. As a result, the elder brother must find a new source of nuts or die of starvation. When the younger brother implores how he might save his brother's life, he asks, "Where did you get the chestnuts? Let me go and seek some for you!" The elder brother answers by stating that he got them from a tree far, far away, where their "forefathers gathered chestnuts long ago." He warns him, however, that the tree is guarded by a white heron and several women with war clubs who prohibit anyone from gathering the nuts. Undaunted, the younger brother makes the journey, finds the tree, and "[fills] his bag with chestnuts." Escaping the white heron and the all-female sentries, he returns home to find his brother still alive. "You have done me a great favor," states the elder brother. "Now I shall be well, and we will be happy."[44]

Chestnuts are mentioned in other Iroquois folktales, including ones involving "flying heads"—large mythological creatures possessing wide mouths, stringy hair, and rows of sharp teeth.[45] According to one version, when the Iroquois roasted chestnuts, they ate them very slowly, "making loud exclamations of how delicious the nuts tasted."[46] The flying head, upon seeing the chestnuts, would then swoop down to grab a mouthful of nuts, but instead would eat the red-hot coals, causing it to burst into flames.[47] In another version, a woman roasts chestnuts over an open fire with her small dog. After she pulls chestnuts from the fire and eats them, the flying head, who believes the woman has swallowed the coals, becomes terrified and flies away, never to return (see fig. 4.1).[48]

Although the true meaning of these stories is debatable, there is little doubt they are a reflection of the tree's importance to the Iroquois and Cherokee people. Chestnut gathering was among their most important foraging activities, an annual ritual involving the collection and processing of nuts as well as the making of unleavened bread. Even raw nuts remained in favor among native groups, due to their availability when traveling or hunting. They were also bartered for various and sundry items at local trading posts, a practice continuing well into the twentieth century. A report published by the U.S. commissioner of Indian affairs noted that for the Cherokees of North Carolina "an abundant crop of chestnuts is considered a great boon, the nuts being used very generally as food, both in the natural state and also in the prepared, 'chestnut bread' being a staple."[49]

Many Native American groups used very similar words for both the tree and nuts, a reflection of their universal popularity and appeal. In Cyrus Byington's *A Dictionary of the Choctaw Language*, published in 1915, chestnut is *oti*, or *uti*, and the word for the tree is *otapi*.[50] Virtually the very same word (*otape*) was also used by the Muskogeans, as noted in chapter 3. In the Cherokee language, the word for chestnut is *oot-te* or *uti-ti*, but the tree itself is known as *tili-tlu-gu-i*.[51] However, according to James Adair, who lived among the Cherokees and Catawbas in the mid-eighteenth century, the fires used to boil chestnuts were also called *oo-te* by both groups.[52] The Onondaga Nation used *oh-ell-i-ata* (prickly burr) when referring to a chestnut tree, and, like other indigenous groups, used the phrase for the tree when referring to the individual nuts.[53] However, for the Delaware and Munsee, the word for chestnut was *woa-pim* or *wa-pim*, and a longer term (*woa-pi-min-schi*) referred to the individual tree.[54] Among the Narragansett, a chestnut tree was referred to as *wom-pi-mish*.[55] Curiously, the town of Anderson, Indiana, was called Woapiminschijeck, a Delaware word meaning "chestnut tree place," by its very first inhabitants.[56] Located in the north-central part of the state, Anderson is outside the accepted historic range of the species; however, as linguist Michael McCafferty has argued, there is little doubt the town was once "endowed with an impressive stand of chestnut trees."[57]

Native American place names identify the locations of chestnut stands and groves, although some of these names are no longer in use. Those with documented provenance include *Eutacutachee* Creek (Rankin County, Mississippi), from the Choctaw *oti* (chestnut) and *hohtak oshi* (small pond); *Otapasso* Creek (Pike County, Mississippi), from the Muskogean

FIG. 4.1. Iroquois folktale. A woman eats roasted chestnuts to ward off a mythical "flying head." Henry Row Schoolcraft, *Archives of Aboriginal Knowledge*, vol. 1 (Philadelphia: J. B. Lippincott, 1860). Courtesy of the Library of Congress.

otapi (chestnut tree) and *oshi* (little); *Topisaw* Creek (also in Pike County, Mississippi), from *otipisa*, a corruption of *otapi* and *oshi*; *Wampmissic* (between Yaphank and Manorville, Long Island) from the Narraganset phrase meaning "place of chestnut trees"; *Ganyestaageh* (Perrysburg, New York), from the Seneca phrase meaning "hill of chestnuts"; *Wampecack* Creek (Washington County, New York), from the Narraganset phrase meaning "place of chestnuts"; and *Wombemesiscook* (Hardwick, Massachusetts), from the Delaware phrase meaning "place of the chestnut trees."[58]

The inclusion of chestnuts in Native American language, folklore, and geography is conclusive evidence the trees (and the landscapes they inhabited) had special meaning for indigenous communities. By the second half of the eighteenth century, however, the oak-chestnut forests were themselves changing, largely as the result of the actions of European traders. The peltry trade had a significant impact on wild game populations, causing the extinction of bison and elk in the Appalachians and the elimination of white-tailed deer over large areas.[59] This placed Native Americans in an extremely precarious position, as they had become increasingly dependent on European goods, items purchased largely on credit. Increasing debts, combined with a lack of game to pay off those debts, forced many Native Americans to abandon the hunt for the raising of livestock, corn, and wheat. This was particularly true for the Cherokees, who began adopting European agricultural practices as early as the 1770s.

Initially, the trees benefitted from the absence of deer, elk, and bison, as more chestnuts were able to sprout in the forest understory. Deer ate not only chestnuts, but the leaves of young seedlings and saplings, a problem well documented in twenty-first-century orchards.[60] Later, when cattle, hogs, and sheep were introduced into the backcountry, these mast surpluses were largely eliminated, as happened in New England a century earlier. In the tree's southerly range, the fur trade had a far more detrimental impact, particularly where human-set fires were used to hunt deer or bear. In Alabama and Georgia there is evidence such fires resulted in an increase in pine and other tree species preferring drier environments.[61] Thus, in those areas where southern pines were common and fires more prevalent, chestnuts declined precipitously in numbers.[62]

Globally, however, the American chestnut suffered few appreciable losses during the first half of the eighteenth century and even increased in number in some areas. Witness-tree data provides a fairly accurate picture of their location and prevalence, not only in New England but across the entire range. Some of the findings are surprising, as the species was not as common in places generally believed to have large numbers of trees and was more prevalent in locations thought to have only a few scattered stands. In New York, for example, concentrations were relatively low across the entire state, with most areas having densities of less than 5 percent.[63] At a location near the Shawangunk Mountains west of Poughkeepsie, however, 40 percent of all trees were chestnut.[64] In the Connecticut township of Norfolk, 5 percent of the woodlands harbored chestnut, whereas in Goshen, only ten miles away, 17 percent of all trees were identified as *Castanea dentata*.[65] In the Allegheny mountains section of what is now the Monongahela National Forest in West Virginia, only 4 percent of the standing timber was chestnut, but in the nearby coalfields portion of the forest, 18 percent of all trees were identified as such.[66] And although chestnut densities seldom rose above 5 percent across southern Ohio, it was the sixth most common species, ranking ahead of pine, tuliptree, chestnut oak, red oak, and other unnamed trees.[67]

Despite their uneven distribution across parts of eastern North America, settlers always noted chestnuts' presence, and, if given the opportunity, lived in areas where the trees were most abundant. Backcountry life demanded the use of chestnut timber, even causing some individuals to select homesites based on its local availability. In some places the trees were remarkably abundant and, because they regrew after felling, paid dividends for decades afterward. Although the American chestnut was still an important food source for Native Americans and sustained the wild game upon which they were dependent, for many backcountry residents it was the wood and its byproducts that made the trees so important to frontier life.

the place is all of good fence

At the time of the American Revolution, humans were having a much more noticeable impact on the American chestnut, causing a reduction in numbers nearly everywhere the tree was present. The most visible declines occurred along the Mid-Atlantic, where trees of all kinds were removed for pasture or tobacco, corn, and wheat cultivation.[68] By 1780, backcountry farmers were using chestnut for home, barn, and outbuilding construction, causing appreciable losses in some areas. However, most of the chestnut removal was for fence making, as free-ranged livestock could destroy large crops in a matter of hours. Fence construction required considerable amounts of timber, as no fewer than forty-five hundred rails were needed to enclose a ten-acre field of corn or tobacco, a task requiring the removal of forty large chestnut trees.[69] Even border fences, which were not as high or substantial, required a considerable amount of timber. According to environmental historian John R. Wennersten, in the Chesapeake Tidewater as many as sixty-five hundred rails were needed to build a single mile of fencing.[70]

Most of the enclosures were "wormfences," as the Anglican minister Hugh Jones famously called them.[71] Worm- or snake-fences were made of split chestnut rails stacked atop each other in an interwoven zigzag pattern around cultivated fields or along property lines. Although some chestnut-rail fences held as many as fourteen rails in a stack, the majority, as architectural historian Vanessa Patrick documented, were "five to eight feet high and had six to nine rails, each about eleven feet long."[72] For extra support, two additional stakes were placed in the ground at the end of each stack, crossing diagonally over the top rail. Finally, a larger and longer rail "termed the rider" was placed above the intersection of the two stakes, which, as one eyewitness observed, "locks up the whole, and keeps the fence firm and steady."[73]

Because of the additional wood and labor required in their construction, rail fences were among the most valued assets on the eighteenth-century farmstead. In 1751, when John Gilleylen published a notice in the *Pennsylvania Gazette* advertising the auction of his Southampton farm north of Philadelphia, he boasted "the place is all of good fence . . . done with new chestnut rails and stakes, which were purchased two miles distance."[74] Thomas Jefferson was perhaps the best-known proponent of the chestnut-rail fence, using them almost exclusively at his Monticello estate. In 1769, Jefferson ordered no fewer than eight thousand chestnut rails to be cut and hauled to the north end of the property in order to enclose a forty-five-acre orchard.[75] Incidentally, the "Southwest Mountains" bordering Monticello were then known as the "Chestnut Mountains," and more than one American president would use trees from the area to improve their plantations.[76]

Indeed, prior to the American Revolution, the chestnut worm-fence could be found from Maine to Georgia. They were most common in Virginia, however, especially in the Blue Ridge and neighboring foothills, where they dominated the entire agricultural landscape. Nineteenth-century historian Philip Alexander Bruce went so far as to call the chestnut worm-fence "one of the most familiar features of the Virginia plantation, a monument, like the fence law itself, of the perpetuation of agricultural conditions beginning with the very foundation of the Colony."[77] Chestnut worm-fences did have their detractors, as they required a considerable amount of otherwise arable land, a ten-foot-wide corridor along their entire course. They also removed a substantial amount of timber, a limitation offset by the fact that such fences lasted, where conditions were favorable, "sixty to seventy years."[78] In 1833, an anonymous observer in central New York noted that rails made from the cutting of chestnut timber "fifty-five years ago" were still in good condition.[79]

Besides fencing, the trees were used to build dairy, hay, and tobacco barns as well as corn cribs, smokehouses, and even storage sheds.[80] In Pennsylvania, the Swiss preferred chestnut when making dairy barns and used it to frame the entire structure.[81] Shakers also employed chestnut in barn building, but often mixed it with other woods.[82] When Scottish settlers built barns in New York's Mohawk River valley, they chose chestnut for the roof and ridge beams, hand-hewn timbers thirty feet long.[83] In Massachusetts, chestnut shingles were used to roof barns, as documented by Thomas Fessenden, editor of the *New England Farmer*. According to

Fessenden, when one of his readers was a young boy, he helped his father cut down a chestnut tree that was "converted into shingles and used in covering a barn." Impressed with the ability of chestnut trees to produce new growth after felling, Fessenden noted that thirty years later, the same person removed one of the sprouts that regrew from the original stump and obtained enough shingles "to replace the old ones."[84] While only a small percentage of barns were built *entirely* of chestnut during the eighteenth century, nearly all structures incorporated the wood in their construction and repair.

Larger structures were also built with chestnut, especially if they needed to be durable. A good example is Fort Presque Isle, a massive wood garrison 120 feet wide at its narrowest point. Constructed by the French in 1753 at present-day Erie, Pennsylvania, it was designed to house 150 men. Four additional buildings were erected inside the structure, including an enclosure for storing gunpowder. According to archival documents, the main fort was constructed with "chestnut logs, squared, and lapped over each other to the height of fifteen feet."[85] Including the interior buildings, it is quite possible that four hundred large mature trees were used to build the main structure. This also explains why French commander Chevalier Pierre-Paul Marin ordered troops to clear the forest one hundred yards beyond the main bastions, an area that included twenty acres of additional timberland.[86] Even more chestnut trees were felled to construct the road from Fort Presque Isle to nearby Fort Le Boeuf, a route taking travelers through "a continued chestnut-bottom."[87]

Chestnut timber was also integral in the construction of Jefferson's Monticello, including the stables and slave quarters adjacent the primary residence at Mulberry Row. In a memorandum to overseer Samuel Clarkson dated September 23, 1792, Jefferson ordered carpenters to build five small houses of "chestnut logs, hewed on two sides, and split with the saw, & dovetailed."[88] A similar log structure was built for Bagwell and Minerva Granger, the slave couple who lived along the southern bank of the James River.[89] The original one-room brick home where Jefferson and his new bride set up housekeeping—known as the South Pavilion or honeymoon cottage—also contained chestnut, including "650 feet of inch chestnut plank, and 520 feet of 2¼-inch [chestnut boards]."[90] The Monticello mansion itself was built with large amounts of chestnut trim, including the Palladian orders around the doors and windows. Initially, Jefferson even wanted the main portico supports to be made of the wood, as the original plans called for "eight columns, seventeen feet long, to be made of chestnut."[91] Although he eventually used bricks and mortar to construct the columns, the original roof, completed in 1809, contained thousands of chestnut shingles.[92] According to architectural preservationist John Mesick, the shingles were so intricately interwoven with sheet iron, copper, and lead that it was one of "the most structurally complex rooftops in North America."[93]

James Madison used substantial amounts of chestnut at nearby Montpelier, especially after 1790, when he was directly involved with its management. At the time, Madison made daily notations in his father's account books, which included instructions for overseers, such as when and where to clear land, build fences, and construct outbuildings. In one ledger, the young Madison instructed overseer Lewis Collins to "have the chestnut logs for the principal stable cut and halled [sic] that they may be ready to be put up whenever he is at leisure."[94] Although most trees were turned into fence rails and palings, or hewed into beams for cabin walls, a few were left completely untouched. Madison had a fond appreciation for trees and even admonished American farmers for their "injudicious and excessive destruction of timber [for] firewood."[95] Appropriately, three American chestnuts were allowed to grow and mature near the principal residence, becoming landscape fixtures during his adult life. The largest was located near what he called "the Temple," a gazebo-like structure built north of his home. In 1903, five years before it was inventoried by William duPont, the tree measured "forty-nine feet around its trunk" (fifteen feet in diameter) (see fig. 4.2).[96] It is possible the tree was two centuries old when Madison inherited the estate from his father in 1801.

Evidence for the tree's advanced age is found in a dendrological study conducted by forester Thomas Dierauf, who cored numerous trees in the Landmark section of the Montpelier estate. Dierauf found that several trees at the location had been "released" in 1670, including a white oak and an unnamed hickory. The white oak, which measured only 35 inches in diameter, was 336 years old in 2009, and possessed an average annual growth rate of ten rings per inch. The hickory was much smaller in diameter (30 inches), with an annual growth rate of eleven rings per inch. However, the largest tree, a red oak measuring fifty-one inches in diameter, grew at a rate of four rings per inch, making its birth or release date 1776.[97] Madison's Temple chestnut was more than seven feet in diameter (each single trunk), making it 360 years old in 1908 (using the estimate of four rings per inch of

FIG. 4.2. The "Temple" American chestnut, Montpelier Station, Virginia, c. 1898. From the duPont Family Scrapbook Collection. Courtesy of Montpelier, a National Trust Historic Site.

growth) and its release date, 1548.[98] The double trunk suggests the Madison chestnut was originally cut down by Native Americans or grew back as joined coppice sprouts after the original tree was killed by fire or late frost. Curiously, a forest history of the estate found evidence of land-use disturbance in 1700, just prior to the purchase of the property by the Madison family and a full century after the Temple chestnut had developed into a mature tree.[99]

The elimination of competing species at Montpelier was certainly beneficial to *Castanea dentata*. By 1915, 10 percent of the estate contained chestnut, making it one of the most common trees on the plantation grounds.[100] However, forest disturbance did not always increase their numbers as suggested by the Montpelier example. In most cases, chestnut only became more plentiful if deforestation occurred without additional human interference, such as agriculture or livestock grazing. The author of a forest history conducted at Great Meadows, Pennsylvania, the location unsuccessfully defended by George Washington in 1754, thus rightly concluded that the chestnut increases at Great Meadows indicated that "the stumps on the hillside were neither grubbed out nor burned after initial clearance."[101] If the stumps had been removed or burned, the trees would have declined considerably in number, as happened in the coastal townships a century earlier.

The Montpelier case notwithstanding, the American chestnut became less prevalent in many backcountry locations during the second half of the eighteenth century. Livestock grazing, land clearance, barn construction, and fence building all took their toll on the trees, particularly those closest to human settlements. However, there were other forces removing large numbers of chestnut trees from the wooded landscape, among them the burgeoning iron ore industry. By 1760, nearly every British colony possessed a large furnace or forge that required the use of charcoal. At that time, wood charcoal was the most common fuel used in iron smelting; in fact, it was so closely associated with iron production that the end product was often referred to as "charcoal iron."[102] After 1770, dozens of such establishments operated in the backcountry, drawn there not only by the ferrous ores, but the large amounts of timber needed in the iron-making process.

better coal than green wood

Eighteenth-century iron making required not only considerable amounts of wood, but also skilled colliers, individuals employed specifically to make the charcoal. The process involved removing all timber from around the forge or furnace and then burning the resulting cordwood under tightly controlled conditions. Even a small furnace could consume four hundred acres of mature woodlands in a single year (640 acres = one square mile), although even more trees were needed if they were smaller in size or growing in thin stands.[103] In 1732, when William Byrd traveled to Fredericksburg, Virginia, to educate himself about the intricacies of iron making, he was told two square miles of wood were needed annually to "supply a moderate furnace."[104] Over time, the effects of charcoal making on the surrounding woodlands were calamitous, as the colliers spared few if any trees in the charcoal-making process. When touring the Union Iron Works property in New Jersey, the German-born Johann David Schoepf learned the facility had exhausted a forest of twenty thousand acres in "twelve or fifteen years," and had lately abandoned it due "to the lack of wood."[105]

Although trees resprouting in cutover forests could in theory be used again for charcoal when they were four inches in diameter, the insatiable appetite of the furnace nearly always outstripped local supplies. In 1750, in order to ensure a steady supply of charcoal to its largest smelters, the operators of the Baltimore Ironworks had to purchase an additional thirty thousand acres of timber.[106] Although colliers favored certain kinds of wood—hickory, oak, or maple—over others, chestnut was always among the mix of trees chosen for the coaling process. In fact, nineteenth-century furnace operators in Connecticut, Massachusetts, Pennsylvania, and Maryland reported chestnut as among their most favored woods, second only to oak in many locales.[107] At that time, the source material was likely third- or fourth-generation "chestnut coppice," saplings specifically grown to replenish furnace stocks after being harvested.[108] Forest historian Gordon G. Whitney has even suggested that in areas where charcoal production was suspended long enough for the successful regeneration of timber, such practices actually increased the prevalence of chestnut.[109] While likely in places where forests were managed specifically for that purpose, in areas where the making of iron was followed by human settlement or agriculture, the species was eliminated from the landscape.

In addition to charcoal production, the species also declined in numbers as a result of climate change, or, more specifically, the Little Ice Age. For much of the eighteenth century, cooler temperatures gripped much of eastern North America, causing the Chesapeake Bay and Boston Harbor to freeze over, deep snows across New England, and late frosts along the entire eastern seaboard.[110] For several decades, the severe winters caused the price of food to rise and many to worry about the availability of firewood. After a severe cold spell in April 1770, the tidewater planter Landon Carter made the following prediction: "We now have full ¾ of the year in which we are obliged to keep constant fires; we must fence our ground with rails, build and repair our houses with timber and every cooking room must have its fire the year through. Add to this the natural deaths of trees . . . [and] in a few years the lower parts of this Colony will be without firewood."[111]

Although the environmental impact of the Little Ice Age on chestnuts remains understudied, evidence suggests the cooling climate did indeed reduce their numbers. This includes a remark made by Thomas Jefferson, who told the Agriculture Society of Albemarle, Virginia, that a May frost at Monticello had totally destroyed "the forest trees at the summit and at the foot of the mountain."[112] The fallen timber was used for Jefferson's most ambitious fence project, a partition of "chestnut pales" three-quarters of a mile long that was placed around his mountaintop orchard.[113] In a memorandum written to overseer Elisha Watkins, Jefferson specifically asked that the material for the palings "be got in the high mountain," implying he wanted dead or decaying trees for the fencing material.[114] Further evidence the palings were sourced from the mountaintop is a letter Jefferson wrote to overseer Edmund Bacon on December 8, 1806. "I think there is a great deal of fallen chestnut on the mountain which will make better coal than green wood," stated Jefferson. "I should think that you might get the rails for the upper end of your long fence on the upper mountain also."[115]

The late-spring freezes of the Little Ice Age not only killed chestnuts but slowed their northward progression, perhaps eliminating some stands in New Hampshire, Maine, and southern Canada.[116] However, trees in the southern part of the range were most affected, as they, like the trees surrounding Monticello, were fully leafed out at the time, thus guaranteeing their demise.[117] The tree death by frostbite did not completely remove chestnuts from the impacted areas, as some trees continued to send up coppice sprouts from the remaining stumps. However, if the trees did survive the colder temperatures, their weakened state and lower mast production would have allowed other species to become more dominant in the forest canopy, especially if the cold snaps were followed by hot or dry summers.[118] This appears to have been the case in southern Alabama, where resident Morgan D. Jones observed fallen chestnuts "five feet or more in diameter" along the hilly slopes near the Pea and Conecuh Rivers.[119] Although Jones had seen dead trees and a few living ones there as late as the 1870s, his stepfather, born in 1812, recalled the biggest chestnuts "were larger and taller than the long-leaf pines nearby, all died in one night as the result of a severe freeze in May." When asked what year the frost occurred, Morgan's stepfather could not exactly remember, but believed it was "in the 1830s."[120]

The American chestnut diminished in numbers during the late eighteenth and early nineteenth centuries, but did so as a result of both natural and human causes. Human activities had perhaps the greatest impact, as geographer Navin Ramankutty and ecologist Jonathan A. Foley documented in 1999. According to Ramankutty and Foley, the amount of land used for agriculture doubled between 1750 and 1800, and many of the woodlands cleared for agriculture contained chestnut.[121] The Chesapeake Tidewater, which had half of its area in standing trees a century earlier, was by 1770 "lacking or scarce in timber."[122] In some locales, township growth had the greatest impact on the species; in others, the iron furnace and forge caused the most visible declines, particularly after the industry became more widespread and its resource extraction methods more intensive. The removal of trees for fence construction also reduced their numbers, although that practice alone did not eliminate them from large areas. In locations where livestock was allowed to continuously graze, or where fields and forests were converted to agriculture, the tree disappeared entirely from the landscape.

This does not mean the American chestnut was scarce or even vanished from large swaths of territory during the eighteenth century. The trees were still common on the streets of colonial townships and remained plentiful in most village woodlots. In 1763, a thirty-acre estate in the heart of north Philadelphia's Chestnut Hill was able to advertise for sale "ten acres of the finest chestnut timber in the county."[123] Even the Chesapeake Tidewater possessed large isolated stands, a fact corroborated by the auction of a plantation near Gloucester, Virginia. On November 7, 1777, the *Virginia Gazette* claimed the property had "sufficient fund of chestnut to supply the plantation with rails forever."[124] As late as 1789, the native trees were still an integral part of Mount Vernon, as Polish playwright Julian Ursyn Niemcewicz noted when touring the estate with George Washington.[125] The trees were also abundant in southern Canada, including the Essex County township of Mersea, where they comprised 16 percent of all trees.[126] Interestingly, the Mersea chestnuts were associated with "good land," meaning their presence signified agriculturally productive soils.[127] And in Maine, the species was found as far north as Conroy Lake near Monticello, a North Woods location near the forty-sixth parallel.[128]

If the range of the American chestnut was incrementally shrinking during the late eighteenth century, its cultural uses were expanding. By the end of the American Revolution, few individuals living in the original thirteen colonies did not daily observe a tree or come into contact with its wood or lumber. In New England, chestnut timber was frequently used in home construction, where it was transformed into flooring, wall paneling, attic beams, and purlins.[129] It was also a favored wood in furniture making, especially for the interior blocking of dining tables, wooden clocks, lowboy chests, and sideboards.[130] By the end of the eighteenth century, craftsmen from Providence, Rhode Island, to Hartford, Connecticut, kept a continuous supply in their home workshops.[131] Chestnut was also a major component in Windsor chairs, as the wood had become the preferred material for their curved seat-bottoms. When the noted Connecticut Windsor chairmaker Ebenezer Tracy passed away in 1803, more than seven hundred board feet of chestnut lumber was inventoried in his private estate.[132]

In addition to supplying wood for furniture and other household items, the tree continued to be valued as a nut producer. In fact, a profitable interstate nut trade had developed well before the Revolutionary War, which included the maritime shipment of chestnuts to the southern colonies. Chestnuts were even among the cargoes confiscated by British Loyalists during the Revolutionary War, as evidenced by a New England sloop impounded in Norfolk loaded with "cider, potatoes, chestnuts."[133] By the war's end, it would not be hyperbole to suggest that most families in the original thirteen colonies consumed a bushel or more of nuts annually.

However, by the first decade of the nineteenth century there were noticeably fewer trees in the largest townships, particularly those along the eastern seaboard. As a result, chestnut shortages were advertised in local newspapers, an indication that demand often outstripped supply.[134] The species also began disappearing from portions of the southeastern United States as the result of a then-unknown pathogen. Although the species would experience a major setback as a result of the deadly disease, the tree remained a common landscape feature across much of its range well into the mid-1800s. And in some areas, it actually increased its presence, becoming more prevalent in parts of New England, the Mid-Atlantic, and southern Appalachians. In many respects, the American chestnut was still a species on the move.

CHAPTER 5
cash will be paid if delivered soon

At the beginning of the nineteenth century, America was a much more urban place, having tripled its population in just two decades. By 1810, the country possessed a dozen townships with ten thousand or more individuals, including Baltimore, Philadelphia, and New York City, which together accounted for two hundred thousand people.[1] The largest townships greatly increased food demands, especially if late-spring frosts or hot and dry summers induced crop failures. Because railroads were virtually nonexistent and canals inadequate for most intercity commerce, large quantities of foodstuffs had to be imported from Europe.[2] Trade with communities in the interior also remained limited due to poor roads and lack of major thoroughfares. With the decrease in contiguous forests near and around coastal seaports, supplies of seasonally favorite commodities like chestnuts were unable keep up with demand.[3] Grocers note their availability as early as 1796, among them John Bartes of Hartford, Connecticut. His public notice in the January 11 edition of the *Connecticut Courant* listed "Walnuts and Chestnuts" among the items sold at the establishment.[4]

Up and down the eastern seaboard, shopkeepers were beginning to advertise their desire to purchase chestnuts, placing weekly or monthly "wanted ads" in local newspapers. Among the earliest notices was one printed in 1798 in the Newark, New Jersey, weekly *Centinel of Freedom*. In their solicitation the two buyers—Joseph and James Brown—stated that they intended to ship barrels of chestnuts to southern ports for the holiday season.[5] A similar notice was placed in the *Providence Phoenix* by William Peckham, a Providence, Rhode Island, storekeeper. The headline of the single-column ad reads "CHESTNUTS WANTED." Peckham's desire to purchase nuts was so great that he was even willing to pay cash for them: "CASH given for CHESTNUTS," announced the notice, dated October 18, 1806 (see fig. 5.1).[6] In larger cities, demand was so great that sellers could ask a more premium price, as the French botanist François André Michaux noted. In the original French edition of his *North American Sylva* (1810), Michaux remarked that "on the streets of New York, Philadelphia, and Baltimore, where there exists a better market for them, they are sold at nearly $2.75 per bushel."[7]

A decade later, American grocers appear to have had less trouble buying and selling chestnuts, as merchants simply listed them among the various and sundry items on hand. In 1816, storekeeper Richard Hosier of Nantucket, Massachusetts, for example, notified the public that he had for sale "Almonds . . . Filberts, Chestnuts, Black Walnuts."[8] Other ads might announce the actual amount of nuts in the shopkeeper's inventory, which, at the beginning of an ad cycle, was substantial. On December 22, 1818, the proprietor of the No. 127 Fly-Market in Manhattan advertised that he had one hundred bushels of chestnuts available for purchase. The ad ran biweekly until January 26, 1819, presumably when stocks ran low or completely sold out.[9]

For larger establishments, such as those owned by wholesalers or shippers, the demand for chestnuts was great enough that inventories often exceeded several hundred bushels. On October 27, 1818, Benjamin Fowler and Sons asked for the delivery of no fewer than three hundred bushels of chestnuts and walnuts to their Hartford, Connecticut, establishment. Their eagerness to acquire the nuts is reflected in the ad's closing line, "cash will be paid if delivered soon."[10] A notice published in the *American Advocate* of Hallowell, Maine, suggests that chestnuts were among the fare consumed during the Thanksgiving holiday, which at that time was celebrated only sporadically after ad hoc presidential degrees. On November 24, 1819, the storekeeper J. M. Ingraham announced he had just received for "Thanksgiving" an assortment of fruits and nuts, including "Citrons—Walnuts—Chestnuts, etc."[11]

By 1820, chestnuts were found in grocery stores and outdoor markets from Augusta, Maine, to Augusta, Georgia, where they were displayed in barrels alongside hickory nuts and black walnuts.[12] In some locations, "Spanish" chestnuts were sold alongside the native variety, further evidence local demand exceeded supply. Spanish chestnuts, however, were often advertised later in the season or with other products imported from Europe. They were also kiln-dried before shipping, although some may have been harvested locally, as they were introduced to the United States as early as 1780.[13] In 1815, when a farm near Newark, New Jersey, was advertised for sale in the New York *Evening Post*, one of its selling features was the more than a thousand native trees on the premises that had been grafted with Spanish chestnut scions. According to the seller, although the trees had

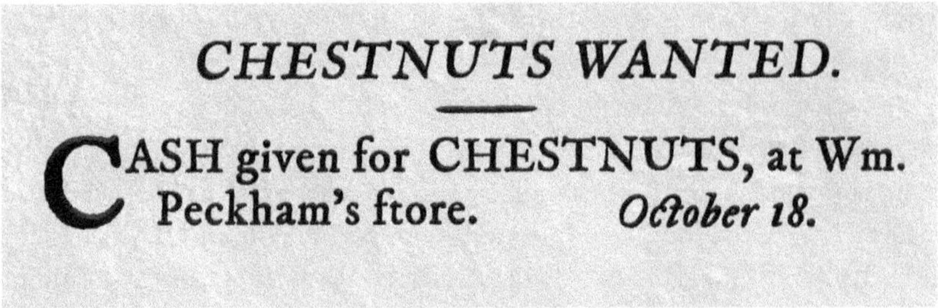

CHESTNUTS WANTED.

CASH given for CHESTNUTS, at Wm. Peckham's ftore. October 18.

FIG. 5.1. "Chestnuts Wanted" advertisement, *Providence Phoenix*, October 18, 1806. Courtesy of the Library of Congress.

been grafted "only 5 or 6 years ago—many of them will probably bear this year."[14]

As demand for chestnuts increased during the first decades of the nineteenth century, supply chains expanded, reaching far into the backcountry interior. As early as 1796, Benjamin Hawkins encountered a Cherokee woman near the north Georgia town of Pine Log who had just returned by foot from Augusta, where she had traded a bushel and a half of chestnuts for a used petticoat.[15] By 1810, most chestnuts were brought to local markets in horse-drawn wagons, usually after a day's journey over poorly maintained roads. John Hogg, who owned a North Carolina grocery store chain, used his Wilmington commission house to procure farm products that were bartered for merchandise or sold for cash. According to historian Lewis Atherton, when wagons arrived at the commission house they were loaded with goods from the backcountry, including "deerskins . . . venison hams [and] chestnuts."[16] The fact that merchants continued to advertise their desire for chestnuts in local newspapers suggests that the bulk of nuts sold in early nineteenth-century markets were supplied by individuals living in the readership coverage area, a distance generally less than forty miles.[17]

The increasing demand for chestnuts, coupled with seasonal shortages in the fastest-growing population centers, also explains why American horticulturalists accelerated their efforts to grow imported varieties. Many believed European chestnuts possessed more commercial promise, as the larger nuts meant greater financial rewards. In fact, George Washington was so enthralled with a Spanish variety that he continued to send nuts to friends and acquaintances well after his presidential term.[18] Thomas

Jefferson placed bets on what he called the *marronier*, a French variety made popular in Paris after 1700. Known in France as the *marron de Lyon* (*marron* in Italy), the nuts were actually from the townships of Savoie and Chablais near the Swiss and Italian borders.[19] In 1813, after failing to succeed in obtaining trees from the French countess Madame de Tessé, Jefferson wrote back to her apologetically, stating that he had given her "a great deal more trouble than . . . intended by my inquiries for the *Marronier*, of which I wish to possess in my own country."[20]

Prior to his death, Jefferson even recommended the marron as one of several chestnut cultivars to be planted on the future University of Virginia campus. That Jefferson wanted marrons, and not common European chestnuts, is implied in his instructions to the resident botanist for procuring them. With regards to "the *Marrionier*," wrote Jefferson in a confident if not covetous tone, "I can obtain from France."[21] Marrons were more flavorful than other European varieties and considerably larger, which astonished Americans seeing them for the first time. An experiment conducted by the president of St. Mary's College in Baltimore in 1827 concluded that European chestnuts from "the south of France" growing on campus grounds produced nuts six times larger than their American counterparts.[22] The results of the experiment were published in the *American Farmer* that same year:

Ten French chestnuts weighed last autumn . . . 8 oz 14 grs
Ten American . . . . 1 ¼ oz
French to American, in the ratio of 120 to 19.

Their comparative volume:
Ten French, made a volume of 6 solid cub. inch.
Ten American. . . . . . . .1 ¾ solid cub. inch.
In volume, therefore, their ratio is as 24 to 7.[23]

Although size certainly mattered to nineteenth-century horticulturalists, their bias against the native tree was not only due to their search for a larger and more marketable nut. The other reason the American chestnut was being overlooked by commercial growers, as Thomas Jefferson himself remarked in *Notes on the State of Virginia*, was a "new theory of the tendency of nature to belittle herself on this side of the Atlantic."[24] Jefferson was referring to a thesis promoted by French naturalist Georges-Louis Buffon, who believed species were in their most advanced evolutionary state when they were domesticated or under the intelligent care of humans.[25] According to Buffon, plants and animals living in the wild were *degeneris*; that is, smaller, weaker, and more prone to disease. As historian Philip J. Pauly, author of *Fruits and Plains: The Horticultural Transformation of America*, explained it, "living things that were *degeneris* had changed from the optimal cultured state in either quality or productivity, generally for the worse."[26] For that reason, Buffon thought plants and animals inhabiting the wilds of North America were biologically inferior, which caused them to suffer in both health and productivity. For American horticulturalists influenced by Buffon's views, European chestnuts were superior specimens, benefitting from centuries of human cultivation. Perhaps it is no coincidence that another Frenchman, Éleuthère Irénée du Pont, was growing marrons at his Wilmington, Delaware, estate as early as 1803.[27]

Although he was critical of Buffon's theory of evolution and took great pains to challenge it in his *Notes on the State of Virginia*, even Jefferson was forced to admit the native nuts were smaller in size.[28] And although the American chestnut was as productive in terms of overall mast production, and larger in height and diameter, it was virtually impossible to convince nineteenth-century nut-growers the tree was superior to its European cousin. The widespread opinion that the American nuts were sweeter in taste also became less important as marrons and other improved cultivars became more widely available. Moreover, in rural settings, the nuts had to be gathered almost immediately after falling to the forest floor—otherwise, jays, crows, squirrels, deer, and turkeys consumed them before they could be harvested, leaving few for local consumption, trade, or export.[29]

Although experiments comparing American and French chestnuts may have in the short term proven the superiority of European cultivars and encouraged their adoption by American horticulturalists, the demand for native nuts hardly slowed. Urban populations on the East Coast doubled in a single decade, with New York City alone possessing more than two hundred thousand inhabitants by 1830.[30] Moreover, improved roads, canals, railroad lines, and riverboat traffic allowed for the delivery of goods over greater distances, so native chestnuts could be more predictably sourced from almost anywhere within the tree's range, including remote parts of the Appalachians. Agricultural historian Lewis Cecil Gray believes chestnuts were among the earliest exports leaving the Cumberland mountains of eastern Kentucky, as records indicate they were shipped down the Ohio, Kentucky, and Cumberland rivers as early as the 1830s.[31]

PLATES

PLATE 1. François André Michaux, *The North American Sylva*, vol. 3 (Paris: C. d'Hautel, 1819). Image courtesy of the Biodiversity Heritage Library.

PLATE 2. Native Americans processing nut-mast, eight thousand years ago. Painting by Greg Harlin (2001). Courtesy of the McClung Museum of Natural History and Culture, University of Tennessee, Knoxville.

PLATE 3. Walton Ford, *Falling Bough* (2002). Passenger pigeon nesting colony. Watercolor, gouache, ink and pencil on paper. Reproduced with permission from the artist. Courtesy of Kasmin Gallery, New York.

PLATE 4. American chestnuts, Poplar Cove, Robbinsville, North Carolina, 1909. *American Lumberman* (January 15, 1910). Courtesy of the Forest History Society, Durham, N.C.

PLATE 5. The author surveying a stand of large surviving American chestnut trees at the Tervuren Arboretum, Tervuren, Belgium. Courtesy of Marc Meyer.

Native chestnuts also appear more frequently in local newspapers, including areas not formerly known for the commercial nut trade. In the October 17, 1831, edition of the *New Hampshire Patriot*, storeowner Asaph Evans requested fifty bushels delivered to his Concord establishment and told potential sellers, "the highest price will be given."[32] Paul Chase, a Brattleboro, Vermont, grocer, lured prospective sellers with cash *and* barter: "the subscriber wishes to purchase immediately at his Temperance Store, 100 bushels of *Chestnuts*; for which he will give a fair price, one half in *cash* and one half in *goods*."[33] A similar tactic was employed by the J. C. Stone & Company dry goods establishment in Guilford, Vermont. In a notice published in the *Vermont Phoenix* on October 12, 1838, Stone offered to pay "fractional bills at par in exchange for goods." Among the requested items include ten thousand feet of hemlock boards and "100 bushels CHESTNUTS."[34] Although it is unclear how much Stone paid for chestnuts in hard currency, retail prices declined as a result of the growing supply, ranging from $1.25 to $1.75 per bushel in East Coast markets.[35]

As a result of buyers like Chase and Stone and improvements to the U.S. transportation infrastructure, native chestnuts continued to dominate the American nut marketplace and would do for another half-century. Although growers did not stop planting or selling European varieties, the sheer volume of nuts produced by the native tree secured its dominant role in the commercial nut trade. What the nineteenth-century orchardists did not know, however, was that the importation of European chestnuts and other nonnative trees was introducing into American soils an invisible organism, a fungus-like pathogen that would eventually eliminate more than one hundred thousand square miles of territory from the tree's historic range.[36]

now scarcely a tree is left

Although not formally identified until 1930, a disease was attacking and killing chestnuts in the United States as early as 1825. Now known as *Phytophthora cinnamomi*, the organism belongs to a group of water molds once classified as fungi, but is now placed in the kingdom Chromista.[37] *Phytophthora* attacks the root collars of trees and shrubs, causing their roots to quickly rot and emit a dark blue-black liquid resembling ink. As a result, the term "ink disease" has often been used to identify its presence in infected plants. The invisible pathogen kills not only chestnuts but avocados, azaleas, Fraser firs, pines, and numerous other tree species. If left untreated, ink disease results in 100 percent fatality among the host plants, especially in warmer and moister climes where the organism is most virulent.[38]

The first *Phytophthora*-infected plants in the United States most likely came from Europe, as the disease was well established there by the late eighteenth century. At least one plant pathologist believes *Phytophthora* entered the European continent via the Azores, arriving on the Iberian mainland sometime during the early 1700s.[39] The Portuguese were heavily involved in the East Indies spice trade at the time and likely imported the infected plants from the Indian subcontinent. In fact, plant pathologists first isolated the pathogen on cinnamon trees in Papua, New Guinea; thus the genus name *cinnamomi*.[40]

The first observations of ink disease among European chestnuts occurred in southern Spain, in the Extremadura region west of Madrid. According to agricultural historian Isabel Azcárate Luxán, the pathogen destroyed large groves of chestnuts near the village of Jarandilla de la Vera in 1726, before spreading westward toward the town of Plasencia. By 1796, the village of Cabezuela del Valle saw 90 percent of its chestnuts killed by the pathogen, resulting in extreme hardship for local residents.[41] The disease was identified among chestnut groves in Portugal in 1838 and afterward in Italy and France. In all four countries, chestnut loss was so lethal and pervasive that authorities encouraged residents not to replant trees.[42]

Phytophthora most likely entered the United States in the late eighteenth century, spreading inland as soil and weather conditions allowed. During the colonial period, when seedlings or scions arrived as ship cargo, they were often planted in the immediate port area before being moved elsewhere after their root systems became more established. Because *Phytophthora* is a slow-moving pathogen, the spread of the disease inland would have taken several years, if not a full decade.[43] Moreover, its reproductive spores can remain dormant up to six years, requiring constant temperatures above 59°F to reproduce.[44] Excessively dry, highly acidic, and well-drained soils also slow the spread of the disease, as does uncultivated and heavily wooded terrain. In fact, one study found the pathogen moves in some areas only six feet per year and even less in hillier, upslope locations.[45]

Because of such facts, several plant pathologists believe ink disease did not establish a foothold in the southeastern United States until the second decade of the 1800s.[46] The disease is first mentioned by Dr. William L. Jones of Riceboro, Georgia, an amateur botanist and regular correspondent to the National Academy of Sciences. In 1825, he reported that all Allegheny chinquapins near his home had perished "in full leaf and with fruit half matured." After examining the trees and finding "no internal cause for their dying," he attributed the die-back to excessive rainfall which, he claimed, caused a "considerable quantity of land not subject to overflow to be covered with water for some time."[47] When reflecting on the problem decades later, however, Jones thought there may have been other reasons for their demise, as the trees continued to perish "up to the year 1845."[48] Whatever the cause, Jones was not very hopeful about the future of the *Castanea* species on Georgia's coastal plain. "If the disease is not arrested, in a few years," he predicted, "I fear it will be entirely exterminated."[49]

Although the pathogen would not be directly linked to American chestnut mortality for another half-century, nineteenth-century commentators had reason to believe a common culprit was, in fact, killing the trees.[50] In 1878, Franklin Benjamin Hough, the first chief of the U.S. Division of Forestry, noted that nearly all chestnut in the North Carolina Piedmont, "down to the country between the Catawba and Yadkin Rivers," had perished "within the last thirty years."[51] In Virginia, the disease reached the interior much later, as forest pathologist George Flippo Gravatt surmised in a 1914 government report. "Throughout the Piedmont section of Virginia, especially in the lower portions," he observed, "there has been for thirty years or more a gradual dying or recession of the chestnut toward the mountains."[52]

Gravatt rightly surmised ink disease had entered the United States through the major southern ports: Charleston, South Carolina, Savannah, Georgia, Wilmington, North Carolina, and Mobile, Alabama. In Georgia, the initial infections likely occurred in Savannah proper before moving deeper into the coastal plain. Later, when infected nursery stock was transported into the Piedmont, even more sites became contaminated with the disease. Cotton plantations were prime locations for the pathogen's introduction, as owners often landscaped their lawns and gardens with European and Asian plants.[53] In Georgia, the Savannah and Ogeechee River floodplains provided natural corridors for the movement of the pathogen, as well as host plants needed for its survival. However, in the upper Piedmont, the pathogen was slowed by hilly terrain, southerly and eastwardly flowing river drainages, and denser forests. Unlike the Allegheny chinquapin, the American chestnut was never common in the coastal plain of Georgia and South Carolina, instead preferring upland soils closer to the fall line.[54] Nevertheless, chestnut trees nearest the southern port cities should have been the first to succumb to the disease, a presumption that, surprisingly, is not always supported by the historical record.

One of the earliest and most comprehensive descriptions of nineteenth-century chestnut decline is found in an 1875 government report by Frederick W. Watts, the former U.S. commissioner of agriculture. Gleaning information from both eyewitnesses and government timber inventories, Watts wrote that the trees of Elbert County, Georgia, were "a mixture of almost all kinds," adding that "chestnut, during the last twenty years, has nearly died out."[55] Watts made a similar observation about chestnuts in Hall County, adding that the trees had once been plentiful there, "but now nearly every tree is dead or dying."[56] In Carroll County, Watts reported that although its original forests contained large quantities of chestnut, "now scarcely a tree is left."[57] Describing the forest health in Walton County, the commissioner simply wrote, "chestnut has all died."[58]

What is interesting about the commissioner's observations is that the four above counties share little in common geographically other than being located in the Piedmont. Elbert County, which had 26 percent of its land in forests in 1870, borders the Savannah River on its western bank and is more than three hundred miles upstream from the city of Savannah. Walton County, one hundred miles to the west of Elbert County in the center of the state, had nearly 36 percent of its area in forests. Carroll County, which borders Alabama along the western edge of Georgia, had 30 percent of its land in cultivation and as much 60 percent of its territory in timber. Hall County, 350 miles from Savannah and the most northerly situated, had the most land in forest (74 percent).[59]

The observations made by Watts thus beg the question, Why did chestnut trees in some Georgia counties escape the northerly and westerly spread of *Phytophthora*, especially those nearest the pathogen's reported point of entry? The obvious answer requires one to accept the likelihood that those four counties were not the only places suffering the ravages of ink disease. If chestnuts in Elbert County were attacked by *Phytophthora* during the 1840s or 1850s, then areas to the south and southeast were impacted earlier and the decline simply went unnoticed. Although the trees were never plentiful in those areas, witness-tree surveys tell us the trees were found as far south as the Florida panhandle during the early 1800s.[60] It is also possible the pathogen had leap-frogged ahead of the disease's inland progression as the result of upland farmers planting infected trees around their homes and gardens.

There is also the possibility the chestnut die-back described by Watts was not due to *Phytophthora* and there is a better explanation for the demise of the trees. In Hall County, for example, the most likely cause for the chestnut decline was a late-spring freeze, a common occurrence during the nineteenth century. As Alabama botanist Roland M. Harper observed, a frost occurring in the spring of 1854 "killed all of the chestnut trees over considerable areas."[61] The cold snap was particularly harmful to chestnuts in northwest Alabama and northeast Mississippi, leaving few survivors. In fact, when soil scientist Eugene Hilgard of Berkeley, California, surveyed a Mississippi forest in 1856, he found chestnuts "both young and old, dead." The largest trees, he observed, were ninety feet in height, but were badly decayed, "the bark dropping off, leaving the trunks bare."[62] In 1911, when Caroline Rumbold submitted a letter to the journal *Science* verifying Hilgard's observations, she added that "chestnuts are still growing in northeastern Mississippi, the epidemic which Professor Hilgard saw did not exterminate the tree in that region." Oddly, both Rumbold and Hilgard believed the trees were killed by an unknown pathogen and not a late frost.[63] However, if ink disease had been the culprit, the die-back would have been far more widespread and permanent.[64] The fact that chestnuts continued growing in the area—including trees farther south, where temperatures were warmer—suggests the decline was due to a late-spring frost.

Carroll, Walton, and Elbert Counties, on the other hand, are closer to the coastal plain, and also held more land in cotton. Upland cotton plantations hastened the spread of the disease, as antebellum plowing and harvest methods caused the organism to move more rapidly over the landscape. Although it is possible those counties were infected with *Phytophthora* before 1850, the first to harbor the disease was likely Elbert County, as it is nearer the coastline and directly upstream from Savannah. Carroll County probably had the pathogen only after 1860 since, as Watts notes, locals witnessed chestnut decline "about ten years ago, and now scarcely a tree is left."[65] Walton County likely became infected with the organism in the 1860s, as residents converted, just prior to the Civil War, much of their woodlands into cotton and corn fields. The Oconee River watershed, which drains Walton County, also possessed the heavier clay soils and moisture needed for the pathogen's survival and spread.[66]

The evidence from Georgia thus provides a much better understanding of how and when *Phytophthora* spread into the uplands during the first half of the nineteenth century. The Georgia example also demonstrates the role that late-spring freezes played in chestnut decline. In South Carolina, the pathogen likely arrived in the port of Charleston during the late 1700s, moving inland as township borders expanded and its population increased.[67] As in Georgia, the spread of ink disease toward the uplands increased with the expansion of cotton and corn cultivation, becoming prevalent in the Piedmont sometime during the 1830s. In 1883, when geologist Harry Hammond reported on forest conditions in the state, he wrote, "chestnut has been dying out for fifty years. In some localities where it once flourished, it has entirely gone, and in others, the large dead stems and stumps are the only vestige of this valuable and stately tree."[68]

In Mississippi, widespread chestnut die-off reached the thirty-first parallel (near Jackson) by 1854, but did not appear to be impacting trees in the northern part of the state, where the trees were still abundant. In fact, that same year, state geologist Benjamin Covington Wailes noted, "chestnut is only found in the interior, and most abundantly in the northern counties. The tree seems to have become diseased in latter years, and is rapidly dying out."[69] In North Carolina, *Phytophthora* likely first entered the state at the port of Wilmington, although it appears cooler and drier weather slowed the inland spread of the disease until the late nineteenth century, when temperatures warmed and rainfall amounts increased.[70] This explains why *Phytophthora* was not reported in either Washington, D.C., or Philadelphia during the antebellum period, even though tens of thousands of nonnative trees and shrubs passed through those two ports before the mid-1800s.[71]

just for their nut-bearing value

As a result of ink disease, late-spring frosts, and the clearing of land for agriculture, the American chestnut declined dramatically in numbers during the first half of the nineteenth century. In retrospect, ink disease did the most harm to the species, as it made the area below the fall line virtually uninhabitable for the trees after 1860. The trees were most impacted along the coastal plain and Piedmont, disappearing almost entirely from the lower-elevation forests of North Carolina, South Carolina, Georgia, Alabama, and Mississippi. Well before the infamous chestnut blight of the early twentieth century attacked a single tree, the species had lost nearly a fourth of its natural range.

Despite appreciable and heavy losses, the influence of the tree on American life hardly waned. In areas where they escaped disease, inclement weather, and the farmer's axe, the trees remained familiar and welcome fixtures on the landscape. The nut grower's fascination with European varieties of chestnuts even appears to have *increased* appreciation for the native tree, or, perhaps more accurately, created the perception that the American chestnut was a separate species with unique and enduring qualities. In fact, as early as 1834 an anonymously published essay in the *American Magazine* stated that although the tree closely resembled those of Europe, "in its general appearance, foliage, fruit, and the properties of its wood, it is treated by botanists as a distinct species."[72] Although this was not entirely true (many botanists classified both trees as the same species for another half-century), Americans everywhere were applauding its uses in and around the homestead, if not embracing the tree as their very own.[73]

Indeed, by the late 1840s many individuals were cultivating native chestnut trees in their own home orchards. In doing so, many believed they could improve their yields, as well as create larger and more improved nuts. In 1847, the noted landscape architect and editor of *The Horticulturalist*, Andrew Jackson Downing, thanked one of his subscribers—D. Tomlinson of Schenectady, New York—for sending him "*native Chestnuts*, of very large size, which we have planted, in the hope of producing still larger varieties." Downing was certain the trees would produce a superior product with successful breeding, adding, "our native nuts have hitherto been kept entirely without the pale of horticultural improvement. There is no doubt whatever, that the size of our Hickorynuts, Chestnuts, Butternuts, &c., may be doubled, and their flavor greatly improved, by selecting and planting the largest and finest seeds of such native specimens."[74]

For those not interested in the promise of native nut-breeding, there was the possibility of allowing the largest and healthiest trees to remain standing in isolated groves near the farmstead. This practice was employed by the father of Theodore K. Long of Perry County, Pennsylvania, who in the 1840s allowed his "finest old chestnut trees to stand . . . just for their nut-bearing value." According to Long, by the late 1850s each of the five or six large trees adjacent his Pfoutz Valley home produced "three or four bushels of chestnuts in a season."[75]

Some individuals were so hopeful about the nut-bearing potential of the species that they transported nuts and seedlings to distant farmsteads, even when relocating outside the tree's native range. In 1840, for example, William Walker planted four chestnuts in the "natural prairie soil" of his McLean County farm near Bloomington, Illinois. Eight years later, Walker publicly shared the results of the experiment, noting the trees varied in height from sixteen to twenty feet, with the largest "measuring 25 inches in circumference."[76] James B. Price, a neighbor of Walker's, planted as many as a hundred chestnut trees that same year and reported they were growing "finely and thriftly" and producing "several bushels of chestnuts."[77] In a letter drafted to the editor of the *Prairie Farmer*, dated July 1848, Walker summed up the results of the two experiments: "my experience is that the chestnut growth and production in Illinois is as rapid as my native state (N. Carolina), a state notorious for the growth of the chestnut. My neighbor James B. Price I hope will send a bushel of chestnuts, the produce of his own farm, to Chicago this fall, as a specimen of what can be done in Illinois."[78]

While many individuals continued to see the nut-producing potential of the tree, others thought the species more valuable for its timber. As noted in chapter 4, the most common use of the tree in the eighteenth century was for building fences, due largely to the rot-resistant quality of its wood. By the nineteenth century, the tree was even more associated with fence making, as Americans everywhere had embraced the chestnut split-rail fence. Indeed, in Noah Webster's *American Dictionary of the English Language*, first published in 1828, the American chestnut is said to be "one of the most valuable timber trees, as the wood is very durable, and forms in America the principal timber for fencing."[79] Eyewitness testimonies about the wood's virtues abound in journals and broadsides of the period, echoing sentiments heard a century earlier in New England and the Mid-Atlantic. As early as 1819, François André Michaux noted chestnut was already the preferred material for fence making in "the Alleghenies," adding it was "strong, elastic, and capable of enduring the succession of dryness and moisture."[80]

By the early 1840s, the chestnut worm-fence could be found as far west as southern Illinois, along the western edge of the tree's historic range.[81] Incidentally, several writers have suggested Abraham Lincoln's nickname—"the Rail-Splitter"—was the result of him splitting chestnut trees along the Sangamon River near New Salem, Illinois.[82] Lincoln was certainly a splitter of rails, but the source material was probably honey locust or black walnut and not chestnut.[83] If Lincoln did split chestnut, it was at his boyhood home in Spencer County, Indiana, which is within the tree's historic range.[84] In Illinois, chestnuts generally did not grow as far north as New Salem, although as late as 1878 a stand was discovered near present-day Rockford, "supposed by some to have been planted by Indians."[85] Pulaski County, near the Kentucky border, had the largest concentration of trees in the state, with a single grove encompassing almost eighty acres.[86] When Reverend E. B. Olmstead visited the stand in 1851, he saw an immensely tall chestnut that had been recently felled and split into rails. After measuring the stump and counting its growth rings, Olmstead determined the tree was 6'2" in diameter and "two hundred and fifty-years old when cut down."[87]

As a result of the tree's ubiquity and the durability of its wood, chestnut fencing remained popular throughout the entire nineteenth century. However, not all individuals preferred the worm-rail fence and by the early 1850s many had abandoned its use. Dairyman George Dennis, for example, who possessed a three-hundred-acre farm in western Maryland, stated matter-of-factly that he could not afford "the ordinary worm or Virginia fence." Dennis and others believed such fences occupied too much arable land and created "a harbor for weeds and shrubs of all kinds." Dennis instead enclosed his many fields with "a post-and-rail chestnut fence," claiming it was a more durable and effective enclosure.[88] In very steep or irregular terrain, both split-rail and post-and-rail fences were replaced by the buck fence, a design using, when and where available, split chestnut rails. Buck fences were constructed by simply driving two stakes into the ground in order to form a saddle upon which a third rail was lain across. As late as 1952, examples of buck, split-rail, and post-and-rail fences made of chestnut could still be seen along several sections of the Blue Ridge Parkway (see fig. 5.2).[89]

Other fences were also made of chestnut during the nineteenth century, including the whitewashed picket fence surrounding homes, large estates, hotel grounds, and government buildings. In the 1838 federal appropriations budget for the U.S. Military Academy at West Point, President Martin Van Buren requested $300 for the purchase of "10,000 chestnut laths for fences."[90] The request was among the most expensive budget items for repairs and improvements, equaling the cost of fifty thousand bricks. The chestnut palings were used to enclose the main campus grounds, including the barracks, officers' quarters, and parade field. If the laths were the standard length and width for the period, the fence stretched a distance no less than 0.92 miles.[91] Most paling fences were not whitewashed, however, and were erected specifically to enclose home gardens. Despite their rudimentary design and construction, paling fences played an important role

FIG. 5.2. "The Fences." Interpretive sign at Ground Hog Mountain Overlook, Blue Ridge Parkway, 1952. An actual split-rail chestnut fence is visible in the background. Courtesy of the National Park Service, Blue Ridge Parkway.

in protecting nineteenth-century food supplies, as without them, rabbits, squirrels, and groundhogs did considerable damage to the kitchen larder.

After 1845, across much of the eastern United States chestnut was used not only for fencing, but a myriad of other purposes, including railroad ties, sleeper cars, flooring, coffins, shipbuilding, and telegraph poles.[92] In Massachusetts, the demand for the trees reached such levels that Henry David Thoreau even worried about the tree's decline from overharvesting.[93] Other Concord residents shared Thoreau's sentiment, complaining "they could find no seedlings for transplant."[94] Thoreau was certainly an astute observer of the species and devoted more space to it in his journals than any other topic.[95] He was also the first to document the natural history of the American chestnut over its entire life cycle, as he wrote about the tree as both a tiny seedling and mature producer of nuts.[96] Although Thoreau's observations were never published in a single volume nor widely available, collectively they tell us a great deal about the influence of the species on both forest ecology and village life both prior to and during the American Civil War.

CHAPTER 6

placed there by a quadruped or bird

In 1845, when Henry David Thoreau moved into the cabin he built at Walden Pond near Concord, Massachusetts, the surrounding woodlands were home to numerous chestnut trees. The small rustic dwelling was itself shaded by a tree that bloomed so profusely, he noted, that its bouquet "scented the whole neighborhood."[1] Several miles from the cabin was a forested area "composed wholly of chestnut" which the writer visited to gather his winter supply of nuts.[2] Located near the town of Lincoln, the "boundless chestnut woods" rewarded him a half bushel of nuts his first winter at Walden Pond.[3] Thoreau's preferred method of gathering chestnuts, prior to the first heavy frost, was to enter the woodlands with a bag over his shoulders and a stick in hand, the instrument he used to pry open the freshly fallen and unopened burrs.[4] To outwit squirrels and jays, his biggest rivals in the nut-gathering competition, he resorted to the less pedestrian, though not uncommon method of gathering chestnuts. "Occasionally I climbed and shook the trees," wrote Thoreau in *Walden, or, Life in the Woods*.[5]

A decade later, the American chestnut had all but vanished from the Concord environs, including the very trees Thoreau had observed during his excursions afield. Most of the trees were cut to build the railroad that connected Concord to Boston and Fitchburg, as noted in a journal entry dated October 17, 1860.[6] Although chestnut timbers were most often used to make crossties or "sleepers"—as they are called in New England vernacular—Thoreau lamented the trees in what is surely an intended pun, writing that they "now sleep their long sleep under the railroad."[7] Despite their visible decline, chestnuts did not disappear entirely from the Concord township and could be sporadically found in woodlots, along fence rows, and within larger forested tracts.[8] However, after 1860, the trees played a far less significant role in local forest ecology, which make Thoreau's writings invaluable to us now. His observations regarding the nut-hoarding behavior of birds and animals offer important insights into the process of forest succession, as well as the migration of trees northward after the last ice age. For Thoreau, squirrels, mice, chipmunks, and jays were not merely consumers of chestnuts: they were, perhaps most importantly, *movers* of chestnuts.[9]

Thoreau's most engaging discussion regarding the American chestnut occurs in a four-hundred-page manuscript entitled *The Dispersion of Seeds*, which he drafted while observing various tree and plant species in and around Concord.[10] Using new material and quotations from previous journal entries, Thoreau paints a portrait of a species in retreat if not rapid decline. By 1860, he is astonished with how difficult it is to find chestnut seedlings and spends nearly two days looking for small saplings to graft or transplant near his Concord home. When he does find seedlings, they are growing in a mixed stand of oaks and white pines, more than "a quarter of a mile from a seed-bearing chestnut tree."[11] After pondering how they arrived at the location, Thoreau was sure they were "placed there by a quadruped or bird."[12] On another outing, the writer found four small chestnut trees near the home of Brister Freeman, the freed slave who twenty years earlier had also lived at Walden Pond.[13] Thoreau was truly "surprised at the sight of these chestnuts," as he knew of no other living trees "within about half a mile of that spot."[14]

Thoreau was confident that rodents had cached the four chestnuts and noted so in *The Dispersion of Seeds*, writing, "I have no doubt that they were buried there two falls ago by a squirrel, or possibly a mouse."[15] For the Concord naturalist, squirrels were the primary movers of chestnuts, which means large numbers of them were needed for the trees to successfully reproduce. Thoreau was openly critical of those who hunted the tree-dwelling mammals and even considered the practice a cause for the trees' scarcity.[16] In the above instance, however, the most likely nut-mover was the common blue jay, as the birds carried multiple chestnuts in their esophagus, as noted in chapter 1. Although it is plausible that a squirrel would take a *single* chestnut a quarter of a mile to secure it safely from competitors, it is doubtful that it would take *four* chestnuts, twice that distance, to the exact same spot. Coincidentally, in 2013, biologist Bernd Heinrich observed blue jays moving chestnuts in much the same manner around his southern Maine farm, "a half mile or so away from the parent trees."[17]

Nevertheless, Thoreau believed mice were important movers of chestnuts, although his own observations suggest they seldom transported them more than one hundred feet. However, on November 14, 1857, Thoreau found evidence that a deer mouse had moved "two fresh chestnuts" at least "twenty rods" (approximately 330 feet) from the nearest nut-bearing tree.[18] When Thoreau observed the two nuts under a microscope, the mammal's incisor marks were clearly visible. The winter nut-stores of deer and field mice could be quite large, making them important players in the process of chestnut reforestation. On January 10, 1853, at a woodlot east of Concord, Thoreau gathered "six and a half quarts of chestnuts," many of which had been placed under the leaves by a single "meadow mouse." In a shallow gallery "under the end of a stick under the leaves," recalled Thoreau, the rodent had arranged "thirty-five chestnuts in a little pile."[19]

Thoreau's observations are also testament to the proliferation of wildlife that invaded chestnut-laden woodlots each autumn. In early October, after the first heavy frost, the trees were assailed from above and below as furred and feathered creatures alike raced to gather the annual nut crop.[20] A trip to even the smallest chestnut grove assaulted the senses, filling the landscape with particular sights, sounds, and smells. On October 6, 1857, when passing through the "chestnut sprout-lands" near the Concord River, Thoreau witnessed vivid chestnut foliage that varied from green to yellow. Upon closer inspection, however, he found the leaves "richly peppered with brown and green spots, at length turning brown with a tinge of crimson."[21] The previous year, Thoreau observed what he referred to as "a rich sight, that of a large chestnut tree" which had lost most of its autumn color. Although the tree and those around it still possessed a few yellowing leaves, most of them, he observed, "strew the ground evenly as a carpet throughout the chestnut woods."[22]

Also appealing to the naturalist were the individual nuts, which he gathered to the sounds of calling jays and scolding squirrels. "It is a pleasure to detect them in the woods amid the firm, crispy, crackling chestnut leaves," wrote Thoreau. "There is somewhat singularly refreshing in the color of this nut—the chestnut color."[23] Thoreau was not the only one gathering chestnuts in New England during the 1850s, as an entire cultural tradition—complete with its own vocabulary, tools, techniques, and folklore—had emerged around the autumn pastime. In fact, the verb "chestnutting" was already in common usage, as many considered it the best way to describe the nut-gathering activity.[24] Thoreau himself used the expression as early as 1856.[25] He sometimes preferred the more colloquial variation "a-chestnutting" when penning his letters or journal entries, as well as in a passage in *Walden* that records his nut-gathering excursions to Flint Pond: "I went a-chestnutting there in the fall, on windy days, when the nuts were dropping into the water and were washed to my feet."[26] Both expressions find their way into literature, poetry, and newspapers of the period, a reflection of the activity's popularity among the general public.

where boys are ranging for nuts

Among the most endearing portraits of New England chestnutting is featured in Harriet Beecher Stowe's novel *Poganuc People*, a collection of stories about life in Litchfield, Connecticut, where the author spent her childhood.[27] Although most of the book is fictional and only loosely autobiographical, chapter 20, "Going 'A-Chestnutting'" is, in Stowe's own words, "drawn from the life."[28] The narrative describes the activities of a parson's family preparing to visit a grove of large chestnut trees on a hillside above town. Before leaving, the parson and his two sons talk strategically about how they will gather the nuts. After one of the boys remarks how "those trees are awful to climb," the father counters, "I won't let you boys try to climb them—mind that; but I'll go up myself and shake them, and you pick up underneath."[29] His response apparently satisfies them, as their chances of bringing home more nuts would improve as a result. When the boys' mother suggests they collect chestnuts in cups or pails rather than baskets, telling them "you won't get more than a bushel, certainly," they quickly reply, "oh yes, we shall—three or four bushels" and "there's no end of what we shall get when father goes. . . . Why, you've no idea how he rattles 'em down."[30] In the end, the outing is deemed a success, after the "doctor" climbs onto the branches of the large trees to release the nuts from their burrs. In the original edition, the accompanying illustration is captioned with an excerpt from the text: "How the doctor climbed the trees victoriously, how the brown, glossy chestnuts flew down in showers."[31]

Harriet's brother, Henry Ward Beecher, also wrote about the chestnut frolic, an event he no doubt experienced as a young boy. In an article published in the *Pittsfield (Massachusetts) Sun*, Beecher called the American chestnut the "very grandfather among trees," adding that "nature was in a good mood when the chestnut tree came forth." His effusive description of the chestnut outing is a reminder the activity served many social functions and was not just about securing food for the pantry. "It was a great day when, with bag and basket, the whole family was summoned to go 'a chestnutting!," remarked Beecher. "There was frolic enough, and climbing enough, and shaking enough, and rattling nuts enough, and a sly kiss or two, but never enough, and lunch enough, and appetite enough. . . . Long live the chestnut-tree, and the chestnut woods on the mountain-side!"[32] For Beecher and his peers, chestnut outings were part entertainment, part social event, and, perhaps more importantly, an outdoor communal exercise that gave participants a memorable bonding experience. It also gave New England villagers access to what many perceived as the disappearing woodland commons, that increasingly scarce parcel of land that allowed everyone—regardless of gender, class, or age—to share in nature's bounty.[33]

The Boston-born Winslow Homer also memorialized the chestnut outing in a woodcut print commissioned by the American magazine *Every Saturday: An Illustrated Journal of Choice Reading* (see fig. 6.1).[34] The black-and-white image depicts two boys and two girls holding a bed sheet under the boughs of a large tree, while a third boy, who sits above them on a lower branch, shakes the highest limbs with a long stick. A basket and cloth sack on the ground is already laden with chestnuts, the result of earlier efforts. In describing the print, the magazine's editors note that "Chestnutting" (also the title of the engraving) "represents one of the pleasant occupations—half work, half sport—which autumn brings to country boys and girls."[35] The description continues with more accolades for the pastime, noting its particular influence on the lives of younger Americans. "Chestnuts are so dear to youthful palates, that had they been forbidden in the first garden, Cain and Abel would surely have fallen in chestnutting time, if Eve had not anticipated them in the apple season. Mr. Homer's picture needs no explanation; the boys and girls are enjoying the process of gathering the chestnuts, as they will enjoy the subsequent process of eating them."[36]

Heavily romanticized in popular literature both yesterday and today, chestnutting had its dark side too, although such occurrences were rare exceptions to an otherwise enjoyable activity. The most adverse outcomes were serious injuries, a fact well documented in the popular press. Indeed, falls from chestnut trees were not uncommon and caused severe scrapes, broken limbs, and even death.[37] On October 27, 1863, the *New Haven Palladium* reported that a New London, Connecticut, boy "had fallen from a tree while chestnutting Friday, and broke his arm in three places, at the wrist, above the elbow and near the shoulder."[38] An even more unfortunate victim was D. W. Beach, a fourteen-year-old boy from East Haven, Connecticut, who fell to his very death gathering chestnuts. According to the *Norwich Aurora*, the young victim, in the company of two other boys and a small girl, went "on a nutting expedition in the North Haven woods."

FIG. 6.1. Winslow Homer, *Chestnutting*, reproduced in *Every Saturday*, October 29, 1870.

After climbing high into a lofty tree, Beach was engaged in shaking down the nuts when a limb suddenly broke, causing him to fall fifty feet to the ground. From eyewitness testimony, Beach landed on his head, and when the others went to his aid they saw "his brains were dashed out, there being two ugly holes in his skull." The unhurt boys ran two miles to the home of a Mr. Philander Robinson, who then raced to the scene in a horse-drawn wagon. Although Robinson transported Beach immediately back to his residence, "he died before reaching there."[39]

Chestnut gathering also involved demonstrative displays of machismo, especially among young men wanting to maximize yields or, if gathering nuts for income, daily profits. This was done by "clubbing" chestnuts down from the tallest trees, as Pennsylvania historian Cornelius Weygandt explained it in a published memoir. For Weygandt, clubbing was a favored technique for gathering chestnuts, even though it required both strength and dexterity to send "a skillfully weighted club true to the top of an eighty-foot tree." With an effective strike, says Weygandt, the nuts would "leap out in all directions and scatter in a shining shower under the tree, followed more slowly by the velvet-lined bits of browning burr." Not all clubs returned to the ground, however, which made walking under the trees hazardous when nut-gathering season was over. "More than once, on windy mornings," recalls Weygandt, "I have had narrow escapes from being hit with some wheel-spoke bearing a fish-plate nut, or some broom handle weighted with lead pipe as I passed on my way to the train."[40] An important part of nineteenth- and early twentieth-century material culture, these weapons of mast destruction would be unidentifiable to almost anyone living in America today. By the early 1930s, they were already rare, causing Weygandt to boyishly muse: "I wish that I had kept all of the chestnut clubs I have found. They would make a display of deadly weapons unrivaled outside of the collections of war clubs in the greater museums."[41]

Clubbing chestnuts could also be done without actually throwing the object into the highest branches, as some trees released their nuts if the main trunk was struck with considerable force by any large object. Stones often served this purpose and, like wooden clubs, left scars on the tree's outer bark. In Thoreau's unfinished manuscript *Wild Fruits*, the author complained that chestnut trees in and around Concord had been "bruised by the large stones cast upon them in previous years and which still lie around."[42] According to Thoreau, the sound created by the impact of the stones was even loud enough to allow one to follow the direction of the activity from a considerable distance. On October 24, 1857, Thoreau wrote that he heard "the dull thump of heavy stones against the trees from far through the rustling wood, where boys are ranging for nuts."[43] The sound was not very pleasing to Thoreau, as he believed the practice was injurious to the trees. His concern for their health caused him to not only condemn the nut-gathering technique, but to have stopped doing it himself. "I sympathize with the tree, yet I heaved a big stone against the trunks like a robber—not too good to commit murder. I trust that I shall never do it again," promised the writer. "It is worse than boorish, it is criminal, to inflict an unnecessary injury on the tree that feeds or shadows us. Old trees are our parents, and our parents' parents, perchance."[44]

For Thoreau, chestnutting was a privilege and an art, allowing one to intimately engage nature as the season changed and the cooling earth readied for its winter sleep. His deep fondness for the practice is reflected in a journal entry dated December 9, 1852, which he penned after foraging for chestnuts near Smith's Hill, a small knoll southeast of Concord. "I love to gather them," wrote Thoreau, "if only for the sense of the bountifulness of nature they give me."[45] A decade later, Thoreau was unable to gather chestnuts, as his ailing health kept him indoors if not fully bedridden, affording him fewer opportunities to observe the natural world. At the time of his death, America had been at war for an entire year, and although the worst of the battles occurred in the southern states far from New England, the conflict caused the naturalist considerable grief.[46]

For those participating in the Civil War, the trees and nuts were, in some instances, the difference between life and death, providing warmth, shelter, and caloric sustenance. This was especially true in major campaigns where battles rose to their greatest intensity. Although the typical Civil War soldier did not rely on chestnuts for daily rations, in autumn and winter they were important dietary supplements.[47] The trees themselves were used to build fortifications and defensive structures, including *cheveaux-de-frise*, the jacks-like wooden stakes that kept charging cavalries at bay. Large living trees would also have protected soldiers from the line of fire in the heat of battle, absorbing the lethal blow of mini-balls and preventing what otherwise might have been certain death.[48]

thrashed out by the bushel

Although the American chestnut receives only occasional mention in newspaper accounts during the Civil War, soldiers very often refer to the trees or nuts in their personal correspondence. In fact, according to Louisiana naturalist Kelby Ouchley, the author of *Flora and Fauna of the Civil War*, "Chestnuts and chinquapins were the most common wild flora mentioned as food by Confederate soldiers."[49] Northern soldiers also made reference to the nuts, among them Private Allen M. Geer of the 20th Illinois Volunteers. In a diary entry written at Jackson, Tennessee, on October 9, 1862, Geer noted, "Went out on a chestnut excursion[,] gathered chestnuts from the trees for the first time."[50] When Lieutenant Colonel Charles F. Johnson of the 81st Pennsylvania Volunteer Infantry wrote home to his wife from northern Virginia, he happily told her, "I have upon my table a pile of Chestnuts and two cups of Parsimmons [persimmons] brought me by different parties of our scouts, who gathered them while out last night and this morning."[51]

Chestnuts were also directly mailed to encampments, usually by relatives of soldiers who desired to augment their daily rations. After receiving a parcel from his family in Henry County, Georgia, Private William R. Stilwell of the 53rd Georgia Volunteers wrote, "You don't know how much good them apples and chestnuts done me. They were so good and then they come from home."[52] Sometimes chestnuts were sent to soldiers via benevolent benefactors, as evidenced by a letter to the editor of the *North Carolina Standard*, a semiweekly Confederate newspaper published in the state capitol. According to the missive, Ann Grinton, who was living with "the family of Dr. Calloway," often sent "articles of apparel and food to the soldiers of this county." The letter writer apparently witnessed the packing of the parcel and provided an exact list of the items shipped. Among the tallied inventory: "1 bed quilt, 1 bed tick, 3 pair socks, 1 pillow case, 1 pound feathers, 4 pair woolen gloves . . . 1 bushel chestnuts."[53]

Berry G. Benson, a Confederate scout who escaped from a Union prison in Elmira, New York, survived the ordeal by consuming chestnuts as he made safe passage back to the Mason-Dixon Line. On October 11, 1864, after walking through Canton, Pennsylvania, and entering what is today the McIntyre Wild Area, the Georgia soldier encountered a tree near the roadside "speckled with ripe nuts." Tired and hungry, Benson was "up the tree in a minute" and shook numerous chestnuts to the ground. He vividly recalled their sound as they hit the earth, noting it was like "twenty drums beating the long roll." Afterward, Benson filled his overcoat pockets with the fortunate find, remarking they were so full "that nuts dropped on the ground as I walked."[54] The very next day, near the town of Ralston, Benson left the main path near a rocky hillside and, after properly concealing himself, "made a little fire and roasted Chestnuts."[55] The following morning, below Williamsport, Pennsylvania, the Confederate climbed over the north end of Bald Eagle Mountain, passing through chestnut woods before making camp above the west branch of the Susquehanna River. The next day, he hiked slowly down to the riverbank, where he commandeered a small rowboat. To avoid detection, Benson withdrew the oars and floated southward toward the town of Lockhaven. Content with his newfound anonymity, Benson lay on his back in the craft for hours, "drifting down, eating chestnuts."[56]

Perhaps the most notable example of chestnuts providing nourishment to Civil War soldiers is found in the memoir of Thomas H. Mann, a Massachusetts infantryman assigned to the Army of the Potomac under General George McClellan. In the autumn of 1861, Mann and eighty thousand Union soldiers were stationed at Fort Corcoran, an encampment at Arlington Heights near the nation's capital.[57] According to Mann, in the early days of October the Federal army's defensive lines were pushed west, causing his regiment—the Massachusetts 18th—to occupy Hall's Hill near High View Park. For several months, says Mann, he and fellow soldiers were engaged in "leveling the magnificent forests of chestnut and pine for several miles in front [west] of this range of hills." According to Mann, the troops were sent out daily with axes, until more than "30 square miles of forest were leveled with the ground." Because it was also chestnut harvest time, large old trees were brought to the ground so the chestnuts could be "thrashed out by the bushel." In the end, recalled Mann, "thousands of bushels were thus gathered by the army." In fact, so many were gathered, he added, "that they were a drug in the camps of the men who munched them all winter, roasted, raw and boiled."[58]

Occasionally, soldiers' desire to obtain chestnuts brought them into direct contact with local residents. When the editor of the Hendersonville *North Carolina Times* went chestnut hunting with Colonel "B" and Major "S" in East Tennessee, they encountered a "large chestnut tree inside a field."[59] Finding no ripe chestnuts on its branches, the trio paused momentarily to contemplate their next move. Immediately, they heard a hostile female voice: "What are you doing here—plundering my property, you Rebel thieves?" After a few moments of reflection, Colonel "B" responded by saying, "We were only looking at that chestnut tree, to see if there were any chestnuts on it." The woman's response was no less threatening, as the Union sympathizer wanted no part of sharing the bounty with a Confederate officer. "And you was going to steal my chestnuts, was you?—get out o'here, or I'll thrash thee yet with your carcasses!" After more threatening words, the trio retreated, as their adversary, noted the editor, "had adopted a decided disposition to assume the offensive."[60]

Incidentally, this was not the editor's only encounter with either chestnuts or unsympathetic locals. A few days earlier, while visiting an ill soldier, the same newspaperman observed a "small chestnut tree, quite full of chestnuts."[61] Escorted there by Private "H," the editor asked him to fell the tree, but because they had no axe, the private was forced to borrow one from a local resident. When he returned with the news that the occupant would not let him have it, the pair went to make the plea together. After knocking on the door several times, a young lady finally replied, "What business have you got here, sir?" "Only to borrow an axe an hour, to fell a chestnut tree," responded the editor. "Give him the axe over the head—he's a chicken-stealing Rebel," added another female voice from inside the home. After several minutes of trying to persuade the ladies they meant no harm, the men were threatened with a fireplace poker, abruptly ending all negotiations. Reflecting on his visit to Unionist East Tennessee, the newspaperman concluded, "Lincolnite women, down on the river, are fighting stock; and I must say I would rather do without chestnuts a great while than to come into contact with one of them again."[62]

Large quantities of chestnuts were also consumed by Union soldiers under the command of William Tecumseh Sherman, especially during the autumn of 1864. Although it is unclear if they obtained the nuts from households or by foraging, it is obvious in Sherman's letter to General Henry Halleck—written from Summerville, Georgia, on October 19—that his men were greatly benefiting from them. "When the rich planters of the Oconee and Savannah see their fences and corn and hogs and sheep vanish before their eyes," wrote Sherman, "they will have something more than a mean opinion of the 'Yanks.' Even now our poor mules laugh at the fine corn-fields, and our soldiers riot on chestnuts, sweet potatoes, pigs, chickens, etc."[63] That chestnuts are listed first among other major foodstuffs says a great a deal about their importance to the local citizenry. Had chestnuts been scarce or of little value, Sherman would have omitted them entirely from the correspondence.

Soldiers on both sides of the conflict no doubt encountered slaves gathering chestnuts, as plantation owners required them to do so as part of their daily chores. However, slaves might also collect nuts independent of their master's orders, as they were generally allowed to eat what they grew, captured, or gathered on the plantation premises. Moreover, if a slave was lucky enough to have an afternoon off or a brief holiday reprieve, many used the time to explore the fields, streams, and forests of their environs. This was not done in a purely recreational sense, but as a way to augment highly rationed, and perhaps truly wanting, food supplies.[64] Although there are numerous references to chestnut gathering in the slave narratives compiled by the Federal Writers' Project of the 1930s, the transcription styles of the interviewers are often offensive as they grossly caricature the dialect of the interviewees.[65] Additionally, many of the ex-slaves in the collection were born after 1850, which means their memories of the period are, at best, from their early teenage years.[66] Nonetheless, the narratives provide substantial evidence that slaves ate, collected, and roasted chestnuts in the same fashion as their white neighbors, despite having far fewer freedoms to do so.

Henry Walker, for example, a former slave who grew up on a plantation nine miles south of Nashville, Tennessee, recalled that each autumn "pigs ran out to eat the acorns, and the children—white and black—to pick up chestnuts, scaly barks, and hickory nuts."[67] On a northern Mississippi plantation near Abbeville, Polly Turner Cancer remembered, "the master let us gather chestnuts and hazelnuts on Sundays and den when de wagons was going' to Memphis, we wud put our sacks on dem and dey wud sell dem fur us and [he] let us have de money."[68] Hannah Fambro, who lived in Monroe County, Georgia, recalled going to the woods to gather chestnuts on Christmas Day. "We had dat day off, and I 'member we'd go up to the woods a-hunting fo' chestnut and chinkapins," she recalled.[69] Frances Willingham, another Georgia native, believed that the plantation of slaveholder Elisha Jones was even profitable as a result of its expansive nut orchards. "He had grove after grove of pecan, chestnuts, walnut, hickor'nut, scalybark, and chinquapin trees," recalled Willingham. "When the nuts was all gathered, Old Marster sold 'em to the big man in the city. Dat is why he was so rich."[70] Not all chestnuts were gathered for profit or daily fare, however. Lucinda Lawrence Jurdon remembered that when she and other female slaves started courting, they went to the woods hunting chestnuts. After gathering enough for a necklace, she remembered, "us would "string 'em 'round our necks an' smile at our fellows."[71]

the axe, and the plow, and the cattle

Besides offering a source of sustenance to slaves, soldiers, and civilians, the tree provided additional benefits during the Civil War. The wood, bark, and leaves were also utilized, often in important ways. As noted in previous chapters, the wood possessed exceptional water-resistant properties, and was easily split, greatly increasing its status among trees. Soldiers appreciated this quality and often put the trees to good use. When William J. Clarke, the commanding officer of the 14th Regiment of the North Carolina Volunteers, submitted a field report to the *North Carolina Standard*, he noted the harsh conditions his men faced at their Meadow Bluff, West Virginia, encampment. According to Clarke, for several weeks in October 1861 his regiment was exposed to cold mountain rains without tents, including downpours that completely extinguished their campfires. When proper shelter finally arrived, many soldiers continued sleeping on the wet ground until they could "split chestnut logs and floor their tents."[72]

Even in places where the trees were relatively scarce, soldiers cut them down for fuel, as Private Wilbur Fisk of the 2nd Vermont Volunteers explained in a letter written from Lewinsville, Virginia. In the correspondence, Fisk noted his regiment was encamped in a wooded grove of second-growth pine, "and the remainder chestnut and oak." He added that most of the trees in the camp had been "used up long ago and since fuel has become scarce, so scarce that hardly a sound stump can be found in this vicinity." Fisk surmised the soldiers were permitted only to cut select trees inside the campground, including oak and chestnut, due to the "anticipation, I suppose, of warmer and calmer weather."[73] In warmer months, the bark of the trees might even serve as a barrier from the morning dew, as Jacob Ritner of the 25th Iowa Infantry stated in a letter to his young wife written from Dallas, Georgia, and dated June 2, 1864. "There is no house near here where we can get rails or boards. So last night I skinned a big chestnut tree and got a wide piece of bark, lay down on it and spread my gum over me and slept first-rate."[74]

For warmth, soldiers frequently burned chestnut fences, thereby limiting destruction to nearby living trees. On September 29, 1861, after Confederate soldiers occupied her Bowling Green, Kentucky, farm, a young

Josie Underwood wrote about the ordeal, "the whole brigade of infantry marched straight through our yard to pitch their tents in our barn lot—and the Pond Orchard—and there they are camped for God only knows how long and already the new chestnut rail fence, around the orchard, is fast vanishing and being burnt in their camp fires."[75] On September 21st, 1862, artilleryman Thomas D. Christie of the 1st Minnesota Battery noted that when he and fellow soldiers approached Glendale, Mississippi, near the northeastern corner of the state, they passed "a clump of chestnut trees loaded with their green burs."[76] Afterward, he told his sister that even though the regiment had bivouacked in a pouring rain and were "wet to the skin," they were able to encircle themselves around "enormous fires of chestnut rails, that evaporated the water as fast as it fell on us."[77] Chestnut fences were even torched as an act of revenge. When Sheridan's Union cavalry were sent to Tobacco Creek, Maryland, just after Lincoln's assassination, the soldiers dismantled and burned no fewer than seventy-nine thousand chestnut rails.[78]

If soldiers fell ill or were wounded, the trees also provided physical comfort, as their leaves and roots could be made into tinctures, salves, or medicinal teas. Physicians promoted their use in hospital clinics, which is not surprising, as a century earlier Europeans were recommending the leaves and roots for a variety of ailments.[79] In the book *Resources of the Southern Fields and Forests*, medical botanist and Civil War surgeon Francis P. Porcher advised Confederate physicians to make a tea from the tree's root, to be used for "diarrhoea by our soldiers in camp."[80] A colleague of Porcher offered an additional remedy: boiling the bark or root "as a substitute for quinine in intermittent and remittent fever."[81] As early as 1862, a tincture made from the leaves of American chestnut had gained reputation for treating whooping cough, although it was never fully embraced by physicians on either side of the conflict.[82] Because chestnut leaves were used as a folk remedy to stop minor bleeding or to treat surface abrasions, they were likely stocked in Civil War pharmacies or gathered directly in the field by those needing to treat their own illnesses or wounds.[83]

Besides providing remedies for common ailments, the largest trees shaded soldiers from the hot sun or hid their precise location from the enemy. During both minor skirmishes and major sieges, groups of chestnuts might even influence tactical decisions, including the direction of assaults and retreats. For example, at Gettysburg, on the third day of the conflict—July 3, 1863—the Confederates' assault commonly known as Pickett's Charge was directed at Ziegler's Grove, a group of oak and chestnut trees rising above the brow of Cemetery Ridge.[84] The grove hid the Union artillery from view as well as provided important shade for horses and soldiers in the extreme heat prior to battle. Although some Civil War historians have debated the species makeup of Ziegler's Grove, author John T. Trowbridge, who visited the location in August 1865, found the hillock comprised of "oak chiefly, but with a liberal sprinkling of chestnut."[85]

Civil War soldiers were even eulogized under the shadows of chestnut trees, as was the case with Jesse L. Reno, the Union commander of the Army of the Potomac under General George McClellan. When reconnoitering enemy forces during the Battle of South Mountain, near Boonsboro, Maryland, Reno was struck in the chest by a sharpshooter's bullet. According to eyewitnesses, the commander died that evening as a direct result of his wound.[86] Four years afterward, when author Benson J. Lossing visited the spot where the general had fallen, a stone marker had been placed there by Daniel Wise, the father of the property owner. Standing over the marker was a large chestnut tree whose outer bark had been severely damaged by intensive cannon and infantry fire. In the *Pictorial Field Book of the Civil War*, published in 1874, Lossing includes an illustration of the lone chestnut, which bore, in his words, "the scars of many wounds made during the battle" (see fig. 6.2).[87]

A more macabre example of the tree bearing witness to the death of Civil War soldiers was observed by Massachusetts infantryman Thomas H. Mann in August 1862. While filling his canteen away from camp, Mann stumbled upon the tattered remains of a Confederate uniform "partially covering the skeleton that had put it on while flesh and life was there." The decaying bones and uniform "were lodged between two large chestnut trees that grew from a common root." Judging from the degree of decay and the position of the bones, Mann concluded the soldier had fled to the location to avoid further injury or capture during "the battle of Bull Run of the year previous."[88] Regardless of why he chose that specific spot to hide or rest, the incident is a reminder of the profound influence the natural environment had on Civil War soldiers. The fields and forests of the conflict directly influenced their level of comfort, what they ate and drank, where they camped and slept, and even specific movements during assaults and retreats.

The Civil War's impact on the wooded landscape, and by extension the American chestnut, was no doubt significant. In and around the major

FIG. 6.2. War-damaged American chestnut at the grave of Union general Jesse L. Reno, near Boonsboro, Maryland. Benson J. Lossing, *Pictorial Field Book of the Civil War*, vol. 2 (Hartford, Conn.: T. Belknap, 1874).

battlefields, on both northern and southern soils, trees of all sizes were removed as part of the general clearing process.[89] In many instances, the impact was intensive and long lasting, as those same areas were very often subjected to additional disturbances, including intentional and unintentional fires, soil impaction caused by mules, wagons, and artillery, and the digging of trenches, canals, and roadbeds.[90] For this reason, the conflict accelerated the process of deforestation caused by market agriculture earlier in the nineteenth century. This was especially true in the southern states, where the number of acres in corn, cotton, pasture, and orchards grew relatively unabated during the period of Reconstruction.[91] The rebuilding process obviously required the cutting of tens of thousands of additional trees, as chestnuts were used for fencing, roofing shingles, and barn and home repair. They were also needed to reconstruct the many miles of railroad lines damaged as a result of Union and Confederate blockades.[92]

While there is considerable evidence chestnut trees were felled in great numbers during the Civil War, not all were removed from the surrounding landscape. At Missionary Ridge, a major battlefield near Chattanooga, Tennessee, some chestnuts even survived the ordeal, even though much of the surrounding area was rendered entirely treeless. Evidence for this comes from the testimony of a dog breeder who owned a kennel immediately adjacent the hilltop location. Forty-four years after the war, the breeder boasted that his animals could "romp to their heart's content, and when fatigued seek the shade of grand old oaks and chestnut trees which escaped the merciless ax of the opposing armies who spared little timber that memorable winter of 1863."[93]

In the Northeast, chestnut decline was not linked to the Civil War, but to railroad construction, as Thoreau had noted in *Walden*. In New England, chestnut had become *the* preferred wood for crossties, which were used for both repairing older tracks and building new ones. Because chestnut crossties were also lighter in weight than alternatives and generally needed replacing only after seven years, railroad construction and chestnut timber even became synonymous.[94] In fact, as early as 1858, at least one railroad advocate suggested the trees be planted in double rows on both sides of the tracks to establish "an unfailing supply of the best material for sleepers."[95] Not all chestnut trees in New England were removed for the purpose of railroad construction, however. A significant number were also felled for agriculture, township expansion, and industry. Even before the war, Atlantic and New England shipbuilders were using large amounts of chestnut in the hulls of schooners, riverboats, and steamships.[96] In fact, the largest vessels, which often exceeded 250 feet in length, were almost always framed with chestnut and oak beams. Of the twenty-nine steamships mentioned in a *New York Times* article published in 1865—featuring steamers engaged in "the American-China trade"—twenty were constructed with chestnut timbers.[97]

As a result of such activities, one would be hard pressed to find trees in New England sprouting directly from nuts and even fewer specimens one hundred years old or more in age. Although chestnut-dominant forests were legendary for their ability to resprout, the species was not indestructible and vanished from large areas if unfavorable conditions prevailed. Thoreau himself was keenly aware of the possibility and even broached the issue in one of his last journal entries. "In the wildwood," he warned, chestnuts and other trees suffered greatly not only from fires, insects, and blight, but from "the axe and plow and the cattle."[98] It was the latter three things that impacted chestnuts the most, he believed, and if suffered in combination, could cause the trees to become "extinct under our present system."[99]

If axes and plows alone did not remove chestnut trees from large portions of the New England landscape, then the introduction of livestock certainly completed the process. Left unattended, cattle and hogs consumed young chestnut leaves and shoots, trampled seedlings, and ate viable nuts. Sheep also consumed chestnuts, leaving few to resprout the following spring. Thoreau witnessed the phenomenon firsthand, as livestock commonly grazed in the fields and forest clearings in and around the Concord township. Agricultural historian Brian Donahue noted that foraging livestock were especially destructive to hardwood seedlings, so if and when New England pastureland reverted back to woodlands, the resulting forest no longer contained chestnut, but instead was replaced with white pine, red cedar, and juniper.[100]

This does not mean the American chestnut was rare or uncommon in New England after 1870. Although large numbers were removed from woodlots in the heavily populated townships, the tree was still abundant in many parts of Connecticut, Massachusetts, and across much of southern New Hampshire.[101] While in visible decline, the tree maintained its presence not only in the Northeast but across much of its entire range (except areas impacted by *Phytophthora*). Large trees were observed in nearly every county in New Jersey, Pennsylvania, and Maryland, as well as in forests impacted by intensive charcoal production.[102] Even in central Ohio, where nearly two-thirds of all standing forests was converted to farmland before 1870, "second growth" chestnut was still visible in the Coshocton County uplands less than fifty miles from the city of Columbus.[103] In 1871, a chestnut tree four feet in diameter was reported as far south as Sparta, Georgia, suggesting some southern trees had survived not only ink disease, but the cotton and corn planter's plow.[104] In the mountains of West Virginia, Tennessee, and North Carolina, individual trees continued to reach widths of seven feet, with some specimens exceeding nine feet in diameter.[105] Even at Concord, Thoreau had observed a chestnut stump "eight feet five inches" in diameter as late as 1852, a tree that, he recalled in his diary entry of October 28, 1860, had "been cut, as I remember, but a short time,—a winter, perhaps two winters, before."[106]

In the southern Appalachians, the trees dominated entire mountainsides, comprising as much as 40 percent of the forest in remote areas.[107]

Consequently, the species continued to fuel the engines of industrial commerce and gave character to a landscape none living today have witnessed. By the late 1870s, wood from the trees comprised much of the nation's built environment, including the interiors of factories, train depots, ocean liners, and residential dwellings. Items used in home interiors were also increasingly made of chestnut, including baby cradles, broom handles, dowel rods, storage crates, packing excelsior, picture frames, barrels, shelving, pianos, and kitchen cabinetry. The ubiquitous nut, celebrated in poetry, song, and popular literature, was even changing the American spoken vernacular. All of these things gave the American chestnut unique status among trees, making it as important to the average U.S. citizen as it was to North American forest ecology.

CHAPTER 7
along all prominent thoroughfares

By 1875, the Northeast had recovered from the economic downturn caused by the American Civil War. In cities along the East Coast, foodstuffs of all kinds were in great demand, especially in Manhattan, where the population would reach a million individuals by the end of the decade.[1] In autumn, when temperatures cooled and the domestic trade in fruits and berries slowed, another popular food item appeared on New York City streets—roasted chestnuts. The trade had become such a familiar institution that as early as 1872, the *New York Times* used a full half-column to discuss its impact on daily commerce. The writer noted the chestnut roasters were found "along all prominent thoroughfares," including numerous street corners on Broadway. Those locations commanded better monetary rewards, as foot traffic was heavier and the visibility of vendors greater. Profits were not insubstantial. For those selling roasted chestnuts from street corner stands, the return on investment was "150 percent."[2]

Exactly how many chestnut vendors operated on the streets of Manhattan is unknown. The same *New York Times* writer claimed the exact number would be difficult to determine, but added they were found at "almost every half-dozen corners" in the borough, "almost any hour of the day or night."[3] In the daytime, the vendors were continuously at their posts, crying "Roasted chestnuts!" until prospective buyers approached.[4] Each season, hundreds of bushels were sold at such stands, a small brown paper-bagful at a time. Sales by the "night-birds" were also profitable, if not more so, as those individuals secured higher prices for their efforts. The ultimate success of the sellers depended on a number of factors, including the location of the stand, how cheaply the roaster purchased chestnuts from wholesalers, and the time of day or night they were sold. The *Times* reporter placed the vendors in one of two categories: (1) those "comfortably settled in life" due to making "comparatively large" profits, and (2) those who failed to secure anything beyond "the means which afford their families a support." What percentage fell into one category or the other is not offered, although it appears all roasters were able to make a surplus "upon which to base a bank account" (see fig. 7.1).[5]

FIG. 7.1. Chestnut vendor, Baltimore, Maryland, 1905.
Courtesy of the Library of Congress.

In most instances, the vendor's profit was determined by what people were willing to pay at the point of sale. During the 1870s, after bulk shipments reached Manhattan, wholesale dealers generally sold a one-bushel gunny bag for five dollars.[6] If purchased by the quart at an indoor establishment, they might cost twenty cents; a small paper-bagful on the street corner, a nickel.[7] The price was also partly determined by the quality of nuts, which was often a topic of complaint with some clients. Buyers in different sections of the country sometimes included wormy or decayed nuts in their shipments, making the quality of the entire sack or barrel suspect.[8] In some cases, deliberations regarding the worth of the chestnuts ensued until the advertised price was ultimately lowered. In parts of the American interior, the price of chestnuts was noticeably lower, as fewer middlemen were needed to get them to market. On October 20, 1871, residents of Nashville, for example, reported that one thousand bushels were available in the city for the "ruling price" of $3.25 per bushel.[9]

It was also during the 1870s that Italian immigrants became associated with the sale of chestnuts in Manhattan and surrounding boroughs. Previously, French vendors were the primary sellers in New York City, a tradition dating back to 1828.[10] Italian chestnuts were imported there as early as the 1850s, when shopkeepers begin advertising their local availability. On February 6, 1856, Henry Maillard announced in the *New York Daily Tribune* that he had received a "splendid lot of Italian chestnuts" at his Broadway and Mercer Street establishment, nuts sold "at a moderate price."[11] A decade later, chestnuts were being imported from several Italian cities, including Genoa, which sent eighty barrels of chestnuts in a single February shipment (a barrel contained 3.3 bushels, or 165 pounds of chestnuts).[12] The Italian merchant Francesco Cuneo is said to have done the most to improve the Italian chestnut trade, as he made the nuts available nearly year-round. In 1879, after a decade of roasting and selling chestnuts on Manhattan street corners, Cuneo opened his own grocery store and nut-import business. As documented by the Works Progress Administration (WPA), the establishment operated well into the twentieth century, gaining him considerable publicity. "The great profits made from chestnuts," claimed the WPA writer, made Cuneo "one of the wealthiest Italians of New York."[13]

The importation of Italian chestnuts hardly slowed the demand for the native variety, as East Coast cities continued to receive large quantities of domestic nuts. Most came from the U.S. interior, where buyers and sellers had developed a supply network along major transportation hubs. In 1878, as many as two thousand bushels of chestnuts were shipped from Johnstown, Pennsylvania, to Philadelphia by railcar, a process beginning in early October and continuing until supplies were exhausted.[14] By 1881, New York City wholesale dealers were turning over ten to twelve thousand bushels of chestnuts per year, nuts sold largely by the "pint or quart, either roasted, boiled, or raw."[15] Despite their late-season scarcity, many New Yorkers preferred American nuts and paid dearly for them, especially those shipped from southern states like Virginia. "Virginia nuts," remarked Charles Helfrich to a *New York Sun* reporter in 1883, "which are the best, are sold at $11 and $12 a bushel, and State [New York] nuts at $10 to $11." Italian nuts sold for ten or twelve cents a pound, or $6 and $8 a bushel.[16] The apparent increase in the price of chestnuts is found in Helfrich's opening byline: "Chestnuts are scarce this year, and will be dear."[17] Indeed, scarcity was the best predictor of pricing, especially late in the season when the native nuts were hard to find at most retail establishments.

Native chestnuts were usually sourced from areas where the trees were most plentiful, including sections of western North Carolina, West Virginia, Pennsylvania, North Georgia, East Tennessee, and southwest Virginia. The penetration of railroads into those areas greatly benefited the chestnut trade, lowering prices as far away as Boston. In Georgia, Gilmer, Fannin, and Union Counties were legendary for their chestnut harvest and remained so well into the twentieth century. In 1870, the *Atlanta Constitution* reported that "fifteen hundred dollars' worth of chestnuts" were gathered for export in Gilmer County alone, shipments weighing forty thousand pounds.[18] The majority were delivered to Atlanta, where they were sold for immediate consumption or transferred by rail to neighboring cities such as Chattanooga, Charlotte, Nashville, or Columbia, South Carolina. Some chestnuts even arrived at the Georgia capital from surrounding states. In 1883, the *Atlanta Constitution* reported "it is not an uncommon thing for chestnut wagons to come here all the way from western North Carolina."[19]

Chestnuts were also shipped to Atlanta as baggage cargo during chestnut season, which usually began the last week of October. The following account by an Atlanta baggage handler is evidence this method of transport was indeed used in the trade. After observing a large trunk "going to Gainesville empty and coming back full," the bellman initially thought it contained illegal whiskey. After learning the town of Gainesville was located in a dry county, the handler closely monitored the trunk as it made numerous subsequent trips. When it finally broke open upon unloading, he observed its true contents, as "three bushels of chestnuts were exposed to view." Reflecting on the episode, the handler wryly told the local reporter he had smuggled "forty bushels of chestnuts to Atlanta in that trunk."[20]

By the mid-1880s, Appalachian communities were known as major suppliers of chestnuts, with some shipping entire carloads to wholesale distribution centers in Wheeling, Richmond, Baltimore, Philadelphia, and New York City.[21] The supply chain included everyone from the small and large gathering parties who collected the nuts to local buyers, railroad shippers, wholesale merchants, and finally the urban grocer or chestnut roaster. Along the way, each buyer and seller made a profit, the amount determined by the established market. To date, few researchers have documented these exchanges, especially those occurring during the late nineteenth century, when transactions along the supply chain went largely unrecorded.[22]

An exception is a legal complaint filed against the Virginia Railroad Commission by J. W. Rangeley of Stuart, Virginia. In the complaint, Rangeley asked the commission to reimburse him for losses sustained when shipping chestnuts to Richmond, Virginia. According to Rangeley, a train car containing his chestnuts had been scheduled to arrive in Danville, Virginia, on November 22, 1887, but was too late "for [connecting] trains of that date." After leaving Danville the next day, the nuts were miscarried to West Point, Virginia, arriving there on November 25th. On November 29, when the waybill for the cargo finally reached West Point, they were subsequently found and returned to Richmond that same day.[23] Upon arrival, the consignee refused the ruined shipment, voiding the entire sale. In response to the complaint, the claims agent sent a recommendation to the railroad commissioner, stating the delay of "eight or ten days in the delivery gives them a valid claim for the value of the chestnuts." However, rather than honoring Rangeley's original request for reimbursement, the agent recommended paying the value of the nuts "at the time they should have been delivered."[24] As a result, Rangeley received $107.80, or 3.5 cents per pound, for the chestnuts. The commissioner also recommended an additional payment of $19.71 for the prepaid freight charges and $3.30 for sacks, a total amount of $130.81.[25]

The reimbursement figures tell us a great deal about profit margins in the chestnut trade as well as shipping and handling costs for the major suppliers. In many instances, the seller simply paid what the next buyer in the chain was willing to offer, a price often based on local conventions and norms. For the grocer purchasing a gallon of nuts from schoolboys, the payment generally did not exceed two cents per pound. If a produce company like Rangeley's purchased large quantities of chestnuts from a local supplier, the price might be 3.0 cents per pound or $1.71 per bushel. Regarding the misplaced shipment from Stuart, Virginia, Mr. Rangeley likely paid 3.0 cents per pound for his chestnuts, as their resale value at the time of delivery was 3.5 cents per pound, or $2.00 per bushel. So if a Richmond wholesaler purchased the chestnuts at that price, he would need to sell the nuts at 4.3 cents per pound ($2.45 per bushel) to absorb the additional 21 percent shipping and handling costs. A review of advertised prices for October 1887 reveals that wholesalers in the Northeast sold chestnuts at 5.2 cents per pound, or $3.00 per bushel, a markup of 23 percent.[26]

Such figures are illustrative of the enormous profits that could be made by those working in the chestnut trade, particularly wholesalers and shippers moving the largest quantities. The son of a station master who shipped chestnuts from southwest Virginia at the end of the nineteenth century remarked that his father's commissions "were just fantastic," since nuts were considered perishables and could be billed at higher rates.[27] And because most chestnuts were shipped to cities on the East Coast, the longer routes also meant higher freight charges.[28] Whoever absorbed the bulk of the shipping charges—buyers or sellers—the supply chain was certainly an effective way of moving chestnuts from forest to table.

In fact, the annual trade became such a predictable part of American life that chestnuts were, in many areas, de facto currency. Individuals used them to purchase everything from coffee, milk, and tea at the crossroads store to local attorney services. In 1886, the *North Georgia Times* even accepted a half-bushel of chestnuts as payment for a newspaper subscription.[29] In Dahlonega, Georgia, the monetary return on chestnuts was so great that local residents often used the funds to pay their annual property taxes. As a reporter for the *Atlanta Constitution* stated it, "we heard an old citizen remark that the chestnut crop alone would be sufficient to pay the taxes for the present year in Lumpkin County, and with some of the mountaineers it is a saying that either chestnuts or blockade whiskey must pay the taxes. The trees up in the mountains are literally breaking with their burdens."[30]

Outside the Appalachians, chestnut gathering for profit was not uncommon, but varied in intensity depending on the region or community. In some New England townships, the annual chestnut harvest was so important it involved dozens, if not hundreds, of local residents. In the village of Hamburg, Connecticut, entire families often participated in the annual event. In fact, the additional income was so important that young children were taken out of school in order to gather the nuts. According to a *New York Times* report, Hamburg families made as much as $12 to $15 a day during chestnut season, an amount equivalent to the weekly salary of a skilled laborer.[31] The method of gathering was earnest and deliberate and involved first shaking the trees, and then carefully raking the burrs from underneath them. Next, the individuals literally crawled on the ground to pick up the fallen nuts or to remove them from the burrs. In anticipation of the annual harvest, merchants placed wooden bins on their shop verandas to store them until properly dried.[32] They were then dumped into wooden barrels and shipped by rail to New York or Boston, where they were sold to wholesalers and street roasters. In 1892, Hamburg residents gathered 321 bushels, or eighteen thousand pounds of chestnuts, a harvest valued at $900.[33]

As lucrative as the chestnut trade was for some Americans, it was not without problems or financial risks. As noted above, chestnuts were perishable freight, and their transport could not be delayed more than four or five days in warmer weather. Such delays allowed the larvae of the chestnut weevil—which lay dormant inside the nuts—to become active and emerge.[34] Because the larvae also consumed portions of the edible nutmeat and looked like maggots, their presence could nullify an entire sale. By the mid-1880s, more individuals were complaining of the weevil problem, especially in areas where the nuts were most plentiful. A report published by the West Virginia Agricultural Experiment Station in 1887 found that of the one million pounds of chestnuts shipped out of state, 44 percent were wormy.[35] Indeed, by the early twentieth century the damage done by chestnut weevils was so widespread that entire shipments had to be seized and destroyed.[36] Although the U.S. Bureau of Chemistry was responsible for these confiscations, the U.S. Marshall generally disposed of the infected chestnuts, an action requiring entire lots to be burned or incinerated.[37]

According to one report, between 1911 and 1915 nearly five hundred bags (each containing sixty to one hundred pounds of chestnuts) were confiscated by the Bureau of Chemistry. In addition, 280 large barrels of chestnuts were retained by the agency, all containing "worms, worm excreta, worm-eaten chestnuts, and decayed chestnuts."[38] In 1916, only seven violation notices were issued, however, including two to Joseph Wallerstein of Stuart, Virginia. According to Public Notice no. 5344, on October 25, 1916, the United States attorney for the Southern District of New York filed "for the seizure and condemnation of 21 bags of wormy and moldy chestnuts, remaining unsold in the original unbroken packages at New York, N.Y."[39] The New York attorney claimed the bags contained a "filthy, decomposed, and putrid vegetable substance, to wit, wormy and moldy chestnuts."[40] On November 14, 1916, after no claims were made on the property, a condemnation and forfeiture judgment was entered and the court ordered the U.S. Marshall to destroy the nuts.

If the weevil problem was a deterrent to those entering the chestnut trade, it did not stop individuals from attempting to profit from the seasonal bounty. Shipments of native chestnuts to urban markets continued well into the twentieth century, with the trade peaking sometime during the 1920s. To hedge their bets, some dealers simply purchased more nuts, anticipating losses incurred by the weevil attacks. Others resorted to boiling the chestnuts in iron cauldrons, which killed the eggs and larvae of the weevils but left the nutmeat unharmed.[41] In parts of the southern Appalachians, some residents continued the Native American practice of burning the leaves around the fallen chestnuts, which prohibited the growth of the weevil larvae.[42]

Although weevil infestations certainly diminished the economic value of native chestnuts, Americans continued to receive both income and daily nourishment from the yearly harvest.[43] As a result, the tree's status hardly suffered, as European chestnuts were not immune to the infestations, although did appear to be less impacted by the weevil attacks. Moreover, as native chestnuts became more widely and predictably sourced, their availability was no longer limited to only one or two months per year. Moreover, purchasing a few bad chestnuts was not necessarily a terrible inconvenience, given their relatively low cost and widespread availability. No longer mere holiday snack fare, chestnuts were by the early 1880s key ingredients in main courses, soups, side dishes, and desserts. Cookbooks of the period devote considerable space to their preparation, an indication of their growing influence on American life and cookery.

something that has been heard before

Perhaps the most popular chestnut recipes of the period were found in the widely read *Mrs. Lincoln's Boston Cook Book*, first published in 1884.[44] These included chestnut stuffing and chestnut gravy, two dishes already associated with Thanksgiving dinner. To make stuffing, instructed Lincoln, a quart of large chestnuts had to first be shelled and boiled in "salted water or stock till soft." The chestnuts were then passed through a sieve, folded into a cup of cracker crumbs, and seasoned with salt, pepper, and chopped parsley. To make chestnut gravy, a pint of hot water was poured over the drippings of the cooked bird and thickened with flour browned in butter. The remaining chestnuts were added to the gravy, as were salt and pepper to taste.[45]

The nuts were also used in main courses, including two popular dishes known as chestnut balls and chestnut croquettes. The instructions for making croquettes required boiling one and one-half pints of chestnuts in a quart of water and then adding a chopped celery root, one sliced onion, and a bay leaf. After becoming tender, the chestnuts, celery, and onion were drained and mashed together. To that mixture, onion juice, salt, butter, and a pinch of cayenne pepper were added, the whole then formed into patties, dipped in egg and bread crumbs, and fried in hot fat.[46] Chestnuts balls were not as popular as croquettes, but could be served as a "sweet entrée" for lunch or dinner. Their preparation required the blending of a cup of boiled "sieved" chestnuts, two egg yolks, one-fourth cup of cream, two beaten egg whites, and a tablespoon of sugar. Afterward, the entire mixture was formed into round balls, dipped into a batter of egg yolks and breadcrumbs, and fried in lard or fat.[47]

Chestnut soups were usually eaten before the main course or simply served alone. In Almeda Lambert's regionally popular cookbook, *Guide for Nut Cookery*, five recipes are featured, including one recipe—"Chestnut Soup No. 3"—which contains *only chestnuts*.[48] The most frequently mentioned chestnut desserts in late nineteen-century cookbooks are chestnut pudding, chestnut custard, and chestnut purée.[49] Chestnut purée was prepared using recipes first made popular in Europe or altered slightly to suit American tastes. In Mary Ronald's comprehensive *The Century Cook Book*, published in New York in 1895, the recipe calls for a pound of chestnuts that are boiled until tender. The nuts were then passed through a purée sieve, sweetened with sugar and cream, and flavored with vanilla extract. Next, the entire mixture was heated in a saucepan and stirred over low heat until firm. Once the mixture cooled, it was pressed through a colander directly onto the serving dish. Finally, the purée was very lightly topped with whipped cream, so as not to destroy the "vermicelli-like form the colander has given it."[50]

Perhaps the pinnacle of chestnut cookery guides is the *Nut Menu*, a widely circulated pamphlet published by the Glen Orchard of Mount Joy, Pennsylvania (see fig. 7.2). Although the Glen Orchard was known for its

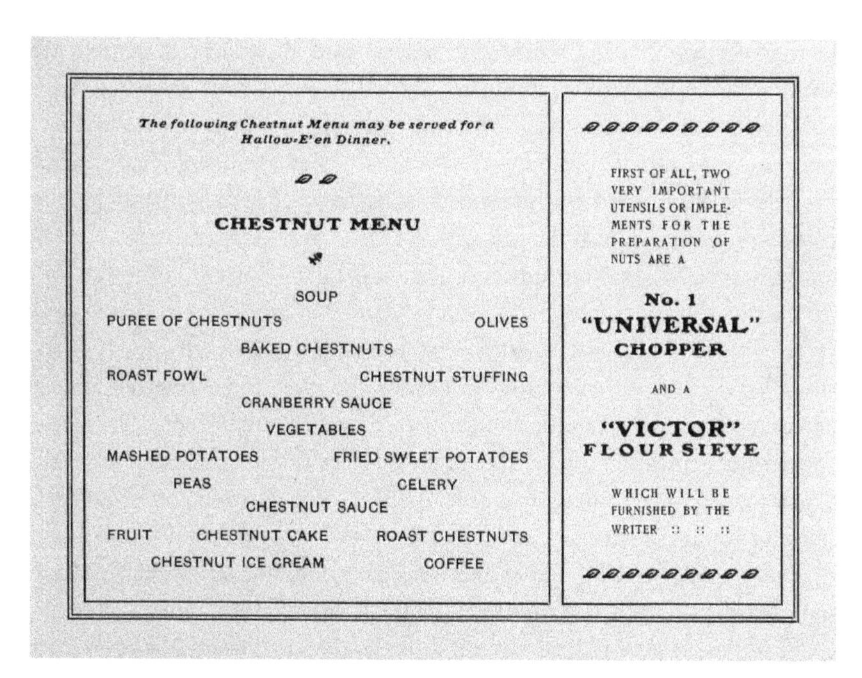
The following Chestnut Menu may be served for a Hallow-E'en Dinner.

CHESTNUT MENU

SOUP
PUREE OF CHESTNUTS — OLIVES
BAKED CHESTNUTS
ROAST FOWL — CHESTNUT STUFFING
CRANBERRY SAUCE
VEGETABLES
MASHED POTATOES — FRIED SWEET POTATOES
PEAS — CELERY
CHESTNUT SAUCE
FRUIT — CHESTNUT CAKE — ROAST CHESTNUTS
CHESTNUT ICE CREAM — COFFEE

FIRST OF ALL, TWO VERY IMPORTANT UTENSILS OR IMPLEMENTS FOR THE PREPARATION OF NUTS ARE A

No. 1
"UNIVERSAL"
CHOPPER

AND A

"VICTOR"
FLOUR SIEVE

WHICH WILL BE FURNISHED BY THE WRITER :: :: ::

FIG. 7.2. Addison S. Flowers, *Nut Menu: A Treatise on the Preparation of Nuts for the Palate* (Mount Joy, Pa.: [Glen Orchard Hotel], 1903). Courtesy of the Library of Congress.

Paragon chestnuts, a European cultivar, native chestnuts were also grown on the premises.[51] By the late 1890s, a hotel on the orchard grounds even served chestnut-themed dinners, including such dishes as golden chestnut salad, chestnut cake, and chestnut ice cream. According to author Addison S. Flowers, the preparation of golden chestnut salad required blanching a dozen chestnuts and then placing them on a bed of lettuce. Before serving, the lettuce and chestnuts were smothered with French dressing as well as two boiled eggs that had been passed through a sieve.[52] To make chestnut cake, a pound of boiled chestnuts was pressed through a colander and then mixed with a pound of flour and a teaspoon of baking powder. In a separate bowl, six ounces of butter and sugar were then blended with three eggs and a teaspoonful of vanilla. The chestnut-infused flour and a gill (roughly half-cup) of milk were folded into the mixture as well as the remaining boiled chestnuts. Finally, the resulting batter was poured into a cake pan and baked in a moderately heated oven for two hours.[53] Chestnut ice cream was made by simply adding the nuts to the following base ingredients—cinnamon, lemon peel, vanilla, egg yolks, and cream—and then hand-churning the entire mixture in a wooden freezer.[54]

As chestnuts took a more prominent role in American cuisine and became more visible in the home and kitchen, they also began influencing everyday speech. By 1880, the word "chestnut" referred not only to the color of a horse, but also the color of one's hair and eyes. Variations for eye color included dark chestnut, chestnut, and light chestnut, although any dark-brown iris encircled by another color—including green, hazel, and yellow—could be labeled as such.[55] By the end of the nineteenth century, the color evoked an entire range of shades and hues and applied to almost any object that was deep maroon, burnt umber, or dark brown.[56]

Another expression linked to the native tree was the term "chestnut coal," a popular heating source after 1870. In the United States, chestnut coal warmed homes, businesses, schools, and factories for nearly a full century. Smaller than egg coal but larger than pea coal, chestnut coal was, as the name suggests, the size of native chestnuts.[57] Legally standardized in 1869, chestnut coal had to be screened in a wire mesh no less than "five-eighths of an inch and no greater than an inch and one-eighth."[58] Although it remained one of the most common coal grades of the late nineteenth century, it decreased in popularity with the later introduction of more efficient heating stoves and furnaces. After 1950, chestnut coal was a scarce commodity.

The expression "chestnut"—in reference to a stale joke, cliché, or oft-repeated phrase—was in common usage by the early 1880s.[59] Although there is disagreement about when the term was first used, most explanations link its origin to *The Broken Sword*, a play dating from 1816 but performed in the United States after the Civil War. In the original script, written by English playwright William Dimond [*sic*], two principal characters exchange the following dialogue:

> CAPTAIN XAVIER: At the dawn of the fourth day's journey, I entered the wood of Collares, when suddenly from the thick boughs of a cork tree——
>
> PABLO: (*Jumping up.*) A chesnut, Captain, a *chesnut*.
>
> CAPTAIN XAVIER: Bah! you booby, I say, a cork!
>
> PABLO: And I swear, a chesnut—Captain! this is the twenty-seventh time I have heard you relate this story, and you invariably said, a chesnut, till now.[60]

According to several sources, in 1867, when theater manager Martin W. Hanley took a production of the play to Rochester, New York, the actor playing Xavier—comedian William Warren—replaced the word "cork" with the word "hickory," presumably to appease American audiences. The Pablo character followed the original script, however, saying, at the appropriate moment, "No, chestnut; I tell you chestnut." Afterward, when one of the actors was telling a familiar anecdote in the theater dressing room, the entire troupe interrupted him with "shouts of 'Chestnut.'" According to Hanley, the expression "clung to the company all the season" and was soon adopted by the entire profession.[61]

A lesser-known but equally probable origin for the term involves the American painter Edwin Austin Abbey (1852–1911), a noted illustrator for *Harper's* magazine. In the late 1870s, Abbey received the nickname "Chestnut" after telling a story about the tree and nuts in a most unique and compelling way.[62] As related by friend and colleague Earl Shinn in 1886, the story focused listeners' attention on chestnuts in a specific tree, and then strung listeners along so that a conclusion about their state and number was indefinitely prolonged. Apparently, the tale could be told by Abbey "for a good part of an hour, without ever releasing the attention or satisfying the expectation."[63] After accepting the nickname, Abbey used the symbol of a chestnut burr encircled by the word *chestnut* on a seal he would impress in wax on all his correspondence.[64] Later in life, Abbey himself claimed responsibility for adding "a word to the English language," at least the one signifying "an old stale story—something that has been heard before."[65] He also encouraged his English literary friends to use the expression, although the term was not commonly written or spoken in England until the early twentieth century.[66] Whoever was responsible for popularizing the expression, all facts verify its country of origin, suggesting Americans found a greater logic in its use. To call something commonplace or overly abundant a "chestnut" made perfect sense in a place where the trees and nuts were as plentiful and inescapable as overused expressions.

Besides influencing written and spoken language, the trees and nuts were also celebrated in popular music, further evidence they were becoming an inescapable part of American life. Within a span of two years, three songs with the word chestnut in their titles were copyrighted, with the first—"Opening the Chestnut Burr"—published in 1884. The song paid direct homage to the best-selling book of the same title by Edward Payson Roe, which had been published a decade earlier.[67] Written by American lyricist E. Clark Reed, the song uses the chestnut burr as a metaphor for the unexpected difficulties faced in life, as well as those joyful occasions

FIG. 7.3. Mulbro, *Chestnuts on the Brain* (W. F. Shaw, 1885). Courtesy of the Library of Congress.

that have bittersweet endings. The chorus even uses a chestnut frolic to make that very point.

> For they take me back to Childhood,
> And the many hours we spent
> In the pleasant autumn time,
> When we a-nutting went,
> All day the toil seemed pleasure,
> But at night we weary were
> With feet all worn and fingers torn
> from opening the chestnut burr.[68]

Another song, "Chestnuts on the Brain," indicates just how much the trees and nuts had permeated the American consciousness, as well as the growing popularity of the expression referring to stale jokes and overused slogans (see fig. 7.3). The comic ditty, published by William F. Shaw in 1885, makes its focal point the biblical Noah, who, after "finding pairs of animals of every kind," gathers up "chestnuts sweet, and puts them in the Ark." Noah does this to give his family "something sweet, to work on after dark." The song ends with the warning that if anyone should think about telling a funny joke or pun, beware, because even "if it starts quite well," someone will most assuredly interrupt and yell, "Oh! chestnuts, got 'em bad."[69]

In addition to popular music, chestnuts and chestnutting were celebrated in short stories and poems, many of which were published in America's most prominent serials and newspapers. Artists featured the trees and nuts in still lifes, landscapes, prints, and drawings. James W. Lauderbach's wood engraving "Gathering Chestnuts," published in the January 1878 edition of the *Art Journal*, for example, commemorates the annual nut-gathering ritual in Philadelphia's Fairmount Park.[70] Native Americans also paid homage to the tree during the period, mostly in the form of published oral histories. Collectively, these documents attest to the tree's indelible influence on American culture during the second half of the nineteenth century. True, the American chestnut was a primary source of sustenance and timber, but it was also the subject of story, song, and art.

Perhaps no other tree species had influenced American life in this way.

PART THREE

Chestnut Decline

CHAPTER 8

the wonder and admiration of all

In the history of American commercial chestnut growing, the year 1896 was an extremely important one. In that year alone, three publications touted the horticultural advantages of Japanese and European chestnuts, giving orchardists and nurserymen a wider public forum to promote their importation.[1] By that date, their popularity had grown to nearly a fever pitch, especially among nut growers, who used hyperbole and even artifice to encourage their cultivation.[2] Perhaps the most notable was *Nut Culture in the United States*, a document published by the USDA's Division of Pomology. The monograph features a lengthy section on foreign chestnuts, including descriptions of twenty-two named varieties.[3] One of the most touted was the Paragon, a hardy and productive cultivar developed in Lancaster County, Pennsylvania.[4] It was discovered in the late 1840s by William L. Schaeffer, who believed the tree had "American blood."[5] Upon closer inspection, it proved to be European in origin, and produced a larger nut much like the Italian marron. It was made commercially available in 1888.[6]

The introduction of Japanese chestnuts into the United States was due mostly to New Jersey nurseryman William Parry, who established chestnut groves along the east bank of the Delaware River in what is today Cinnaminson, New Jersey.[7] As early as 1882, Parry promoted the Japan Giant cultivar, a tree that produced nuts as much as two inches in diameter.[8] However, the first importer of Japanese chestnuts was Samuel B. Parsons of Flushing, New York, who grew them at his Kissena nursery in 1876. Although the original Parson's Japan cultivar eventually fell out of favor, Japanese chestnuts were sold at the nursery well into the twentieth century.[9]

In 1887, Joseph Williams, one of Parry's close neighbors, set out seventy-five hundred American chestnut seedlings on his adjoining property, initiating a curious trend among commercial chestnut growers. Two years later, Williams whip-grafted both Japanese and European chestnut scions onto the American saplings, producing a seamless union between the two trees. By 1896, the whip-grafts had grown into young saplings, creating, in effect, a neat and orderly orchard of Japanese and European chestnuts.[10] In Meriden, Connecticut, nut grower Andrew J. Coe grafted Japanese chestnut trees onto eighteen acres of native chestnut sprouts, making him the first New England resident to grow them commercially.[11] Coe obtained the trees in 1895 from Luther Burbank, the plant breeder responsible for creating the Shasta daisy and July Elberta peach, among other noted cultivars.[12] Burbank spent a considerable amount of time cultivating chestnuts at his California nursery, and was the first to import Japanese chestnuts into the Golden State. In 1884, Burbank received a box of twenty-five "monster chestnuts" from Japan, which he used in various breeding experiments.[13]

It was during that same period that Burbank imported Chinese chestnuts into the United States, making him perhaps the first individual to do so. In a handwritten account of the experiment, Burbank claimed that in addition to Japanese chestnuts, other Asian varieties were growing on his premises, including ones "which came from China."[14] In 1898, Burbank told George H. Powell, the noted Delaware horticulturalist, he had several hundred hybrid chestnuts "beginning to bear"—crosses of Japanese, European, Chinese, and American chestnuts and chinquapins.[15] Well before that date, Burbank had shipped thousands of Japanese chestnut trees to various growers along the East Coast, including John H. Hale of South Glastonbury, Connecticut. Hale, a celebrated peach orchardist, had also developed a keen interest in growing and selling chestnuts and needed nursery stock for that purpose.[16]

By 1900, the cultivation of Japanese chestnuts had advanced to such a degree that thirty different cultivars were being propagated in the United States and Canada. With names like Giant, Mammoth, and Jumbo, Americans were enamored with these new varieties, even if the trees themselves were smaller in size. The largest number of Japanese trees were growing at the Albion Chestnut Company of Clementon, New Jersey, whose premises spread over 150 acres of farmland ten miles west of Philadelphia.[17] Adjoining that property was the Mammoth Chestnut Company orchard, which, like the Albion establishment, possessed several thousand Japanese and European chestnuts.[18]

John H. Hale was so enthralled with Japanese chestnuts that he planted ten acres of the trees on his two-thousand-acre farm in Fort Valley, Georgia. Hale—who by then was known as Georgia's "Peach King"—believed the trees offered nut and fruit growers a truly unique commercial opportunity. However, Hale's operation was also heavily dependent on cheap southern labor, a topic of keen national interest. In 1901, the Connecticut-born Hale was even asked to testify before Congress at a hearing investigating the labor practices of absentee landowners. He told the committee members he had grown Japanese chestnuts with the idea of selling them for "10 cents a pound [or] $5 per bushel."[19] Among those in attendance was New York congressman John Farquhar, who questioned Hale about the advantages of the imported variety. When Farquhar asked about their particular size and flavor, Hale responded by holding up a black-and-white photograph of the nuts and stating proudly, "They are as sweet as our native American chestnuts, yet as large as horse chestnuts."[20]

What is interesting about Hale's claim is that his "Japanese" trees were most likely hybrids, possessing an American chestnut parent, grandparent, or great-grandparent. Although he undoubtedly exaggerated their sweetness, Luther Burbank, who sold trees to Hale, bred chestnuts for flavor, which generally required American chestnut or Allegheny chinquapin parentage. The trees may have even had some Chinese chestnut heritage, as they were growing in close proximity to trees imported from the Chinese mainland, making cross-pollination a possibility. And if cross-pollination did occur (either intentionally or accidentally), not only was the size and flavor of the nuts altered, but other characteristics would have been transferred to the trees as well.[21]

Of course, not all Japanese chestnuts grown in the United States could trace their origins to the California gardens of Luther Burbank. As noted above, several nurserymen imported trees from Japan during the 1890s, including John T. Lovett of Little Silver, New Jersey.[22] Lovett's establishment—known in the trade as the Monmouth Nursery—was located fifteen miles south of Brooklyn, near what is today Middletown, New Jersey. The Monmouth Nursery annually published *Lovett's Guide to Fruit Culture*, a handsomely illustrated catalogue containing hundreds of different kinds of ornamental flowers, berries, and shrubs—all available via mail order. Among Lovett's offerings was the Japan Giant chestnut, which, as he described it, had such large nuts, it was "the wonder and admiration of all."[23]

Although most growers had abandoned the cultivation of American chestnuts for commercial profit, amateur orchardists had not done so, particularly those living near productive trees or groves. As noted in the USDA volume *Nut Culture in the United States*, such trees were scattered across much of the eastern United States. Occasionally, if a tree produced nuts of large size or exceptional flavor, it even became a named variety. The Griffin, for example, was marketed and sold by J. L. Harris of Griffin, Georgia.[24] Equal in size to the Griffin was the Dulaney, named after William L. Dulaney of Bowling Green, Kentucky. Another cultivar known for exceptional size and flavor was the Murrell, whose namesake was George E. Murrell, a Virginia resident living in Coleman Falls near Lynchburg (see fig. 8.1).[25] For the commercial chestnut grower, however, such trees were exceptional anomalies, causing them to ultimately reject the idea of growing native trees in orchard settings. Of course, it was not just the size of the nuts that mattered to commercial orchardists: it was also the perennial problem of spraying taller native trees for harmful insects like chestnut weevils.

Moreover, with the growing popularity of chestnuts in American kitchens, demand often exceeded supply. Cookbook authors even began recommending the use of European nuts in recipes, as they were much easier to prepare. As a result, even larger quantities of chestnuts were imported from Europe, particularly from Italy, France, Spain, and Portugal. From 1890 to 1893, an estimated 6,442,908 pounds of chestnuts were shipped to the United States, nuts valued at $235,976.[26] By 1906, that figure more than doubled, with four Italian cities alone exporting $121,251 worth of chestnuts to America.[27] With demand outstripping supply, it is little wonder U.S. growers were so adamant about growing foreign varieties. And even though there were still those championing the American chestnut as a valuable nut producer, native trees were becoming scarcer in urban areas where demand was greatest. That fact alone caused many orchardists to promote the exclusive growing of European and Japanese chestnuts.

Some even suggested commercial chestnut cultivation include *no* American chestnuts. As New England forester Ernest A. Sterling bluntly stated it, “nurserymen are not slow to follow up a line of work which promises even mediocre returns; hence, the mere fact that our [native] chestnut has received little attention from them is in itself proof that its cultivation as a shade or nut-bearing tree is not exceedingly profitable. This refers only to the wild native chestnut, and not to the improved or acclimated foreign varieties.”[28]

A similar view was shared by Pennsylvania grower Nelson L. Davis, who recommended removing all native trees when establishing a chestnut orchard. In his opinion, the ideal chestnut grove should contain “seventy-five to one hundred [native] trees per acre,” the exact number of stumps required for grafting.[29] According to Davis, each stump should be grafted with two or more European sprouts, so that in five or six years, they “will be

FIG. 8.1. American chestnuts: 1. wild, unnamed variety; 2. Murrell; 3. Hulse; 4. Excelsior; 5. Ketcham; 6. wild, unnamed variety; 7. Watson; 8. Otto; 9. Dulaney; 10. Griffin. European chestnuts: 11. Numbo; 12. Ridgely; Japanese chestnut: 13. Japan Giant. William P. Corsa, *Nut Culture in the United States: Embracing Native and Introduced Species* (Washington, D.C.: GPO, 1896). Courtesy of the USDA National Agriculture Library.

transformed into a . . . chestnut grove that is giving returns."[30] While some followed the advice of Davis, others were more reluctant to do so, instead grafting trees on already established American chestnut saplings. In both cases, the Japanese and European trees survived as biological symbiotes, receiving water and vital nutrients from the established root systems of the American stock, while the American "trees" received energy and nutrients from the imported grafts through photosynthesis, a process they no longer performed. The imported and native trees together, as a single plant organism, shared not only the same space, but molecules and even DNA.[31] If one species was exposed to a fatal disease, both succumbed.

By 1903, European and Japanese chestnut orchards were established in a dozen U.S. states and twice as many townships, encompassing nearly ten thousand acres of farmland.[32] Although this represented only a tiny fraction of the geographic area occupied by native chestnuts, the orchards were hardly inconsequential. When growers abandoned native trees to make way for foreign varieties, the American chestnut was doomed as a vital, self-reproducing species. Flourishing on the North American continent for millennia, the species soon faced a foe more devastating than insects, late frosts, hungry livestock, forest fires, and industrial timbering. Humans were responsible for this unfolding chain of events, as twentieth-century nut growers continued to embrace eighteenth-century prejudices against untamed wild nature. In their minds, native chestnuts needed further improvement if they were going to truly benefit humankind. A well-managed orchard of foreign chestnuts—void of pests, disease, and unpredictable yields—was superior to even the stateliest grove of wild trees.

classed with the most destructive parasites

When Hermann W. Merkel surveyed the grounds of the New York Zoological Park (Bronx Zoo) in the summer of 1904, he did so as its first chief forester. This was not an easy task, as the park encompassed some 260 acres and several bodies of water, including the Bronx River, nearly two hundred feet wide. It was also home to many large trees, as two-thirds of the park was covered with a mature forest of oak, chestnut, tuliptrees, beech, and hickory.[33] On an otherwise uneventful summer day, Merkel investigated several diseased American chestnut trees that had caught the attention of groundskeepers.[34] Thinking they had been attacked by a common fungus, Merkle sprayed the infected trees with fungicide and continued to monitor their health as cooler weather approached. The following spring, Merkel realized the problem was far more serious, as nearly every chestnut in the park showed signs of infection. In fact, the situation was so grave he asked William A. Murrill of the New York Botanical Garden—whose grounds bordered the park at its northern end—to inspect the dead and dying trees.[35] Murrill, who served as assistant curator of the Botanical Garden, was also a noted mycologist specializing in North American plant fungi. Upon closer inspection, Murrill was unable to identify the fungus, but recommended spraying all 438 trees with a "Bordeaux Mixture." He also asked Merkle to remove, as much as humanly possible, the infected limbs and branches.[36]

Knowing the importance of identifying the pathogen, Murrill took samples from the trees and began growing the fungus in the Botanical Garden greenhouse. In 1906, after months of studying the pathogen's life cycle, he named the fungus *Diaporthe parasitica*, believing it to be a new, unnamed species.[37] Over the next decade, most Americans referred to the fungus as "chestnut bark disease" or, more commonly, "chestnut blight" (it was renamed *Endothia parasitica* in 1912 and *Cryphonectria parasitica* in 1978).[38] To the layperson, the fungus appeared not as a single pathogen, but as two or three different organisms. Initially, when it penetrated the bark and cambium of the tree, the surface of the infected limb or trunk turned a pale brown. Weeks later, small yellow-brown pustules emerged from the outer bark, giving it a slightly warty appearance. In a matter of days, these nodules formed reddish-brown masses that emitted millions of "summer" spores, which, as Murrill noted, were disseminated "by wind and other agencies, such as insects, birds, squirrels, etc."[39] Finally, the so-called "winter" spores, or ascospores, emerged, small dark-colored sacs with strings extending from a circular apex. Both summer and winter spores were capable of reproducing the fungus, although each required specific ranges of moisture, temperature, and sunlight to do so.[40]

In most cases, the infection caused the bark to form large cankers, usually at the location where the fungus first entered the tree.[41] Once the fungus girdled a limb or branch, the healthy leaves above it drooped and turned brown, a sign water and other vital nutrients were no longer traveling to that part of the tree.[42] Easily seen from a distance, observers were able to determine if a tree was infected without visiting each and every specimen. It was also a sign the death of the tree was imminent, as nothing appeared capable of stopping the fungus. Writing in the spring of 1906, Murrill was not at all optimistic. "In its effect on the host, this fungus may be classed with the most destructive parasites," he wrote. "The attack [is] so vigorous that young trees often succumb in one or two years, and older ones soon lose branches of such size that the vigor of the entire tree is materially impaired and its beauty and usefulness practically destroyed."[43] After combatting the illness for more than a year, Merkel was also not very hopeful about the trees' chances for survival: "it is safe to predict that not a live specimen of the American chestnut (*Castanea dentata*) will be found two years hence in the neighborhood of the Zoological Park."[44]

To the shock of New Yorkers, Merkel's prediction came true. By the autumn of 1906, the forester reported all the chestnuts in Brooklyn's Forest Park were either dead or dying, and many of the trees in Prospect Park were also seriously affected.[45] By the following spring, the fungus had also destroyed large numbers of trees on the grounds of the New York Botanical Garden. According to Nathaniel L. Britton, the director-in-chief of the Botanical Garden, many of the trees had already been heavily pruned, but he conceded most would need to be completely cut down.[46] In the spring of 1908, Murrill told a *New York Times* reporter the Botanical Garden had lost a total of three hundred trees and was "waiting for Winter and workmen at liberty to cut down as many more." In New York City alone, the value of the dead and infected trees was estimated to be between five and ten million dollars. Murrill did not find this an exaggeration, claiming a single mature tree could be valued at a hundred dollars, and "a single large chestnut tree in a favorable position on an estate may be worth, and would increase the value of the property, as much as $1,000."[47]

If the precise value of a chestnut tree was debatable in 1908, everyone was in agreement that the blight had spread well beyond the New York City boroughs. As early as August 1907, the fungus was reported as far north as Poughkeepsie, New York, and as far south as Trenton, New Jersey. Chestnut blight and numerous dead trees were observed on western Long Island and along the Connecticut shores as far east as Fairfield. Curiously, if one drew a straight line along the known points of infection, the result would be an isosceles triangle with the Bronx Zoo at its center.[48] After learning this, New York residents began asking questions about the origins and subsequent spread of the disease. "Where exactly did the fungus come from?" "Did it arrive first in the Bronx?" "Has it been observed elsewhere in the United States?"

Although it is unlikely chestnut blight entered New York City at a single location or via a single shipment of infected trees, if that indeed were the case, then the Bronx or Queens was the initial point of contact.[49] Of course, knowing the exact point of entry was critical in stopping the disease, as that information might disclose its true source as well as a potential cure. Although many speculated about who was responsible for bringing the fungus to North America, those closest to the outbreak's epicenter were perhaps most qualified to solve the biological riddle. William A. Murrill was unable to point his finger to a specific source, but did observe, in the autumn of 1906, a Japanese chestnut with blight symptoms along the eastern boundary of the Botanical Garden.[50] Murrill, like others at the time, was under the impression Japanese chestnuts were immune to the blight, meaning they could be carriers of the fungus but not die from the disease.[51] For that reason, he was truly surprised when, a year later, the tree had developed a severe canker at the base of its trunk. In fact, the fungus growth had advanced to such a degree that Murrill was unwavering in his prognosis: "the tree will doubtless succumb soon after the next season opens."[52]

Murrill never publicly mentioned the Botanical Garden's Japanese chestnuts as a possible source for the disease, which is understandable, given the enormous liability issues had that been true. Moreover, Murrill's official duties at the Botanical Garden did not start until 1903, so the provenance of the Japanese trees was likely unknown to him. Had he inquired about their origin he would have learned Japanese chestnuts were growing at the Botanical Garden as early as 1897, when plant inventories place several in the greenhouse nursery and along the borders of the main grounds.[53] The trees were likely purchased from the Yokohama Nursery Company of Yokohama, Japan, one of several large nurseries selling rootstock and scions to U.S. buyers. They may have also been acquired locally, as establishments such as the Monmouth Nursery in Little Silver, New Jersey, sold trees directly to the public, as did other area nurseries.[54]

Did the New York Botanical Garden's Japanese chestnuts carry the pathogen prior to 1904? Unfortunately, the question cannot be answered with certainty as the Japanese rootstock may have had the fungus without showing signs of possessing the disease (in its dormant phase, it is invisible to the naked eye).[55] And some Japanese trees harboring the pathogen could have even slowed or stopped its spread at the point of infection. Severely infected specimens, however, would have spread the disease to other nursery stock or nearby adult native trees. In fact, as early as 1898, horticulturalist George H. Powell—who worked exclusively in Delaware and New Jersey—reported a "blight" on Japanese chestnut nursery stock in one of those two states. In a published government report, Powell noted it was more virulent on the southern and western sides of trees, causing the bark to split and collapse inward.[56] A photograph of the infection reveals the same cankers associated with *Cryphonectria parasitica*, as well as the scarring of the outer bark that often results from the growth of its summer spores. Powell noted both American and European trees appeared susceptible to the disease, and seemed surprised by its lethalness. "In a lot of one thousand Imported European seedlings," he observed, "nine hundred and fifty . . . died soon after setting out."[57] If Powell did, in fact, witness *Cryphonectria parasitica* on those trees, he did little to increase public awareness of the pathogen. In the report, he simply labeled the disease a "sun scald," a generic term used for any number of common plant ailments.

Evidence of chestnut blight does not surface again until 1902, when mycologist Frederick D. Chester reported seeing an infection on Japanese chestnuts in Delaware.[58] Although Chester believed the pathogen to be a "*Cytospora*," he was unsure of its identity and rendered a detailed drawing in order to make, in his words, "a record of the fungus at this time since it may be of possible interest in the future" (the drawing shows a remarkable resemblance to *Cryphonectria parasitica*).[59] In 1903, a disease with characteristics of chestnut blight also surfaced in a chestnut orchard in Bedford County, Virginia, a location more than four hundred miles from New York City. And in 1905, only nine months after the disease was observed at the Bronx Zoo, blight symptoms were seen in chestnut orchards in Warren County, New York, Lancaster County, Pennsylvania, Greensboro, North Carolina, and British Columbia, Canada.[60] When Dr. Haven Metcalf, the chief plant pathologist at the USDA, later speculated on how the fungus could be found in such remote places, hundreds of miles from the supposed point of entry, he matter-of-factly concluded they "contained chestnut orchards with Japanese chestnut trees."[61]

In fact, it was Metcalf who first promoted the idea that imported Japanese chestnuts were responsible for bringing chestnut blight to North America, a theory he introduced as early as 1908.[62] Metcalf based his belief on the relative immunity of Japanese chestnuts to the blight, but also the knowledge that they were introduced and cultivated "on Long Island and in the very locality from which the disease appears to have spread." Both of these things together, wrote Metcalf, suggested "the interesting hypothesis that the disease was introduced from Japan."[63] Despite overwhelming evidence that his theory was accurate, it was debated for nearly a decade and did little to influence U.S. blight-prevention policies.

Metcalf's hypothesis was partially confirmed in 1914 when Dr. Johanna Westerdijk, a well-known Dutch botanist, observed a pathogen "identical with *Endothia parasitica*" (*Cryphonectria parasitica*) on Japanese chestnut trees near Nikko, Japan.[64] Her specimens were lost at sea, however, so it would take another year before North American scientists would be convinced of the blight's origins. The official pronouncement resulted from the travels of plant hunter Frank N. Meyer, who in 1915 found blight on trees inhabiting an exposed mountainside above Nikko.[65] Days later, Meyer also observed the fungus on wild trees surrounding the grounds of the Yokohama Nursery Company in Yokohama, providing the smoking gun that linked chestnut blight to the importation of Japanese chestnut nursery stock. Upon his return to Washington, D.C., Meyer presented the infected limbs to USDA plant pathologists Cornelius Shear and Neil Stevens, who reproduced the fungus in cornmeal and potato agar.[66] Afterward, the pair confirmed the cultures were identical with those made from "typical *E. parasitica* [*Cryphonectria parasitica*] collected in this country," adding that the "mycelial and spore characters of this fungus as well as its cultural characters are so distinctive as to leave no doubt as to its identity. The fungus collected by Meyer at Nikko is unquestionably *Endothia parasitica*."[67]

easy prey for their natural enemies

If the route taken by chestnut blight to North America was obvious to some observers, not everyone was convinced Japan was the source of the disease. As late as 1912, nearly a decade after the blight's initial discovery, Connecticut botanist George P. Clinton argued the fungus was native to North America and referenced past chestnut die-offs to support his claim.[68] Clinton, who was Metcalf's biggest critic, believed widespread chestnut mortality was due to other factors, including forest fires, harsh winters, late frosts, severe droughts, and harmful insects. Any or all of those things, argued Clinton, would have weakened the trees and made them prone to disease outbreaks.[69] He also believed the blight was the same disease that had eliminated chestnut populations in North Carolina, Georgia, and Alabama during the nineteenth century. In Clinton's opinion, the pathogen killing trees in the southern Piedmont was not ink disease, but a relative of *Endothia gyrosa*, a fungus observed in Europe in 1862.[70] A supporter of Clinton's view was Harvard mycologist William G. Farlow, who also believed "our American chestnut tree fungus does not appear to be new but to have been known on chestnuts in Italy fifty years ago."[71]

Ultimately, Clinton was unable to defend his position regarding the origin of the blight fungus. However, the notion the trees were physically weakened as a result of natural and human-induced stresses continued to hold weight among both experts and the general public well into the twentieth century. In 1905, William A. Murrill suggested the repeated coppicing of chestnut had made the trees prone to the fungus, as the practice "cannot fail at length to impair the vigor of each generation of sprouts and render them peculiarly liable to speedy infection and vigorous attack."[72] A similar view was held by government forester Raphael Zon, who in 1904 found that 90 percent of all chestnut trees in eastern Maryland were coppiced.[73] Zon was certain repeated coppicing weakened the trees, making them even more susceptible to disease. "It must not be forgotten," he stated forcefully, "that a chestnut stump can not [*sic*] go on coppicing forever. With each new generation of sprouts the stump becomes more and more weakened, and hence gradually loses its capacity to produce healthy and vigorous sprouts."[74]

Although Zon admitted it was impossible to know exactly how many generations of coppice sprouts could be grown from a stump before it loses vitality, he was certain the practice had increased the number of dying chestnuts in the state.[75] Although he offered no direct evidence that chestnut blight had entered Maryland, his concluding remark was, if not prescient, an indication of how prone the trees were to external threats. "The immediate cause of their death can nearly always be traced to attacks of either insects or *fungi*," argued Zon, "yet the prime reason is their decreased vitality, which makes them easy prey for their natural enemies."[76]

If coppicing caused chestnut blight to spread more rapidly, then injurious insects also might. One such insect was the periodical cicada, which in the eastern United States includes five species of the genus *Magicicada*.[77] Locally known as locusts or "jarflies," periodical cicadas reproduce in extremely large broods every thirteen or seventeen years.[78] In the past, many adult females laid their eggs on the smaller branches of chestnut trees, leaving noticeable scars and entry points for fungi and pathogens.[79] In 1902, seventeen-year cicadas did such extensive damage in New Jersey, Pennsylvania, New York, Maryland, and Connecticut that trees lost entire branches as a result of strong winds (the outbreak was caused by what entomologists refer to as "Brood X" or the "Great Eastern Brood").[80] Although wind damage also made trees more susceptible to the blight fungus, the heavy scarring caused by the cicada's egg-laying activities was more widespread. In 1910, when Bucknell University biology professor Nelson F. Davis observed cicada damage at a chestnut grove near Shamokin, Pennsylvania, several years after the trees had been "stung" by the adult females, many large branches were beginning to heal, but the smaller ones "still possessed wounds through which the spores of the fungi could enter."[81]

Another insect weakening the vigor of the American chestnut was the two-lined chestnut borer, a dark-colored beetle that deposited its eggs in the crevices of the tree's outer bark. After hatching, the larvae tunneled into the inner bark, making meandering channels in the underlying cambium and sapwood layers. The extensive feeding galleries slowed if not prevented the flow of water to the tree's upper branches, causing them to wilt and later die.[82] Tens of thousands of chestnut trees perished annually from the attacks, sometimes over large areas. In 1893, the insects did serious damage to large stands in Fairfax County, Virginia, killing "about 75 percent of the chestnut trees . . . in that and adjoining counties."[83] Of all the insects attacking the tree, the two-lined chestnut borer was by far the most destructive. A 1912 report in the *Macon Telegraph* matter-of-factly proclaimed, "450 species of insects inhabit the chestnut. While all of these are not destructive . . . the so-called two-lined chestnut borer is directly responsible for the death of more timber than all of the others combined."[84]

The chestnut timber worm also did considerable damage to chestnuts, although its presence generally did not result in the death of trees. Chestnut timber worms are actually small brown beetles that, in their larval stage, create long, narrow channels in the tree's heartwood. Because they only make small holes as they enter and exit, timber worms usually do not adversely impact the structural integrity of the wood. They also received less attention from entomologists and lumbermen of the period.[85] However, the holes and channels of the timber worm were clearly visible after the wood was milled, making identification of the infestation an easy task. Initially viewed as a defect, the wood later became both aesthetically desirable and even commercially profitable.

Indeed, by the late-1890s, the wood was in such demand that a separate grade of lumber—"sound wormy chestnut"—was approved by the commercial lumber trade (see fig. 8.2). Both coffin and piano makers used it as a foundation for high-grade veneers, as the worm holes proved beneficial to the gluing process. Home parlors and living rooms were also paneled with wormy chestnut as the look became desired by late nineteenth- and

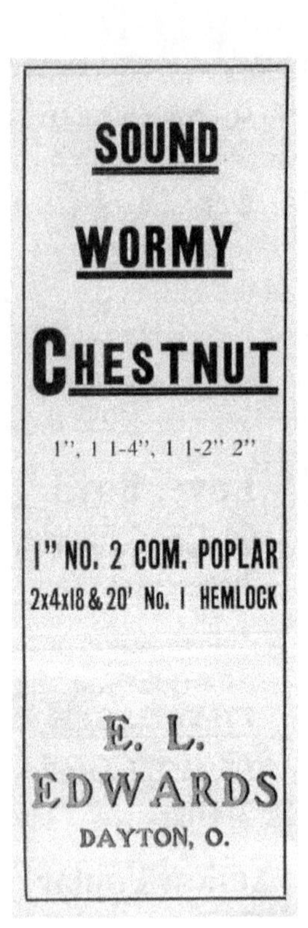

FIG. 8.2. Sound wormy chestnut advertisement. *Southern Lumberman* (June 10, 1907).

early twentieth-century homeowners.[86] In some instances, however, the damage from the tunneling larvae was so pervasive the wood was rendered unusable. In some areas, timber worm infestations were so widespread that very few trees escaped their attacks.[87] In 1908, when Theodore Roosevelt's National Conservation Commission surveyed timber worm damage in the eastern United States, including locations far removed from the ravages of chestnut blight, it found nearly every tree of merchantable size affected by timber worms and a large percentage so "seriously damaged that the product [was] reduced to the lowest grade."[88]

Although Americans later associated the sale and proliferation of wormy chestnut lumber with the blight fungus, thinking it actually encouraged the insect attacks, the reverse was more likely true, at least for the first two decades of the twentieth century. Indeed, if timber worms, wood borers, and periodical cicadas did not directly encourage its spread, they certainly gave the fungus the upper hand in attacking trees. The entry and exit holes created by the insects allowed the blight to penetrate the trees more easily, and that, along with their weakened state, ensured its growth and mobility. Some entomologists argued that the trees were in poor health even before the insect infestations—as a result of droughts, excessive rainfall, lightning, and frequent fires.[89] Fires were particularly detrimental to chestnut health; in fact, there was a reported relationship between fire and insect damage, as well as with the growth and spread of chestnut blight.

Prior to the blight's arrival, chestnuts nearly everywhere had been injured by natural and human-set fires.[90] One of the hardest-hit states was New Jersey, which saw two hundred thousand acres of its woodlands destroyed in 1894 alone.[91] Although most of the fire damage occurred in the southern part of the state, by the beginning of the twentieth century the interior and northern highlands were impacted as well. In 1904, a highly destructive fire in Sussex County, in the Montague township, damaged, if not "killed outright," 3,100 acres of oaks and chestnuts.[92] Near Hibernia, also in Morris County, a fire ignited by hunters burned 2,500 acres of chestnuts and oaks, causing an estimated $10,000 in damage.[93] In Tabor, a fire starting from the sparks of a locomotive burned three hundred acres of timber, "badly injuring the chestnuts."[94] A locomotive-started fire in Old Bridge destroyed two hundred acres of oak and chestnut, creating timber losses valued at $1,200.[95] In fact, of the eighty-one fires reported in New Jersey in 1904, eighteen of them burned chestnut trees. In total, on the 41,539 acres suffering fire damage in the state that year, chestnut was the first- or second-most common species, representing 28 percent of all affected trees.[96]

Such fires extended well beyond New Jersey and surrounding states, impacting woodlands as far away as the southern Appalachians. Many were started by individuals working for the railroad operators, who had been employed by the very timber companies engaged in removing trees from the forest.[97] The scorched timber provided a perfect medium for the spread of the fungus, as experiments conducted at the time revealed it was much more virulent on the bark of "fire-killed" chestnut.[98] Resprouted coppice wood in the burned-over areas also appeared to attract the blight, a view shared by the noted forest pathologist Samuel B. Detwiler. In 1913, when doing field observations in Pennsylvania, Detwiler noted the infections were mostly "scattered through the young shoots growing up after fire."[99] In Connecticut, the relationship of blight to fire injury was even embraced by local residents, including Avon native Samuel W. Eddy, who "reported that he found the fungus abundant on the cut wood and fire-injured trees, but scarce on the perfectly healthy ones."[100] In Massachusetts, State Forester Frank W. Rane also promoted this view, writing in 1912 that "there is an unbalanced condition again where forest fires have run through the state year after year, and the trees are abnormal, and only half alive anyway. There you find the disease seems to travel more rapidly than it does where the trees are under normal conditions."[101]

absolutely no remedies against it

The impact of fire on the growth and spread of the *Cryphonectria* fungus was also investigated by the Pennsylvania Chestnut Tree Blight Commission, the government agency assigned the task of solving the blight problem. Operating for nearly three full years (1911–13), the commission was composed of scientists, foresters, public officials, and volunteers. It also hosted a major national conference and published numerous scientific studies regarding the growth and spread of the disease, not only in Pennsylvania but across the entire United States.[102] Although some of the commission's findings were shared at the Conference for Preventing the Spread of the Chestnut Bark Disease held in Harrisburg in February 1912, most documents were not publicly available until 1914, after the blight had spread across much of Pennsylvania and into more than a dozen states.[103]

Initially, as a result of commission policy, Pennsylvania landowners were told to cut down and remove all infected trees as well as burn any remaining stumps, regardless of their size or the severity of the infection. However, when coppice growth returned afterward, the fungus also attacked the young sprouts, often with greater intensity. In response, the commission required the removal of all bark from the charred stumps, which prevented the growth of coppice sprouts as well as completely killed the trees.[104] They also endorsed the removal of healthy trees—"all trees, both young and old, that stand within half a mile of diseased trees"—as well as burning any discarded limbs and young trees too small for harvesting.[105] By burning and removing all trees, the species no longer provided ecosystem services, including nesting cavities for birds, food for insects, or organic matter for forest soils. Although a small number of the trees might have temporarily resisted the blight, or even fought it off entirely, the commission was unwilling to take that risk. Instead it embraced a blight-eradication program that resulted in the immediate and permanent removal of hundreds of thousands of healthy native chestnut trees.[106]

The commission also did little to investigate the blight's country of origin. Despite Haven Metcalf's insistence that it had originated in Japan and arrived in the United States on imported nursery stock, the agency never officially endorsed the assumption, which made the control of the fungus difficult if not impossible. As result, no effective interstate quarantines were established for Japanese chestnut trees, which allowed them to be shipped, without real consequence, anywhere within the continental United States.[107] And even when the U.S. government attempted in May 1915 to establish a more stringent quarantine for nursery-grown trees, the initiative fell short of Congressional approval.[108]

The commission's inability to slow the spread of the disease is not surprising, considering the role America's largest nursery owners played in controlling the blight narrative. They were, after all, wealthy businessmen with much to lose if they could no longer ship trees, shrubs, and plants across state lines.[109] To admit they had caused the catastrophe would have been, in their own minds, economic suicide. Many of the same nurserymen had been responsible for introducing another foreign pathogen into the United States and all saw their businesses suffer as a result. The earlier epidemic involved San José scale, a tiny, scale-like insect that arrived in the eastern United States on imported Japanese plum and pear trees from San José, California.[110] The parallels between the two outbreaks are quite remarkable and expose an industry in abject denial about the role it played in the introduction and spread of nonnative species. As early as 1896, San José scale was considered one of the most destructive insects in North America and was commonly known as "pernicious scale." First identified in fruit orchards in the Northeast in August 1893, pernicious scale was responsible for destroying, in less than a decade, tens of thousands of pear, peach, apple, and plum trees and as many tons of marketable fruit.[111]

In 1894, the introduction of pernicious scale on the East Coast was linked to two prominent nurseries, one on the Delaware River and the other along the Atlantic shoreline.[112] The Delaware River establishment was owned by none other than William and John Parry, the father-and-son team who in the previous decade had established a Japanese chestnut orchard at the same location. The second nursery was operated by another grower and seller of Japanese chestnut trees, John T. Lovett, who owned the Monmouth Nursery in Little Silver, New Jersey. In 1894, when the younger Parry answered an official query from USDA entomologist Charles V. Riley about the source of his scale-infected trees, he replied the scale had arrived on Kelly's Japan plums that had been shipped from California.[113] Parry also told Riley that he preferred that the name of his nursery "not be published in connection with this pest," before adding, "we realize it would ruin our business."[114]

In response, Riley acknowledged Parry's request to remain anonymous, stating that "no particular good could be accomplished" by revealing the nursery as a distributor of the pest. However, Riley warned Parry the incident would have to be officially recorded, and bluntly told him that he would need to "make every effort to destroy the insect in your nursery."[115] John T. Lovett's connection to the spread of pernicious scale eventually became public, however, causing him not only to move his nursery operations, but to provide evidence to buyers that his stock no longer carried the infestation. In 1900, the opening page of Lovett's mail-order catalogue included a notice of certification from New Jersey Board of Agriculture entomologist John B. Smith assuring readers that the Monmouth Nursery was "free from San José Scale and other dangerously injurious insect pests."[116]

Despite numerous letters and warnings from the federal government, few inspections were actually done in the field and reprimands or fines were almost never levied on violators. In fact, most of those duties were delegated to state agricultural inspectors who let growers do their own monitoring and disposal of infected trees. This in effect allowed the diseased plants to be shipped across state lines with little government oversight. As a result, orchards in a dozen states reported cases of the scale as early as 1896, with these infestations coming from just two New Jersey establishments.[117] A third establishment, the Parsons family's Kissena Nursery of Flushing, Long Island, was also implicated in the spread, an important fact since it was also among the first importers of Japanese chestnuts.[118] With the arrival of the chestnut blight fungus, the pattern of infection was doomed to repeat itself, as few state inspectors understood the growth and lifecycle of exotic fungi. And for those plant pathologists who became fully educated about chestnut blight, publicly revealing the source of the fungus in infected nursery stock was a bit like telling the naked emperor he had no clothes.[119]

Regarding the release and spread of *Cryphonectria parasitica*, the most obvious sources were nurseries selling Japanese chestnuts in closest proximity to the Bronx Zoological Park. Meeting both criteria were the Monmouth Nursery owned by John T. Lovett, the Pomona Nursery operated by William and John Parry, and the Kissena Nursery belonging to the Parsons family of Long Island. It is likely all three nurseries received shipments of blight-infected chestnut stock after 1896, possibly from the same source.[120] Earlier importations would have resulted in the fungus being detected on native trees before 1898, and there is little evidence of that beyond a few anecdotal observations. The infected shipments most likely arrived from the most prominent Japanese nurseries—the Yokohama Nursery Company or the Tokyo Nursery Company—both of which exported trees known to have possessed chestnut blight on their premises.[121] After arriving in the United States, the infected saplings or scions may have remained isolated for a year or two, although they could have been planted adjacent to domestic chestnut stock destined for orchard plantings.

However, it would have taken numerous growing seasons before the infected saplings would have developed fertile ascospores, allowing the disease to spread more rapidly outside the nursery. Unlike the heavier and stickier summer spores, ascospores are capable of traveling far greater distances with heavy rains and strong winds. In fact, a study published in 1914 demonstrated that a strong prevailing wind could transport them "over 380 feet from the nearest chestnut tree."[122] Thus the most ideal meteorological event for spreading the blight both widely and uniformly was a tropical storm or hurricane, and several passed over Long Island and southern New England prior to 1901.[123]

With the aid of wind, moisture, insects, and fire, *Cryphonectria parasitica* moved out of the commercial nurseries of western Long Island or northern New Jersey sometime after 1898, infecting native trees in the nearby woodlands. George Powell's observation of a "body blight" on Japanese, European, and American chestnut nursery stock that year is perhaps evidence the fungus was on the verge of spreading beyond the New Jersey border. For those trees infected at the Bronx Zoological Park, the Kissena Nursery in Flushing, Queens, was the most likely source of the fungus, since the blight spores needed to travel only nine miles to reach park boundaries. However, to accomplish that feat, the spores had to be blown several miles, including across a large body of water, the East River, which in places is several miles in width. The most likely catalyst for moving the blight toward the Bronx was Hurricane Nine, a tropical storm that passed over New York City in the autumn of 1899, delivering sustained winds of fifty miles per hour across the Atlantic States and southern New England.[124] The severe winter of 1903 also hastened the spread of the fungus as ice storms and low temperatures damaged large numbers of trees across the metropolitan area.[125] Because the trees identified by Merkel in 1904 had advanced stages of the disease, the spores causing the infections likely arrived in the park before 1903. The fungus may have also spread from the Monmouth Nursery at Little Silver, New Jersey, although that is a less likely scenario as it would have traveled a longer and more easterly route to do so.

Regardless of the specific route the blight took to reach the Bronx, after 1905 the fungus was found across the entire New York City area, as well as large portions of western Long Island. This suggests that there was not a single location for the blight infection, but many. It also means the arrival of chestnut blight to the Bronx Zoological Park was due not to a single group of imported trees, such as those growing at the New York Botanical Garden, but to several delivery agents, including heavy winds, excessive rainfall, and even flocks of birds.[126] The pervasiveness of chestnut blight was undoubtedly

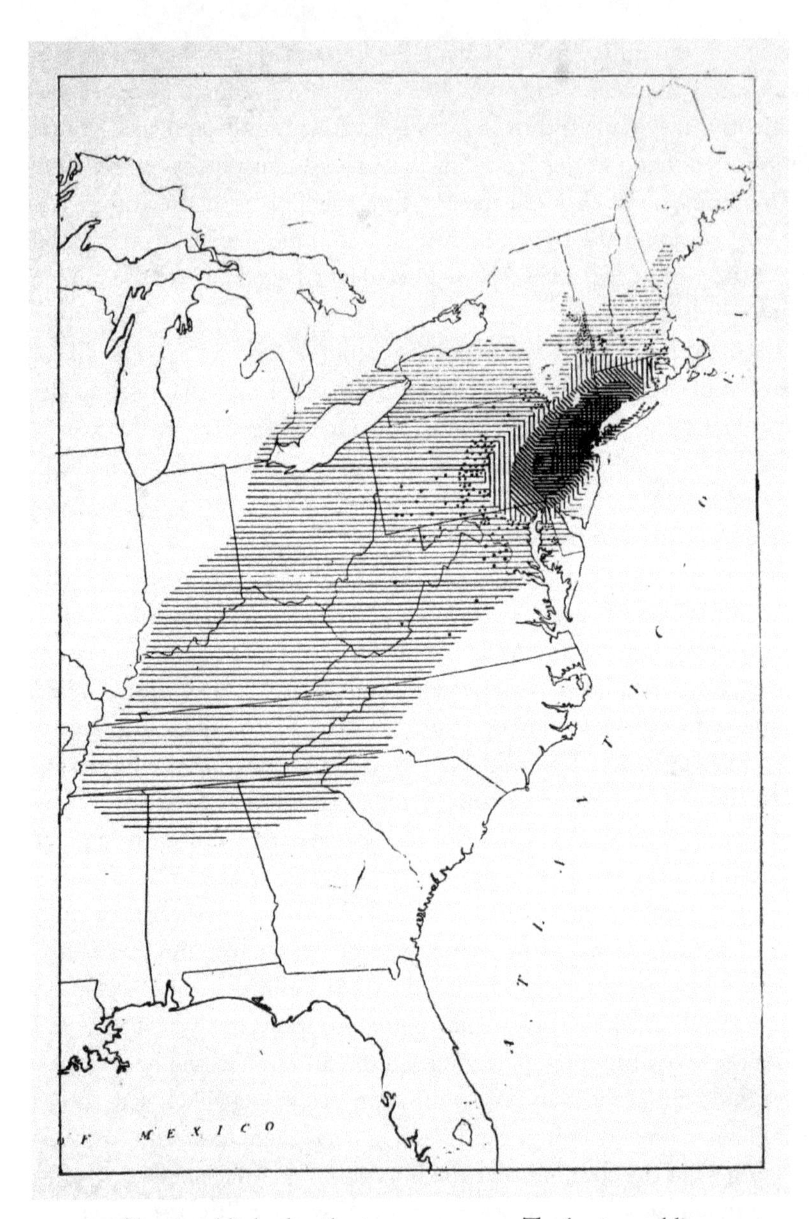

FIG. 8.3. Chestnut blight distribution map, c. 1911. The horizontal lines depict areas where no blight had been reported. The single dots indicate isolated infections of one or two trees and the heavier cross-hatched lines indicate "varying degrees of infection culminating in an area of infection about New York City in which all chestnut trees are dead." Pennsylvania Chestnut Tree Blight Commission, *The Publications of the Pennsylvania Chestnut Tree Blight Commission, 1911–1913* (Harrisburg, Pa.: W. S. Ray, 1915).

compounded by a tropical hurricane that moved up the eastern seaboard in September 1904, a category 1 storm that dumped ten inches of rain on northern New Jersey and damaged thousands of trees from Long Island to New Haven, Connecticut.[127] The storm certainly ensured a wider diffusion of the blight ascospores, condemning more trees to an early and untimely death. By 1906, so many chestnut trees had died in the metropolitan area that the *New York Observer*, a publication edited by the Presbyterian minister John Bancroft Bevins, likened the fungus to the plagues of Egypt and suggested the disease was evidence "we live in a world that has been cursed with sin" (see fig. 8.3).[128]

For less apocalyptic observers, the fungus was still stoppable, but only by some yet to be determined treatment or prophylactic. As long as scientific inquiry moved faithfully forward, they reasoned, a cure for the blight might still be found. Some avoided the discussion altogether, describing the pathogen as invincible, including William A. Murrill, who stated as early as 1908 that there were "absolutely no remedies against it."[129] Others waxed more philosophically, arguing that the fate of the American chestnut was linked to the overdomestication "of the vegetable and animal world." This view was advanced in a *New York Times* editorial in 1909 after it was made public that sixteen thousand chestnut trees had died in Brooklyn's Forest Park, a loss estimated at more than five million dollars.[130] The writer firmly believed the disease was introduced via nursery stock from Japan and applauded the USDA's recommendation to create an inspection service for future importations.[131] However, the author of the editorial was uncertain if such measures would solve the problem, adding, "the new graftings of Japanese chestnuts are valuable. But the very act of human progress becomes self-destructive unless accompanied by additional care and artifice." In the writer's view, the principal cause of the deadly disease was human interference, the result of a near obsession with creating "domesticated organisms."[132] In the end, the nursery owner and commercial orchardist had not actually improved the natural world, they simply made it in the image they themselves thought it to be. For the *New York Times*, at least, such actions were not only detrimental to Nature, but to human civilization as well.

CHAPTER 9

to maintain the balance of nature

When the U.S. Congress passed the Plant Quarantine Act in 1912 to limit and monitor the importation of foreign plants into the country, it was too little, too late for the American chestnut. Shipments of diseased nursery stock continued unabated across state lines, infecting chestnut trees as far away as South Carolina.[1] The devastation had an immediate economic impact, as the annual contribution of the trees to the U.S. economy was estimated at $25 million.[2] However, the annual timber harvest was only a small percentage of the total number of trees, leading many to predict even greater economic losses as a result of the blight. In 1912, when Samuel B. Detwiler estimated the monetary value of the American chestnut in Pennsylvania—using a formula that converted a portion of the trees into railroad ties, poles, and cordwood—the entire stand was believed to be worth $55 million. The remaining nut crop, orchard, park, and shade trees, added Detwiler, were worth another $15 million, putting the total value at $70 million ($2.2 billion in today's currency).[3] That figure hardly represents the true economic benefit of the trees, as chestnut lumber produced numerous value-added products, including paneling, coffins, flooring, and furniture.

The Northeast was the first U.S. region to witness significant chestnut losses, although it would take a full decade for the trees to be rare or absent in most communities. However, many trees were removed immediately upon infection as part of state-endorsed blight prevention programs. Urban landowners were also quick to cut and haul away trees, citing safety or aesthetic reasons for the hasty salvage operations.

Perhaps the most notable individual engaged in chestnut removal was Theodore Roosevelt, who was observed cutting down blight-infested trees on his Long Island estate in July 1910. Among the witnesses were members of an African American delegation sent there to ask the former president to deliver a speech at the Colored State Fair in Richmond, Virginia.[4] Upon arrival, they found Roosevelt wielding a heavy axe in order to take down a large chestnut tree at the main entrance. According to a *New York Times* report, the tree was a mature older specimen, three feet in diameter. The reporter noted, however, that its upper limbs "had died of a peculiar blight which has been attacking chestnut trees all along the north shore of Long Island for the last three years."[5] After several minutes watching Roosevelt, who supposedly "was wielding the axe like a pioneer," the delegation made their request. When Roosevelt said that he would indeed consider their invitation, a member of the group then asked him if they could take some chips as souvenirs, to which he boisterously replied, "Help yourself!" Within minutes, added the reporter, "there wasn't a chip left. Each man had an armful and their pockets bulged."[6]

After tens of thousands of individuals repeated Roosevelt's actions across both New England and the Mid-Atlantic, it became increasingly apparent the loss of the American chestnut was a fait accompli. This did not stop speculation about who was responsible for the devastation, as many continued to blame lack of government oversight for the relentless spread of the disease. Others pointed their fingers at nursery owners, particularly those importing blight-infected saplings. Both government agencies and commercial nurserymen were obviously responsible for the catastrophe, but so, too, were the thousands of individuals purchasing imported trees. In 1925, when writer Oliver Peck Newman, a former D.C. councilman, reflected on the disappearance of the tree across the Atlantic seaboard, he faulted both weak government regulations and the personal whims of individual chestnut growers. "The chestnut-bark disease might have been excluded if the present quarantine policies had been in force fifteen years ago," wrote Newman in a critical if not facetious tone, "when the blight is believed to have been brought in with a shipment of Japanese trees imported to complete a collection of chestnuts of the world."[7]

Regardless of who caused the epidemic, very few individuals were willing to accept responsibility for the devastation it was causing to our nation's forests. Most commercial nurserymen refused to address the issue publicly and professional foresters seldom embraced blight-control strategies that did not include the removal of each and every tree. Others were more circumspect, including Samuel B. Detwiler, who suggested the blight was a necessary evil, reinforcing, in his words, "the fact that man has greatly changed and disturbed the conditions under which our forests have grown, and which are necessary for their proper development."[8] For Detwiler, the ignorance and indifference of forest landowners encouraged the blight's spread, just as their actions were also responsible for the loss of birds, rodents, snakes, and other animals needed "to maintain the balance of nature."[9] If Detwiler thought chestnut blight a warning sign that turn-of-the-century forest management was woefully inadequate, he also believed the catastrophe might encourage a more rigorous response to the larger, and perhaps more important, issue of forest health. "We invite calamity, and the chestnut blight, if its warning is heeded, may prove a blessing in disguise," surmised Detwiler.[10]

With a few notable exceptions, Detwiler's plea for better forest management fell on deaf ears. And despite efforts in Pennsylvania to stop the spread of the blight by cutting, burning, and hauling away all trees, it became more apparent with each passing year that the tree was a doomed species. Timber companies also did little to stop or slow the removal, as they saw the infection as simply another excuse to increase harvests. Although the annual production of chestnut lumber peaked in New England and the Mid-Atlantic states in 1909 (as it did for all wood types), the logging of chestnut trees continued largely unabated until 1915.[11] At the end of that year, in states where blight was present, the largest timber salvage operation in our nation's history had been successfully completed. Only ten years after the blight was first reported in New York City, no fewer than 2.1 billion board feet of chestnut lumber, worth a total $36 million, had been processed in New England and the Middle Atlantic states alone.[12] Connecticut contributed five hundred million board feet, or 23 percent of the total harvest. An equal number of trees were also taken for poles, crossties, charcoal, cordwood, and fencing, so the amount of land disturbed for chestnut removal exceeded 250,000 acres, roughly 8 percent of the state's total land surface.[13]

Elsewhere, chestnut-dominant forests continued to be logged at historically high rates, causing the removal of hundreds of thousands of trees ahead of the range of infection. Among the commercial enterprises removing large amounts of chestnut timber was the telegraph industry, which became economically important during the second half of the nineteenth century. It did so by sending millions of coded messages over suspended electrical wires, creating what one scholar rightly called "the Internet of the Victorian Age."[14]

After 1870, the American chestnut was as important to the telegraph industry as the telegraph was to American life. In most locations, the wires were suspended atop chestnut poles twenty-five to thirty-five feet long, placed in the ground at two-hundred-foot intervals. The industry obviously increased demand for such trees, especially those with straight and slightly tapering trunks.[15] In 1873, the Western Union Company alone owned no fewer than sixty-two thousand miles of poles, of which fifty thousand miles were cedar and chestnut.[16] In many cases, the poles were second- or third-generation coppice growth, as is documented in industry reports. Coppiced chestnuts not only grew straighter, but were more durable than trees grown from nuts.[17] The faster-growing coppice sprouts were also less likely to bend or warp due to wind, ice, or snow.[18] If no preservatives were applied to their surface, chestnut poles had a lifespan of twelve years; if treated with creosote or other preservatives, they lasted twice that long.[19] By the early 1880s, with the introduction of the telephone in many American cities, more and larger chestnut poles were needed to satisfy the growing demand. The necessity for multiple lines required longer and stronger poles as well as older and more mature trees. In 1885, there were 150,000 miles of telegraph and telephone lines stretching across the United States, resulting in the removal of some 1,350,000 chestnut trees.[20] As telegraph and telephone companies merged and consolidated, and made even greater profits, lawmakers began holding public hearings regarding their day-to-day operations. And because chestnut poles were among their greatest investments, the hearings often made reference to the trees, including the very stands of timber from which they were harvested.

On November 11, 1887, at a hearing regarding the operations of the New York and Pennsylvania Telephone Company, lawmakers in Albany were told the annual earnings of the firm were $151,957.88 and its stock value one million dollars.[21] When manager William N. Estabrook was asked about the number and value of his company's telephone poles, he told the committee they owned 1,792 miles of poles and 840 miles of additional wire. When questioned about the type of poles the company possessed, he responded by saying they "were generally chestnut, but a few were of cedar and others of hemlock." Estabrook added the poles "did not last half as long as people generally imagined," suggesting they needed replacing after seven years. Doubting the veracity of Estabrook's statement, the chair of the committee, Daniel E. Ainsworth, noted "his father's farm was inclosed by a chestnut fence that was in place when he was born, and it was still there." After chastising Estabrook for the remark, Ainsworth concluded the longevity of chestnut poles ultimately "depended upon the character of the soil and the atmosphere."[22]

Remarkably, at the beginning of the twentieth century there were some eight hundred thousand miles of telegraph and telephone lines crisscrossing the United States, with wires suspended atop some nine million chestnut poles.[23] The Western Union Company alone owned 190,000 miles of poles and wires and used them to transmit no fewer than 68 million messages annually.[24] Similar poles were needed for streetlamps, streetcar cables, electric utility lines, and signposts, further increasing demand for chestnut timber. According to one report, of the 3,574,666 new poles purchased in the United States in 1906, 988,804, or 28 percent, were chestnut.[25] In Massachusetts, however, it was not uncommon for entire telegraph and telephone grids to be supported on chestnut poles. In the following townships, they comprised no less than 95 percent of the total wire infrastructure (actual number in parentheses): Chicopee (636), Marblehead (1,001), North Attleboro (1,301), Peabody (731), Westfield (584), Fitchburg (1,366), and Northampton (1,234).[26]

During the last quarter of the nineteenth century, telegraph and telephone companies were responsible for removing no fewer than ten million chestnut trees from North American forests. Although the trees were not large, and generally coppice growth, their removal did not go unnoticed. However, railroads had an even greater impact on the species, as larger trees were needed for crossties, fencing, trestles, corduroy roads, and depot construction.[27] By 1880, there were more than ninety thousand miles of railroad track in the United States, with some forty-eight thousand miles located within the tree's range.[28] In New England, railroad companies purchased some 1.5 million chestnut crossties annually, a figure representing 70 percent of the total amount. They also maintained 144,000 chestnut fence posts on adjoining properties, wood valued at some $18,000.[29] Other parts of the country used less chestnut, as oak and cypress had become the preferred wood for crossties across much of the southern states. This was due in part to the lack of adequate rail lines into southern Appalachian forests, as well as the availability and lower cost of crossties made from other trees.[30] Elsewhere, demand for chestnut crossties continued to rise, growing almost exponentially in the Northeast and Midwest.

a special interest in forest preservation

By 1887, there were 140,000 miles of railroad track in the United States, a figure representing an extraordinary amount of timber.[31] As forest historian Michael Williams has argued, the construction of railroads removed more trees from the American landscape than did all other activities, including agricultural development and urban growth.[32] The tripling of crosstie demand in only two decades even caused many to worry about the future of U.S. timber supplies. Among the more notable responses was a report by Bernhard E. Fernow, the chief of the Forestry Division of the USDA.[33] In the introduction to his *Report on the Relation of Railroads to Forest Supplies and Forestry*, Fernow lamented that in order to construct and maintain the present railroad system, "more than one hundred million acres, or one-fifth of our present forest area, were stripped during the last fifty years." He blamed the industry for current shortages and seemed puzzled it had offered few viable solutions. "While, therefore, directly and indirectly, railroad enterprises have contributed largely to a considerable reduction (if not destruction) of forest supplies," wrote Fernow, "it might be presumed that . . . railway companies would feel a special interest in forest preservation." He also predicted the industry was doomed to failure if it ultimately refused to embrace more prudent and sensible uses of America's forests.[34]

Despite Fernow's pleas for forest conservation, the harvesting of chestnut for the industry continued well into the twentieth century. In 1905 alone, railroad companies purchased no fewer than 4,717,604 chestnut crossties, a figure representing hundreds of thousands of mature trees. Ninety-five percent came from North Atlantic states, with the remainder divided among central and southern regions.[35] A year later, when the government began publishing specific uses for crossties, a total of 6,588,975 were acquired by the industry, with 5,355,500 going to the construction of steam railroads and 1,942,213 for streetcar lines.[36] In southern states, chestnut had fallen almost completely out of favor, with white oak, pine, and cypress becoming the most popular woods for crosstie production. Whereas in the Atlantic states chestnut sleepers represented 43 percent of the cut, in the southern and central regions they accounted for less than 1 percent of the total harvest.[37]

The growing demand for railroad ties, as well as the perceived shortages, also increased the number of creosote-treatment facilities in the United States. When applied to crossties, creosote oil greatly extended their life, as it also did with telephone and telegraph poles. In 1900, there were eleven such plants in operation; a decade later, more than one hundred.[38] The growing popularity of creosote oil, as well as the use of other kinds of trees for crossties (and wood imports from Canada), eventually appeased those worrying about the future of U.S. timber supplies. Chestnut did not always fare so well in the prognosis, however, causing some in the Northeast to propose alternative sources. In a letter to the editor published in the August 26, 1899, issue of *Scientific American*, George E. Walsh suggested the substitution of other woods for chestnut had already taken place. "Originally," noted Walsh, "the chestnut was considered the finest tree for supplying railroad ties, but forests of chestnuts are scarce in all parts of the country."[39] By "all parts of the country," it is unclear if Walsh means the Northeast or the eastern United States. Although the comment would not be accurate in terms of reflecting forest inventories elsewhere, it did represent a reality endorsed by most New England timbermen: *chestnut timber is scarcer today than it was yesterday and will be more so tomorrow.*[40]

Although the pole and crosstie industries had a visible impact on the prevalence of chestnut and accounted for a large portion of all wood harvested, commercial markets for milled lumber were equally destructive.[41] By the early 1890s, dozens of industrial and home products were made of chestnut, including tool handles, mining timbers, threading spools, shuttles, crates, food barrels, dowels, office furniture, plaster lathing, pulleys, beehive supers, poultry houses, window blinds, camera tripods, picture frames, and kitchen cabinets.[42] To make these products, chestnut timber harvests rose from 206,688,000 board feet in 1899 to 243,537,000 board feet in 1904.[43] Those numbers were not inconsequential. A stand of second-growth chestnut, forty-five years of age, yielded approximately five thousand board feet of lumber per acre.[44] This means that as much as fifty thousand acres of chestnut was annually processed at American sawmills, no small amount given the number of trees also taken by the railroad, telegraph, and telephone industries.[45]

A portion of the lumber was even used in the manufacturing of musical instruments, a trend more commonplace after 1908.[46] In that year alone, 300,000 board feet of chestnut was purchased in Massachusetts, wood used in the construction of pianos, organs, and phonograph cabinets.[47] In 1911, 15.2 million board feet of chestnut was used by New Jersey manufacturers, with 2.7 million going toward the construction of musical instruments.[48] By 1913, forest products expert Albert H. Pierson estimated Connecticut manufacturers were using chestnut more than all other hardwoods, as it had become the preferred construction material used for upright and standard model pianos.[49] Regarding the use of chestnut in piano construction, very little was actually visible in the finished product, as the wood was usually concealed underneath the veneer panels.[50] As much as 110 square feet of chestnut went into the making of a single instrument, no small amount, since more than three hundred thousand pianos were manufactured domestically in 1917.[51] In 1919, fifty-eight New York factories used seven million board feet of chestnut to make keyboard instruments, although only 4 percent of that total was harvested within state boundaries.[52] Across the entire country, thirty-eight million board feet of chestnut was annually used for making pianos and organs, enough wood to construct forty-two hundred typical American homes.[53]

In the Northeast, considerable amounts of chestnut lumber went into the making of coffins and caskets, although after 1910 the wood was largely purchased from states outside the region. Forty-eight million board feet was yearly purchased during the World War I period, making the industry among the largest consumers of milled chestnut lumber in the United States.[54] This was due to the fact that chestnut boards held veneer glue extremely well, repelled moisture, and were relatively inexpensive. Some wood was even used for coffin or casket lids, although such construction obviously required "higher-grade chestnut stock."[55] Ohio manufacturers consumed 3.1 million board feet of chestnut in 1911, wood used primarily for "cloth-covered caskets."[56] However, government forester Carroll W. Dunning noted even the most expensive coffins could be made from chestnut, particularly if treated with "a varnish finish."[57]

In New York, chestnut contributed more than 50 percent of the raw material to the casket industry, an amount equal to 15.7 million board feet of lumber.[58] The Spanish flu of 1918–19 also greatly increased demand for coffins and caskets, as 675,000 Americans died of the disease in those two years alone.[59] In New York, nearly all of the twenty-one coffin manufacturers purchased chestnut from the southern Appalachians, a region that was beginning to develop its own burial-box industry.[60] North Carolina casket manufacturers consumed 7.8 million board feet of chestnut in 1919 and paid $340,345 for chestnut lumber, or about $6.2 million in today's currency.[61] Although uses for chestnut varied little from state to state, in North Carolina large amounts of wood were transformed into home furniture, especially during the second and third decades of the twentieth century. In 1919, 11.2 million board feet of chestnut was purchased by furniture makers—54 percent of the total—with the rest consumed by the following producers: coffin and casket makers, box and crate manufacturers, planing-mill owners, sash, door, and window-blind craftsmen, picture and mirror-frame makers, and wooden-fixture builders.[62]

Whatever the final product, commercial timbering reduced the number of healthy trees in the forest, even before the arrival of the blight fungus.

It also encouraged coppice growth and the spread of harmful insects, and injured both mature trees and saplings not scheduled for immediate removal. Without question, the overharvesting of chestnut timber, along with repeated fires and insect damage, hastened the blight's spread across the eastern United States. Although it is doubtful the absence of these factors would have stopped the blight from killing trees, the outcome might have been different if they had been in a more ecologically balanced state.[63] At the very least, more trees showing natural resistance to the disease could have been identified and thus included in future breeding programs. Only a handful of commentators followed this line of reasoning, however, believing the fungus, if left unchecked, would doom the species to extinction. In the short term, the better solution was not to improve the health of the forest, but to aggressively eliminate the fungus from it, including by cutting down and burning all infected trees and, in many instances, uninfected ones as well.[64]

a perfectly practical proposition

With the help of piano, casket, and furniture manufacturers, chestnut harvests reached record levels during the second decade of the twentieth century. However, because of the wood's economic importance and the continuous spread of the blight southward, efforts to find a cure for the seemingly unstoppable disease hardly waned. The earliest and most common strategies for preventing chestnut blight involved the use of prophylactics; that is, the application of various organic substances used to shield or inoculate trees from the harmful effects of the fungus. By 1914, orchardists, plant breeders, as well as the lay public were experimenting with dozens of such remedies, including spraying trees with numerous fungicides, the application of mud or pine tar on infected areas, exposing saplings to sulfur fumes, and the injection of trunk and roots with various chemical solutions.[65]

Initially, chemical injections showed considerable promise; so much so that the USDA's Bureau of Plant Industry funded many of the experiments. By 1918, USDA employees had injected no fewer than fifty-six organic and inorganic substances into living trees, including ones already infected with the blight fungus.[66] In one study, the chemicals were administered using three glass bottles fitted with long rubber hoses connected to glass tubes. The bottles—which held various liquids such as lithium carbonate or hydroxide—were hung from wires fastened to the limbs of the infected or soon-to-be-infected trees. The tubes pierced the outer bark and sapwood and were held in place by large sprocket chains clamped around the main trunk. The unusual configuration allowed for the gradual release of the chemicals into the sap, limbs, and leaves of the tree. While some of the chemicals delayed or stopped the growth of the fungus, nothing appeared to be a magic bullet that fully protected trees. In 1920, Caroline Rumbold, the plant pathologist pioneering the experiments, concluded that the chemicals evaporated too rapidly after treatment, which made the trees subject to reinfection.[67] Although she truly believed "giving medicine to trees" was an important field of research, the experiments only offered, in her own words, "a temporary check against the disease."[68]

While a handful of researchers continued to experiment with chemical injections as a way of thwarting the fungus, others focused on breeding blight-resistant trees. Several horticulturalists believed that hybrid crosses between American and Asian chestnuts could be vigorous enough to fully repel the blight. Aware that such hybrids might not resemble the native tree in height and stature, they concluded it would be better to have a reasonable facsimile of the American chestnut in the forest than no chestnut trees at all. Among the first professional plant breeders to approach the problem was Dr. Walter Van Fleet, who was growing hybrid chestnuts for the USDA's Division of Forest Pathology as early as 1909. Although Fleet's primary goal was to develop a blight-resistant chestnut for tannin, timber, and wildlife, he was also interested in creating a nut-bearing tree for commercial orchards.[69] By 1915, Van Fleet was making national news with his efforts, which included the crossing of Japanese and Chinese chestnut trees with the native variety. He also crossed Japanese chestnuts with Allegheny chinquapins, resulting in a tree that possessed prolific nut-bearing tendencies as well as extremely large nuts.[70]

Van Fleet's experiments were similar to those of Luther Burbank, who decades earlier had crossed native chinquapins with Japanese and American chestnuts. However, Van Fleet incorporated Chinese chestnuts in the breeding program, using varieties that Burbank himself did not possess. Burbank also grew chestnuts from China at his California nursery, but it is unknown if the trees were actually *Castanea mollissima* and not Japanese varieties imported from the Chinese mainland. Several horticulturalists on the East Coast maintain that Chinese chestnuts did not arrive in North America until 1907, when nuts were shipped to the United States by plant hunter Frank N. Meyer.[71] Even if Burbank did possess such trees prior to that date, Meyer brought additional Chinese cultivars to the United States, including trees that purportedly grew "one hundred feet high in their home forests."[72] After their arrival, the trees were shipped to growers like Van Fleet, who were given the task of evaluating them for future plantings. Some were even hopeful the new Chinese varieties might replace, as one grower put it, "in some measure our vanishing chestnut stands."[73] In 1917, one lot of nuts was sent to James Hobson of Jasper, Georgia, who grew several generations of the trees at his home orchard. Although the trees were considered blight resistant and were available commercially, the Hobson chestnut eventually lost public favor and its propagation was discontinued (see fig. 9.1).[74]

In the end, Van Fleet's experiments with hybrid crosses gave the public little hope for the future of the species, as the resulting trees remained susceptible to the disease, with many developing large and deadly cankers. When William A. Murrill of the New York Botanical Garden visited Van Fleet's experimental orchard at Glenn Dale, Maryland, in 1917, he saw several Chinese/American hybrids attacked by the blight, "and some even fatally so."[75] Upon returning to New York, Murrill concluded that it was still "a perfectly practical proposition" to grow hybrid chestnuts, implying he still had faith in the government breeding program.[76] However, at the time of his death in 1922, Van Fleet had still not produced a blight-tolerant hybrid closely resembling the native variety. Nevertheless, plant pathologist Russell B. Clapper continued Van Fleet's work at the Glenn Dale facility, and over the next two decades produced thousands of Chinese, Japanese, and American hybrids, including intercrosses of all three species.[77]

Clapper was aided in the effort by R. Kent Beattie, who in 1927 was sent to Japan and China in order to find new and even more desirable chestnut varieties. Over the next three years, Beattie shipped more than seven tons of chestnuts to the United States, including nuts from the Henry's chinquapin (*Castanea henryi*) and the Seguin chestnut (*Castanea seguinii*).[78] Part of Beattie's mission was to find trees with high tannin levels, as leather producers were becoming increasingly worried about the availability of chestnut extract. Beattie's shipments of nuts produced some 320,000 seedlings, which were distributed to federal and state foresters, tannin producers, nurseries, hunting clubs, and other cooperators.[79] In 1930, Clapper was joined in his effort by Flippo Gravatt, who had become the director of the chestnut breeding program at the U.S. Division of Forest Pathology. Their strategy was simple: make crosses between the various *Castanea* species and

FIG. 9.1. The "Hobson" Chinese chestnut, Jasper, Georgia, c. 1960. Courtesy of the Georgia State University Library, Special Collections and Archives. Reproduced with permission from the Associated Press.

then intercross those progeny in countless permutations. All crosses were recorded in notebooks using the genus name of each species: e.g., (*mollissima* x *dentata*) x *dentata* (the progeny of Chinese and American chestnuts crossed again with an American tree).[80]

Although first-generation Clapper and Gravatt hybrids often exhibited blight resistance, the trees did not always possess the desired characteristics. Many of the offspring were shorter in stature than the American chestnut, even after a decade of growth. Nor did the trees grow as rapidly as the native tree. Whereas *Castanea dentata* might grow five or six feet in a single growing season, the intercrossed trees seldom exceeded annual growth rates of three feet.[81] The blossoming times of the hybrids also varied widely, with some trees blooming a week or more earlier than the native variety. Such traits caused many to doubt the ability of the hybrids to repopulate the eastern deciduous forest. As a result, the USDA shifted its focus to exclusively growing Chinese chestnuts, as many believed the variety might, without *any* intercrossing, result in a prolific nut-producing tree for commercial growers. In 1937, only a decade after they were planted, Chinese chestnut trees at the United States Pecan Field Station in Albany, Georgia, were "producing 50 to 70 pounds of nuts each."[82]

For Americans wanting a forest-type tree that produced large amounts of nuts and lumber, the government breeding program could hardly be deemed a success. There were also those who continued to worry about the overall health of the forest, as well as the impact of chestnut decline on other tree species. Some remained extremely concerned about wildlife populations and their possible decline in the tree's absence. Although it was assumed oaks and other mast-bearing hardwoods would eventually replace chestnuts in the woodland canopy, the ultimate outcome was still unknown. At the Bronx Zoological Park, where the blight was first observed by Hermann Merkel in 1904, eastern hemlocks appeared to be benefiting from the species' absence, a fact mentioned in the *New York Times* as early as 1911.[83] The removal of dead chestnut from the Park's Hemlock Forest—"one of the most important natural features of the Bronx"—noted the report, not only eliminated the dense shade that prevented hemlock regeneration, but "gave more air and freedom [to the] soil, with the result that the famous forest is now recuperating itself."[84]

Outside the Bronx, however, the impact of chestnut blight on forest ecosystems was still open for debate. Many commercial foresters were not concerned about the long-term consequences of the blight problem, as they saw it merely as a salvage problem. Remove the host trees, they argued, and the fungus will eventually go away. As a result, discussions about limiting chestnut harvests fell largely on deaf ears, resulting in the removal of even more healthy trees. And because the blight struck when demand for chestnut was as its greatest—as a result of the booming railroad, telegraph, piano, and coffin industries—the voice of conservation had few, if any, champions. Additionally, the growing demand for the wood opened up new markets in the southern Appalachians, which still possessed large and expansive stands of healthy trees. This meant that consumers in the Northeast and Mid-Atlantic never had to directly confront the problem of scarcity, as chestnut products could always be imported from outside those areas.

Eventually, however, chestnut timber would become scarce even in the southern Appalachians. Before 1895, most chestnut removal in the region involved small-scale operations that did relatively little damage to forest ecosystems. At that time, most trees were hand-rived into roofing shingles, split into fencing material, or worked into small bolts for the tannin industry. By the turn of the century, a considerable amount of chestnut was also removed for telegraph and telephone poles, coffins, and tanbark. The construction of narrow-gauge railroad lines in the mountains eventually increased the rate of removal, providing access to trees well beyond the rural townships. Thus, in 1907, for the very first time, more chestnut timber was taken from southern Appalachia than any other U.S. region.[85] West Virginia yielded the most chestnut lumber that year, providing 118 million board feet, or one-third of the total U.S. harvest.[86] In fact, West Virginia lumbermen harvested more chestnut in the first quarter of the twentieth century than any U.S. state, processing some 2.3 billion board feet between 1907 and 1925.[87] Consequently, some chestnut-dominant forests in the Appalachians were depleted even before the arrival of the blight fungus.

However, in many parts of the southern Appalachians, large, healthy stands of chestnut survived well into the 1940s. As a result, the species continued to shape the economy and culture of the mountain region in unique and important ways. Many residents of southern Appalachia still owed their very existence to the tree and could not even imagine life in its absence.

CHAPTER 10

grandfather had lived in a log

If there is a single photograph that illustrates the prominence of the American chestnut in the southern Appalachians during the early twentieth century, it is the image taken by Sidney Vernon Streator in 1909. Appearing in the January 15, 1910, issue of the *American Lumberman*, the black-and-white image features five large chestnut trees in Poplar Cove, in an area above Little Santeetlah Creek in Graham County, North Carolina (see plate 4).[1] Streator most likely captured the image in the summer months prior to its publication, as the leaves on the trees' cascading branches are fully mature. Standing among the large chestnuts are timber agent D. W. Swan and timber warden E. B. King, who possibly worked for the Whiting Manufacturing Company when the photograph was taken.[2]

The Whiting Manufacturing Company was owned by brothers Frank and William Whiting of Philadelphia, who operated large saw mills in Abingdon, Virginia, and Judson, North Carolina. Under the guise of an independent news story, the *American Lumberman* published the image and numerous others to advertise the company's newly acquired timber holdings. The goal was to convince American or British investors to fund the construction of a railroad into the remote area, as the Whitings did not have the necessary funds to do so. Although the position of Swan and King in the image distorts the size of the largest two chestnut trees in the foreground, both appear to be at least six feet in diameter. In the publication, the caption beneath the photograph—likely composed by Streator himself—describes the trees as "large, sound, and free from visible defects."[3]

Streator's iconic image provides documentation of one of the last remaining old-growth stands of American chestnut in the eastern United States. Although the chestnuts were exceptional trees and did not represent the typical stand, they are visual reminders of what a mature grove might look like if afforded proper nutrients, rainfall, and sunlight. They were not the only large chestnuts in the company's holdings, as the Whiting brothers owned some seventy thousand acres of mature timber across four watersheds.[4] In fact, when the same parcel was surveyed several years earlier by

Lemieux Brothers & Company, an independent timber cruising firm based in New Orleans, chestnut was the second most dominant species, accounting for 209,346,743 board feet of lumber.[5] For perspective, that volume is greater than the total amount of chestnut milled in the Mid-Atlantic states in 1909, the historic peak year of production.[6] In the Belding Tract, which included Poplar Cove where the Streator photograph was taken, Lemieux Brothers estimated the parcel contained forty million board feet of chestnut. In fact, the trees comprised 30 percent of all standing timber in the tract—only eastern hemlock was more plentiful.[7]

Although western North Carolina timber had already acquired international fame when Streator took the photograph, not all chestnut trees in the southern mountains were sound or free from visible defects.[8] Because of human-set fires, which damaged the outer bark and made the tree more prone to scarring and disease, the trunks of the American chestnut often became hollow as they matured.[9] However, this also made them beneficial to animals making dens inside the trees, especially bears, bobcats, raccoons, and opossums. Lumbermen generally avoided such specimens, and thus older trees often remained uncut, allowing them to form, over time, even larger interior cavities. As a result, humans often found creative uses for the trees, including temporary and permanent shelter.

One of the most innovative examples of the use of such trees was documented in the *New York Times* in 1904.[10] According to the report, federal revenue agent Thomas Vanderford was summoned to inspect a large chestnut tree in the Pisgah mountains near Asheville. The tree was emitting smoke from its trunk both morning and evening, suggesting a smoldering fire at its base or interior.[11] The smoke was emerging from a hole in the top of the massive tree, a specimen otherwise alive and in perfect health. Prior to Vanderford's visit, several individuals dug around the base of the tree and found it firmly rooted with "no hollow under it."[12]

Upon arrival, Vanderford made a careful examination of the tree and observed no acceptable cause for the smoke. The next day he brought an iron rod to the site and repeatedly thrust it into the ground in concentric circles around the tree. Returning on a third day, Vanderford detected, after additional searching, something odd about one hundred yards from the base of the tree. In the evening, he departed for nearby Hendersonville and the next morning returned with six revenue officers. At daylight, all seven men observed smoke coming out of the chestnut "in full blast." Finding the same spot from the day before, the men dug a large hole with picks and shovels, which led to a long underground tunnel. Armed with carbine rifles, the men moved cautiously toward the interior of the tree, where they discovered "a blockade still running at full capacity." They also found Amos Owens inside the chestnut, who the *New York Times* called "the most incorrigible revenue violator in this State." Owens was apparently asleep when Vanderford found him, but awoke when the officer tapped him on the shoulder. "I supposed you would find me out after a while," Owens told them. "I knew you were prospecting around here."[13]

In another instance, a hollowed-out chestnut tree provided temporary housing for an entire mountain family. Oleta Nelms, who grew up near Little Snowbird Creek, North Carolina, recalled in 2007 that her grandfather, John Denton, had once built a rough cabin that incorporated a "huge fallen chestnut tree" into the structure.[14] Using an axe and other tools, remembered Nelms, her grandfather extended the dwelling "right into that chestnut log."[15] The hollowed-out portion of the tree was so large it allowed him to stand fully erect without bumping his head, despite the fact he was 6'3" tall. According to Nelms, the tree and cabin provided shelter for Denton and his family for several months until a more permanent home could be built. Not surprisingly, the tree remained part of community folklore for decades and even caused the younger Oleta to be teased at school. As she later explained it, her classmates thought it peculiar her "grandfather had lived in a log."[16]

An equally remarkable story was told by Charles Grossman, one of the first rangers employed at the Great Smoky Mountains National Park. On a mountainside above Cosby, Tennessee, Grossman documented a chestnut tree 9'8" in diameter at a point six feet off the ground. "The hollow portion is so large that [an adult] could stand up in it," wrote Grossman after discovering the tree. "The hollow runs more than fifty feet up the trunk and at its narrowest point is not less than three feet," he recalled. "This must be the tree of which I heard. A man lost some stock during a snowstorm and later found them safe in a hollow chestnut tree" (see fig 10.1).[17]

However, the largest American chestnut on record was located at Francis Cove, North Carolina, near present-day Waynesville. According to several published sources and one eyewitness, the enormous chestnut measured "seventeen feet in diameter."[18] Although some have doubted the veracity of the claim, the late Colby Rucker, of the Eastern Native Tree Society, believed the tree likely possessed "the greatest known diameter of any eastern hardwood."[19] Gene Christopher, a native of Francis Cove, recalled

FIG. 10.1. Large decaying chestnut tree, Great Smoky Mountains, East Tennessee, c. 1902. United States Department of Agriculture, *Message from the President of the United States, Transmitting a Report of the Secretary of Agriculture in Relation to the Forests, Rivers, and Mountains of the Southern Appalachian Region* (Washington, D.C.: GPO, 1902). Courtesy of the Library of Congress.

seeing photographs of the tree when it was still alive, and even played in the decaying stump as a young boy.[20] According to Christopher, there were other large chestnuts at the site, including one tree with such an enormous hollow trunk that, after falling on the ground, cattle could not only enter inside, but turn around and exit it at will.[21]

Christopher believed the giant chestnut was felled for firewood in 1915, a full decade before blight reached the Francis Cove community.[22] His use of the term firewood is somewhat misleading, as chestnut remained unpopular for use in open fireplaces due to its tendency to throw off sparks.[23] Chestnut kindling, on the other hand, was desirable for early twentieth-century cookstoves, woodstoves, and locomotive fireboxes. Indeed, in airtight structures chestnut burned hot, evenly, and longer than pine and other woods. As a result, chestnut stovewood was a common heating source for home parlors, community stores, and one-room schoolhouses.[24] Cured chestnut kindling also left fewer ashes and produced less smoke, making it a favorite among moonshiners needing to conceal their illegal distillery operations.[25]

to purchase groceries and other supplies

In addition to being a desirable stovewood, the American chestnut was valuable to residents of Appalachia for yet another reason. The tree also produced large amounts of tanbark, a commodity mountain residents—particularly after 1890—relied on for cash income. Tanbark, it should be noted, is the generic name for any bark that produces tannin, the ingredient used in the production and preservation of leather products.[26] For most of the nineteenth century, hemlock bark was the major source of tannin, especially in the Northeast, where individual tanneries consumed several tons per day.[27] After the Civil War, chestnut oak was also used for that purpose, especially in the Appalachians, where the trees were large and numerous.[28] By 1890, however, both hemlock and chestnut oak were becoming so overharvested that tanbark was a scarce, if not unavailable, commodity.

The scarcity of hemlock and chestnut oak, along with the development of new tannin-extraction techniques, made the gathering and processing of chestnut bark not only possible, but an extremely lucrative enterprise. As a result, many northern tanneries shuttered their establishments and moved their entire operations to the southern Appalachians, where chestnut was still widely available.[29] In 1898, when the Hans Rees & Sons tannery opened its new facility in Asheville, North Carolina, it was the largest belt-making enterprise in the United States. Its twenty-two-acre campus contained thirty separate buildings and processed no fewer than thirty thousand cowhides daily.[30]

After 1905, chestnut cordwood replaced tanbark as the favored source of tannin.[31] Before 1880, leather manufacturers in Europe had successfully rendered tannin from European chestnut, but experiments in the United States failed to produce the desired product. After it was discovered that tannin from trees in the southern Appalachians was comparable to the European product, a new industry was born. Although trees used to make chestnut extract were generally smaller in size, mature specimens were also harvested. To make extract, chestnut cordwood was placed into metal grinders that transformed the bolts into finely ground chips.[32] The chips were then placed into large tanks and steeped in hot water, creating a relatively weak tannin liquor. With additional cooking and straining, and the passing of the liquid from one extractor to another, the tannin obtained the desired strength and density. To create the final product—in both liquid and solid forms—the solution had to be filtered and steam evaporated, a process requiring thousands of gallons of water.[33]

By 1907, chestnut extract was used at 462 different tanneries in the United States, in communities stretching from Pennsylvania to Alabama. Two years later, 475 facilities required 184 million pounds of chestnut extract, making it the most common source of tannin in the U.S. leather industry.[34] After 1915, most North American tanneries no longer produced their own tannins, relying instead on the growing number of extract producers in the southern Appalachians.[35] Although the largest tanneries were enormous industrial complexes sprawling over dozens of acres, most had limited space for storing cordwood or tanbark. Tannery owners preferred instead to operate in close proximity to extract producers, thereby reducing shipping costs and the price of producing leather goods. If transported by rail, chestnut extract was placed in wooden barrels weighing as much as five hundred pounds. For larger purchases, entire railroad tank cars were needed to carry the liquid to its final destination.[36]

Tanneries purchasing the largest quantities of chestnut extract produced the sole leathers for shoes as well as the heavy leathers for making horse harnesses, saddles, boots, and large suitcases. Many establishments, such as the Hans Rees & Sons tannery, also produced the thick leather belts that ran industrial machinery. When blended with other vegetable tannins, chestnut extract improved the firmness and tensile strength of all leather products, which explains its popularity among tannery owners.[37] In fact, by 1919, chestnut extract was used in making nearly half of all heavy leathers produced in the United States. In that year alone, 432 million pounds of chestnut extract were consumed by American tanneries, an amount worth almost $16.3 million dollars ($297 million dollars in 2021 currency).[38]

Before 1921, the state of Pennsylvania consumed the largest quantity of extract in the United States.[39] When hemlock became less available at the end of the nineteenth century, Pennsylvania tanneries were forced to purchase other kinds of tannins, including quebracho, a product imported from South America. In 1912, there were still numerous chestnut-extract producers in Pennsylvania, even though the blight had spread across much of the state.[40] When it was discovered that wood from blighted chestnut did not alter the quality of extract, tanneries were able to continue, if not increase, their production quotas. According to Joseph Shrawder, one of the first chemists to investigate the problem, in twenty experimental tests, all but one "showed a higher percentage of tannin in the infected bark than in the healthy bark of the same tree."[41] The higher tannin content also made the finished leather a "medium dark color," which, he believed, was a more desirable hue. Shrawder thought chestnut extract made from infected bark would become "a matter of much interest," if not eventually preferred by the entire leather industry.[42]

At that time, producing tannin from chestnut bark was already falling out of favor, making Shrawder's claim a largely moot one. Pennsylvania had millions of blight-infected trees with little if any bark (the bark usually fell from the trunk two or three years after infection), as well as blight-free stands in the western part of the state.[43] Needless to say, seeking out and harvesting chestnut tanbark from blight-stricken trees would have been an inefficient, if not entirely impractical, undertaking. Moreover, by 1920, much of the 1.5 million pounds of chestnut cordwood arriving at U.S. tanneries and extract facilities was already debarked in the field.[44] The practice helped to standardize the strength and quality of the tannin as well as streamline production; in fact, most of the cutting and preparation of chestnut cordwood—or "acidwood," as it was often called—followed the same industry protocols. Not only was the bark removed prior to processing and transported separately, but the length and width of the chestnut bolts had also become standardized.

For example, in a wood specifications broadside produced by the Marion Extract Company of Marion, Virginia, woodcutters were told to prepare trees by first sawing them "close to the ground and in full five-foot lengths." Regarding the required widths, bolts nine inches in diameter did not require splitting, but those eleven inches in diameter "should be split once only." Bark removal was so important that failure to do so could cause forfeiture of the entire sale. "All bark must be removed," warned the company, "and if it is not removed it will not be accepted."[45] One representative even communicated the going price for chestnut cordwood on the back of a company broadside. The typewritten letter, addressed to B. Y. Dickey of Sugar Valley, Georgia—a rural community located at the base of Chestnut

Mountain in Gordon County—reflects not only the strict standards of the company but their immediate need for extract material.

> Dear Sir: Until April 30th, 1922, we will pay $5.00 per cord of 160 cubic foot for prime chestnut wood prepared in accordance with the specifications on the reverse side of this sheet and loaded in full minimum carloads at Sugar Valley, Georgia, station on the Southern Railway. We will appreciate your shipments and hope that you will be able to begin shipping at once.
>
> Yours truly,
>
> The Marion Extract Company[46]

As the Marion Extract Company expanded its operations southward, so did America's largest tanneries. By 1925, the leather industry had relocated almost entirely to the southern Appalachians, making it the geographic center for the production and consumption of chestnut extract.[47] Even before that date, there were as many extract plants in the region as there were tanneries, which held true well into the 1940s. Tanneries and extract plants also gave important jobs to mountain residents and were, in many communities, the major employers. For more than three decades, the trees not only provided year-round employment but cash incomes for farmers willing to gather cordwood and tanbark. In 1923, fifty-four tanneries in Kentucky, Virginia, Tennessee, North Carolina, Georgia, and West Virginia maintained nearly six thousand employees on their payrolls, with many facilities employing several hundred or more individuals.[48] Although most chestnut-extract facilities were smaller and required fewer workers, many required enormous amounts of cordwood and armies of suppliers. The Kingtan Extract Company of Kingston, Tennessee, for example, purchased no fewer than three thousand railroad cars of chestnut wood annually.[49]

Taking perhaps the greatest advantage of the forest resources in what was then called "the chestnut belt" was the Champion Fibre Company of Canton, North Carolina. Founded by publisher and paper manufacturer Peter G. Thompson in 1905, the company produced paper pulp at the facility as early as 1908.[50] Although many thought Thompson built the Canton mill to access the enormous stands of spruce trees in what is today the Great Smoky Mountains National Park, Reuben B. Robertson, his son-in-law and manager of operations when the facility opened, stated there was another rationale for the company's move to western North Carolina. "Well, we came into this territory for two reasons," Robertson told a Forest History Society interviewer in 1959, "the first was the spruce and the second was the enormous stand of chestnut."[51] The company's move to North Carolina was made even more attractive when chemical engineer Oma Carr discovered in 1904 a way to extract tannin from chestnut wood without destroying its papermaking fibers. As a result, the same wood chips could be used for making tannin extract *and* finished paper.[52] This allowed Champion Fibre, which purchased the Carr patent, to double its profits. By 1910 the company was one the largest pulp producers in the United States and the country's most profitable producer of chestnut extract.[53]

Indeed, by 1915, the Champion Fibre Company employed more than a thousand individuals and daily received "fifty carloads of wood" at the Canton facility.[54] Half of that total was chestnut, as ninety-five thousand cords were needed annually for the pulp and extract operations.[55] In its first decade after opening, the plant manufactured five million pounds of chestnut extract a month, a product that was also shipped overseas, especially during World War I, when shortages in Europe made chestnut extract a highly sought-after commodity.[56] Some extract made on the premises was used to tan the boots of World War I soldiers, although heavy shoe leathers were often treated with a variety of tanning agents and preservatives.[57]

On the paper production side, chestnut-sourced products included standard newsprint, book pages, cardboard, postcard stock, and the lower-quality paper found in pulp-fiction comic books and magazines. To make such products, Champion Fibre maintained an insatiable appetite for chestnut, processing no fewer than 275 cords each "twenty-four hour working day."[58] Such demand required the company to keep a two-month reserve on hand (fifteen thousand cords), wood piled several stories high across seventy acres of property.[59] Large amounts of cordwood were stored at railroad stations or along major highways in a radius extending a hundred miles beyond the Canton facility.[60] Although some individuals worked full-time supplying chestnut to the mill, most worked seasonally, as a way to supplement farm incomes or pay important bills. Asheville attorney George Henry Smathers, who represented Champion Fibre in legal matters, recalled that most mountain residents used the funds "for payment of taxes or to purchase groceries and other supplies" (see fig. 10.2).[61]

As influential as Champion Fibre Company was to the southern Appalachian economy, it did have its competitors. In 1927, the Armour Leather Company, which purchased annually forty-five thousand cords of chestnut for its tanning and extract operations, convinced the Mead

FIG. 10.2. Chestnut cordwood used in the making of chestnut extract. Champion Paper and Fibre Company, Canton, North Carolina, 1937. Courtesy of Evergreen Packaging, LLC.

Fibre Company to open a cardboard plant adjacent its facility in nearby Sylva.[62] The Mead Fibre Company was one of the largest paper producers in the world, and, like Champion Fibre, used enormous amounts of chestnut in its operations. However, the wood purchased for the Sylva plant was used to make corrugated cardboard, as George W. Mead, the principle owner, found it insufficient for high-quality papers.[63] While the Sylva facility focused on cardboard production, using chestnut pulp as the core material, the company itself invested in additional extract facilities. By the mid-1930s, the newly consolidated Mead Corporation owned eight extract plants in the region, including facilities in Damascus, Virginia, and Knoxville, Tennessee. In 1941, the company claimed to be the largest leather-tannin extract producer in the United States, and even advertised its product in trade magazines.[64] Being the nation's top producer of chestnut extract was no small feat, as the U.S. production of acidwood that year was 571,000 cords, enough wood to manufacture 125 million pounds of tannin for domestic and foreign use.[65]

Tanneries and extract facilities provided important jobs and incomes for Appalachian residents, but did so at the peril of mountain ecosystems. Although the vast majority of chestnut trees would eventually perish as a result of the blight, the intensive and widespread harvesting of cordwood had a visible impact on the mountain landscape, causing erosion, increased water runoff, and a reduction in leaf matter that would have otherwise returned to soil. Watercourses in the Appalachians perhaps suffered the most, as the tannery and extract operations released tons of toxic effluent into streams and rivers, causing major fish and mussel kills and, in some instances, rendering entire watercourses void of aquatic life.[66]

Working conditions at the extract and paper mills were also not ideal, as employees of all ages worked ten-hour shifts in unsafe conditions. At some plants, workers were even paid in scrip that could only be tendered at the company store.[67] If individuals opted out of such work, they had no other choice but to rely on the mountain environs for their livelihood, which required unregulated access to the forest commons. By the second decade of the twentieth century, large expanses of what had been a de facto commons were privatized, so the cutting of acidwood was no longer possible without permission of the principal landowner.[68] This did not stop individuals from entering the forest, however, particularly if property owners were unable to employ full-time caretakers or wardens. As a result, the gathering of chestnuts remained an important subsistence activity in the Appalachians, even after large numbers of trees were harvested for the leather and paper industries.

so that the people are independent

Despite losses incurred by lumber, acidwood, and pulpwood companies, the commercial nut trade continued relatively unabated in the southern mountains well into the 1920s. Many families had their own privates groves or had access to productive trees on the properties of relatives, friends, and neighbors. Remote rural areas where property ownership was unknown or not legally posted also provided nut-gathering opportunities. Chestnut harvesting was also possible on public property, including state or national forest reserves. Nevertheless, some communities fared better than others in the annual harvest. A 1911 report published in the *Daily News* of Frederick, Maryland, noted that six stores in the vicinity of Smoot, in Greenbriar County, West Virginia, had purchased no fewer than 106,000 pounds of chestnuts from local sellers.[69] In western Pennsylvania, twenty tons of chestnuts were shipped from Warren County that same year, with a single dealer contributing nearly two hundred bushels.[70] In southwest Virginia, the chestnut harvest was no less impressive. In his important study "Like Manna from God: The American Chestnut Trade in Southwestern Virginia," Ralph H. Lutts found that the residents of Patrick County gathered as many as 160,000 pounds annually.[71] In 1914, 9,156 pounds were traded at a single country store in the county, nuts destined for Baltimore, Philadelphia, or New York City.[72]

Not all chestnuts in Appalachia were sold or bartered, however. In many households, several bushels were retained for the kitchen larder or reserved for family holidays. In some communities, having a large quantity of nuts on hand was even equated with economic independence. An editorial published in the *Western North Carolina Democrat* in 1913 claimed that Edneyville residents had "all they need for home consumption. Wheat, corn, and potatoes in abundance and the woods full of chestnuts so that the people are independent."[73] Some residents even depended on the sale of chestnuts as their primary source of cash income. According to a 1911 report in the Frederick, Maryland, *Daily News*, an individual living on the western slope of South Mountain who possessed fifteen acres of "unusually fine native chestnut trees" gathered and sold between $400 and $800 worth of nuts annually. The nuts were of significant economic value, claimed the writer, if not "the most important crop on his small farm."[74]

A small minority of mountain residents even relied on chestnuts for their very survival, particularly the poorest citizens possessing little arable land or disposable livestock. The grandfather of Kentucky native Samuel D. Perry, who lived on the Cumberland Plateau at the beginning of the twentieth century, recalled that some families went hungry when there were no chestnuts in the forest.[75] According to the elder Perry, "without chestnuts ... some of those folks, living way out in the woods, with no close neighbors to look out for them, might have starved to death."[76]

More affluent mountain residents consumed chestnuts far less frequently, reserving them almost exclusively for the Thanksgiving or Christmas holidays. The harvest also inspired well-to-do families to organize chestnut roasts, an event bringing young people together in the same way chestnut frolics did in nineteenth-century New England. One such roast, sponsored by a direct descendent of Daniel Boone—Colonel Abner C. Boone—was held on September 27, 1913, beneath Glen Burney Falls in Blowing Rock, North Carolina. As reported in the *Watauga Democrat*, the roasting party consisted of six young couples and several adult chaperones, among them Mr. W. P. Pendley and Mrs. T. H. Coffee of the Watauga Inn. At the start of the roast, participants were given popcorn poppers attached to long poles, allowing them to maintain a safe distance from the hot flames. According to a reporter covering the event, two full bushels of nuts were roasted at the gathering, excluding "one burnt popper full which fell into the fire having slipped off the pole held by I. G. Bell of Savannah, Georgia."[77] Prizes were awarded for the "best popper of chestnuts," "singing the best song," and "best dramatic recitation." All prize winners, concluded the report, received "wooden trays, bowls, rhododendron napkin rings . . . all made by the mountain people."[78]

Outside the Appalachians, chestnuts were becoming increasingly scarce due to both the blight and World War I, which limited the import of Italian and French varieties. Demand for native nuts remained strong during the war years, particularly the week before Thanksgiving and the last two weeks of December.[79] They were also more expensive. In 1916, New York produce markets experienced a diminishing supply of chestnuts as well as a price of $10 per bushel, more than double what was paid at the beginning of the war.[80] A year later, chestnuts sold for $8.00 per bushel in Manhattan and other Atlantic port cities, including Baltimore and Washington, D.C.[81]

The high prices ensured substantial cash incomes for those willing to bring nuts to market, although southern chestnuts, as a result of their proclivity to be wormy or scorched by fire, had fallen out of favor in many northern markets. Consequently, they brought only 80 percent of the price of northern chestnuts, a fact affording them separate listings in market reports and newspaper ads.[82] In 1919, dry October weather and a New York City railroad strike reduced the supply of chestnuts coming from Virginia, Maryland, and Pennsylvania, increasing prices to an all-time high of $12 per bushel.[83] By 1920, even northern chestnuts were difficult to purchase after Thanksgiving and sold, when available, for $15 per bushel.[84]

After 1920, with the market for northern chestnuts in steep decline and markets opening up in larger southern cities, chestnut gatherers in the southern mountains were poised to make substantial profits. They also benefited from the building of roads and highways into remote rural areas, as reported in the *Charlotte Observer* in October 1923. "In former years," noted the editorial, "there was little incentive for mountain people to gather the nuts, as they had no way of transporting them 'down country,' except by an occasional covered wagon, and it would be weeks before the wagoneer could return home and distribute the meager profits among [those] who had contributed to his wagonload." As a result, added the writer, "many thousands of bushels were permitted to rot among the winter leaves, a neglected source of revenue."[85] After roads were constructed into these areas, however, chestnut gatherers could load them into trucks or cars, hastening their shipment to urban markets.[86] One of the largest such hubs was Hickory, North Carolina, where no fewer than twenty-five thousand pounds of chestnuts were sold to Charlotte buyers the first week of harvest. The nuts were gathered in Avery, Watauga, Ashe, and Allegheny Counties "by individuals who will this year take thousands of dollars that formerly went to waste."[87]

As lucrative as the annual nut harvest was for Appalachian residents, it was not a sustainable proposition. With each passing year, fewer trees remained in the forest as a result of the lumber, acidwood, and pulpwood industries. But it was the arrival of chestnut blight that impacted the harvest the most, as the fungus could render entire hillsides void of nut-bearing trees in a single year. By 1920, the blight had moved well into central Virginia, although smaller isolated outbreaks were recorded as far south as the Black Mountains of western North Carolina.[88] There were also reports

FIG. 10.3. Fallen American chestnuts, Poplar Cove, Joyce Kilmer Memorial Forest, Robbinsville, North Carolina, 1975. These are the same trees photographed by Sidney Streator in 1909. Courtesy of Craig Lorimer.

of the fungus killing trees along the North Carolina/South Carolina border, although government officials tracking the blight's movement believed the observations unreliable.[89] In parts of northeast Georgia nearly half of all trees were dead or dying by 1926, with isolated infections occurring as far west as central Kentucky and as far south as northwest Alabama.[90]

It was also in the mid-1920s that the Little Santeetlah Creek watershed was first impacted by the blight, which meant a premature death for the five large chestnut trees photographed by Sidney Streator in 1909. Although the Poplar Cove trees had escaped the lumberman's axe, they would have little defense against the fog of spores that were beginning to spread over the mountain area.[91] The trees in the cove did not die all at once, as the watershed contained six million board feet of chestnut—"with tight bark and some green leaves"—as late as 1935.[92] In fact, the timberlands of Poplar Cove were designated a "virgin forest" in 1936 and consequently offered up for sale.[93] After it was purchased by the U.S. Forest Service, the area was set aside as the Joyce Kilmer Memorial Forest to honor the author of the well-loved poem "Trees" ("I think that I shall never see / A poem lovely as a tree").[94] The government preserve was created not only to pay homage to Joyce Kilmer, but to showcase one of Appalachia's last remaining old-growth forests. Ironically, the trees that once comprised 30 percent of the standing timber in the cove were, by the early 1940s, no longer an integral part of the landscape.[95] At the end of that decade, the only remaining evidence of the American chestnut's former dominance in the watershed were the hundreds of decaying snags and logs that lay scattered over the forest floor.

CHAPTER II

a National calamity

The spread of *Cryphonectria parasitica* into the southern Appalachians had a profound effect on those dependent on the American chestnut for their sustenance and livelihoods. Outside the region, the loss of trees was not as impactful, since chestnut products could always be purchased from areas where the blight had not yet advanced.[1] And although it would take another two full decades to exhaust all timber and nut supplies in the southern mountains, the effect of chestnut blight on those relying on the trees for income was devastating. While the human response varied from community to community, and did not follow the same chronological or emotional trajectory, nearly everywhere there had been chestnuts, there were individuals who mourned their loss. Those living in the mountain region had the most to lose as a result of the blight, but the impact was far reaching and pervasive and affected everyone living within the tree's native range.

In Washington, D.C., the loss of the American chestnut was felt as early as 1917, when the D.C. bureau chief for the *Topeka State Journal*, Frederick Haskin, commented on their scarcity in northern Virginia. "You can walk for miles thru the woods of some sections," wrote Haskin, "and never see a chestnut burr."[2] Haskin also noted the increasing rarity of chestnut roasters on the streets of the nation's capital. "Neither can you buy chestnuts with anything like the ease you could a few years ago," added Haskin. "The sizzling roasting pans with which fruit stands formerly did a flourishing business now rarely roast anything but popcorn." After citing a USDA report claiming the area would see, in just two years, its last living tree, Haskin offered his own personal eulogy to *Castanea dentata*. His words could have been written by almost anyone in America, as individuals everywhere shared his passion for the trees. "Before [the blight], the spreading chestnut tree was known and greatly respected in all parts of the eastern United States. Young ladies who were famous for their home-made candy used to sally forth in the fall and bring back baskets laden with chestnuts which they had picked up in the neighboring woods. Farmers' wives used to store big supplies of chestnuts in the attic to be used off and on during the winter in preparing various savory nut dishes. . . . Farmers used to make their fence posts out of chestnut timber; the railroad companies made chestnut ties, and industry depended on the chestnut tree for a large part of its tannin." Haskin's final remark is telling, as it suggests the level to which local residents relied on the nuts during the wartime years: "The tremendous scarcity in chestnuts this year," wrote the reporter, "is a disappointment to many people who were counting upon them as an aid in reducing the high cost of living."[3]

After World War I, the number of chestnut roasters declined significantly up and down the Atlantic seaboard and continued doing so through the early 1920s. In Bridgeport, Connecticut, a single chestnut roaster—Giovanni Incerto—was seen on city streets in 1921, a fact reported at length in the *Bridgeport Telegram.* Originally from southern Italy, Incerto had been forced to sell hot dogs and ice cream, leaving only a few weeks between Thanksgiving and Christmas to sell the roasted nuts. "Two, three weeks, den good chestnut business all over," remarked Incerto in the illustrated feature.[4] The same article mentions the impending demise of the tree as well as the growing scarcity of chestnuts across the entire state. "Years ago, when native chestnuts were plentiful as peas," noted the writer, "they literally rolled into the warehouses [and] were dumped into great bins and sold by the bushel or sack just like ordinary nuts. Now they are put up in small crepe paper boxes and sold by the dozen."[5] Similar observations were made in 1920 in eastern Maryland, where state forester Fred W. Besley predicted the species "will someday be but a memory." The large groves of trees that were once common in the eastern part of the state, he added, had "all but disappeared."[6]

In the Mid-Atlantic interior, chestnut shortages were experienced much later, although their availability was always a topic of public concern. In 1927, a local buyer in western Pennsylvania told the *DuBois Courier* the nuts were "becoming more scarce and it looks as if there will be hardly any at all this year."[7] A similar observation was made in Oil City, Pennsylvania, a town fifty miles west of DuBois. The blight was considered the cause of the shortage as it had killed numerous trees in the immediate vicinity.[8] In Maryland, there were very few living trees in the western portion of the state after 1926, including Garrett County, along the West Virginia border. In fact, the forestry warden for Garrett County was so confident that chestnuts were a thing of the past, he sent Fred W. Besley a few quarts of nuts with the following note: "Make the most of these as this year's crop is very likely Garrett county's last."[9] The disappearance of chestnuts in the area north of Frederick was even blamed for the out-migration of squirrels from Catoctin Mountain, including the wooded tract that later became known as Camp David.[10] In Virginia, the blight's impact was felt as far south as Patrick County by 1926, causing not only a chestnut shortage, but personal hardship for those who gathered and sold the nuts. After 1928, there were virtually no nut-bearing trees in the entire county, a remarkable fact given that the area had a decade earlier produced tens of thousands of bushels for northern markets.[11]

Although chestnuts did not become scarce across northern Georgia until the 1930s, the economic impact on residents was no less severe. Chestnuts had been a cash crop for mountain families for more than a century and their loss meant the purchase of fewer store-bought provisions and essential goods.[12] Indeed, some Georgia residents made their living bartering chestnuts, as the trade was used to supply families with anything not made or grown on the farmstead. Rabun County resident Noel Moore, who witnessed families delivering chestnuts at a nearby country store, recalled the annual event. "The old man would have a big coffee sack full of chestnuts on his back, and the little fellers would have smaller sacks, and even the mother would have a small sack of chestnuts caught up on her hip," he remembered. "They'd all trek to the store and they'd swap that for coffee and sugar and flour and things they had to buy to live on through the winter."[13] Dahlonega resident Leamon Walden remembered gathering large quantities of nuts during the late 1920s in Chestnut Cove, then hauling them by mule and wagon to the town of Gainesville more than twenty miles away. Walden sold the chestnuts to buyers on the main square, who were not always satisfied with their purchases. When a customer complained to his grandfather about the large number of wormy chestnuts in his offerings, he replied unapologetically, "the man that buys land buys stones, the man that buys beef buys bones, and the man who buys chestnuts buys worms."[14]

For the Eastern Band of Cherokees, the loss of the American chestnut had particular significance, especially for those families maintaining long-standing cultural traditions (see fig. 11.1). As much as 30 percent of the timber on the reservation was comprised of chestnut, although some stands contained twice that amount.[15] During the 1920s, the Eastern Band sold a considerable portion of chestnut timber to the Champion Fibre Company, affording the tribe relative prosperity for several years.[16] By 1930, trees both on and off the reservation had succumbed to the blight, evoking considerable concern among tribal leaders. Along with corn and beans, chestnuts were a dietary staple that were used to make delicacies such as chestnut bread and chestnut dumplings.[17] However, it was not only the nuts that were important to the Cherokees, as the trees were valued for other reasons. In 1959, when anthropologists Raymond D. Fogelson and Paul Kutsche discussed the impact of the blight at a Smithsonian Institution symposium on Cherokee culture, they noted its influence on nearly every aspect of the tribe's existence. "Besides losing an excellent fuel and building material, the Cherokee also lost the chestnut itself, one of the delicacies in their diet. More important," they added, "the chestnut had helped maintain the local game supply, besides serving as fodder for domestic animals who in the past had been left free to forage for themselves in the forests."[18]

In fact, it was the ability of the American chestnut to produce large quantities of mast that made the trees so important to mountain families. The extensive chestnut and oak forests allowed farmers to give livestock free range without expense, so even if they did not benefit from the sale of timber, acidwood, or nuts, they certainly profited from the available forage.[19] Maggie Axe Wachacha, who lived on the Cherokee reservation most of her adult life, remembered the scene around her rural homestead prior to the blight's arrival. "There were about a hundred pigs when I first moved here," recalled Wachacha. "Pigs and hogs were so fat. There was plenty of chestnuts back then. . . . That is what they lived on."[20] The link between chestnuts and successful hog-raising was observed as early as 1854, when Frederick Law Olmsted noted that in areas where chestnuts were abundant, hogs were allowed to fatten on "the mast alone, and the pork thus made is of superior taste."[21] In some locales, chestnut-floured pork hung in the smokehouse all winter, becoming a major source of protein for many mountain families. A southwest Virginia farmer summed up the importance of chestnut mast to the mountain economy when he succinctly exclaimed, "It didn't cost a cent to raise chestnuts or hogs in those days. It was a very inexpensive way to farm. The people had money and had meat on the table too."[22]

Given the many uses for the American chestnut, it is not surprising their loss elicited a deep emotional response from mountain residents. The trees had been a major source of lumber, fencing material, and stovewood, and the nuts provided income and barter opportunities for hundreds of thousands of individuals. Chestnut mast, as already noted, sustained mountain livestock, which in turn provided families with additional sustenance and

FIG. II.I. Large American chestnut tree adjacent the blacksmith shop of John Owl, Qualla Boundary, Cherokee, North Carolina, 1908. Courtesy of the State Archives of North Carolina.

economic reward. Squirrel, turkey, deer, and bear populations were also more plentiful as a result of the chestnut crop, meaning more game on the kitchen table.[23] All of these things increased the cultural importance of the tree, elevating it to almost reverential status in some communities. James Mullins, born in Dickinson County, Virginia, recalled that his grandfather had such sentimental attachment to a large chestnut tree that he asked to be buried inside its stump, which was eight feet in diameter. "We put a six-foot grave through it, you know, about six to eight feet long, and it still had the sides on it," remembered Mullins.[24] Before doing so, several surviving family members cleared the road leading to the burial site, as it was covered in deep snow. They apparently did not mind the effort, acknowledging the importance of honoring their grandfather's dying wishes. "We told him if that is where he wanted to be, then we'd put him [in] it and we did," recalled Mullins.[25]

Many of those who personally witnessed the decline and disappearance of the American chestnut in the southern Appalachians described the event using words like "tragedy," "disaster," "calamity," and "catastrophe." Among the Cherokees, the loss of the tree created a deep psychological void, particularly among older residents who valued chestnuts as "a staple, a given of everyday life."[26] The dead and dying snags were certainly painful reminders that the mountain landscape, and possibly an entire way of life, was irrevocably lost. "Man, I had the awfulest feeling about that as a child, to look back yonder and see those trees dying," recalled Joe Tribble, a native of eastern Kentucky. "I thought the whole world was going to die."[27] Knott County, Kentucky, native Verna Mae Sloan recalled that life without the chestnut tree was indeed unthinkable. "At first we thought they would come back, we didn't know they were blighted out forever," she remembered. "But the chestnut tree was the most important tree we had. We needed those chestnuts."[28] Kentucky filmmakers Nina and Dean Cornett went as far as to subtitle their 2011 documentary about the American chestnut *Appalachian Apocalypse*, believing the loss of the trees subjected individuals in the mountains to unprecedented economic and emotional hardships (see fig. 11.2).[29]

Certainly the timing of the blight could not have been worse for mountain residents, as the onset of the Great Depression occurred just as the trees were losing their reproductive vigor. One study found that in 1929 alone, 150 million board feet of chestnut timber in the Appalachians was rendered unsalvageable as a result of blight, a figure equivalent to one-half

FIG. 11.2. Blight-killed chestnut trees, Skyline Drive, Shenandoah National Park, near the Byrd Visitor Center, Stanley, Virginia, 1935. Courtesy of the Library of Congress.

of the chestnut harvest for the entire United States.[30] Some observers even linked Depression-era poverty to the loss of the trees. After visiting several impoverished communities in the Blue Ridge Mountains of Virginia in the summer of 1929, one commentator offered this opinion about the impact of the blight on the local residents: "Once the great source of revenue was chestnuts, now the dead, white trunks of the chestnut trees, victims of the blight which has swept up the Appalachians, cover the mountain slopes—ghostly symbols of the death and desolation which seems to be coming to the communities themselves."[31] Chestnut blight was even a factor in the federal authorization of the Shenandoah National Park. The widespread destruction caused by the fungus supported the argument for the park's establishment, which required the removal of several thousand individuals. If there were no economic opportunities for those living within the proposed park boundaries, proponents argued, relocation was justifiable.[32]

After 1930, competition to gather chestnuts greatly intensified, in some cases leading to violence. In the Pond Mountain community of western North Carolina, a man named Mel Jones was even murdered as the result of a dispute over chestnuts. The incident took place in the mid-1930s below the summit of Bald Knob, where a truckload of interlopers had camped the night before. "Just before good daylight, somebody killed him," recalled Pond Mountain resident Clara Daugherty. "They shot him with two different guns on a Sunday morning." Apparently, the individual had been there to claim the chestnuts, although he was not the actual property owner. "There were more chestnuts up there than anywhere else, you see," explained Daugherty. "Somebody owned the land that didn't live there and they just thought they'd take them. They were sold for about a quarter a pound, and that was right in the Depression. That was pretty good money."[33]

For residents of rural Appalachia, the loss the American chestnut was indeed the death knell that put an end to their self-sufficient and forest-dependent way of life.[34] Unable to access the forest commons or turn their own trees into needed cash or lumber, many were forced to seek employment in neighboring mill towns or urban centers. Chestnut blight certainly accelerated the decline of subsistence farming in the region, causing measurable population losses in some communities. Some residents left the region entirely. The outmigration had started decades earlier, partly as a result of the purchase of enormous swaths of timberland by absentee landowners who, in some counties, owned 70 percent of the land surface.[35] Transformed into lumber, paper pulp, and acidwood, these vast timber holdings greatly benefited private landholders, but did little to increase the fortunes of ordinary mountain residents. True, the American chestnut was a major component of the upland forest, and generated considerable wealth for those with access to the trees, but not all Appalachians profited from their presence.[36]

Whatever the impact of chestnut blight on mountain farmsteads or communities, the tree's influence on economic and social life in the region was beginning to wane. Outside the region, the spread of blight remained of considerable public concern and continued to be chronicled in the local and national media.[37] The disappearance of chestnuts in one region signified their loss everywhere, so interest in the blight remained high, even as the fungus moved into central Alabama and northern Mississippi. If the passing of the American chestnut was a foregone conclusion by the late 1920s, the problem remained in the spotlight well into the 1930s. North Carolina forester John S. Holmes even described the loss as "a National calamity," believing American citizens everywhere should still be concerned about the tree's demise.[38]

Although federal- and state-funded breeding programs gave the public some hope a chestnut hybrid might be reintroduced into the wild in the near future, a handful of researchers shifted their attention to the blight's larger impact on the forest. Would oaks and other desirable species replace chestnuts in the woodland overstory? How were wildlife populations being affected by the blight? Knowing how the wooded environs were responding to the disease was indeed a growing concern, and finding answers to such questions required firsthand investigations into the blight-ridden areas. As a result, botanists, plant ecologists, and others began surveying the most impacted forests, documenting the subtle and not-so-subtle changes to the wooded landscape. Although it would take several decades before researchers understood how chestnut blight transformed the eastern deciduous forest, some changes were immediately apparent. While the changes were not universal, and differed from region to region, the loss of chestnut trees had an important and lasting effect nearly everywhere they had been present.

by species more xeric than itself

In America's urban areas, the absence of chestnut trees was less impactful, although it did change the overall character of the city landscape. Ecologically, the impact was harder to measure. The trees provided important shade, food, and shelter for birds and mammals in parks and city streets, but so did other mast-bearing species. In rural and other heavily wooded areas, however, chestnut blight changed the surrounding environment in both substantive and perceptible ways. First of all, in places where trees were most plentiful, large gaps formed in the forest canopy. This did not happen overnight, as the largest trees might remain standing for several years, even after the bark fell from the decaying trunks. Over time, other species became more dominant in the understory, including trees not generally associated with oak-chestnut forests. Although some woodlands recovered quickly from the absence of chestnuts, the rate and density of replacement varied from locale to locale.

In southern New England, northern red oak, chestnut oak, and white oak trees returned to the newly formed openings, as well as smaller distributions of scarlet oak, gray birch, and sugar maple.[39] In New Jersey, the replacement species were, in descending order, white oak, chestnut oak, and white pine, with some areas possessing significant amounts of red maple, black cherry, and scarlet oak.[40] In central and western Pennsylvania, white oak, black oak, and sweet birch were among the most common species in the new forest, along with some hickories and maples. In eastern Pennsylvania, tree replacement was most successful in moister areas, but on drier slopes the regrowth was less "extensive and the associated species [were] less desirable."[41] Among the "undesirable" species were flowering dogwood and hop hornbeam, trees considered of little value to period lumbermen. The same study found that species composition changed more dramatically if the woodlands were clear-cut or if the area was damaged by fire after the blight's arrival. Although tens of thousands of chestnut coppice sprouts were observed in the forest, the majority were dead or dying by 1924.[42]

On the Cumberland Plateau, American chestnuts were replaced by tuliptree, American basswood, and red maple, although flowering dogwood and American ash were also present by the late 1940s. Not all trees maintained their dominance in the canopy, as some replacement species yielded to others over time.[43] For example, a stand that in the 1920s contained "71.2 percent chestnut" by the mid-1950s was comprised of tuliptree, northern red oak, American basswood, and American butternut.[44] When the same stand was surveyed in the year 2000, American basswood had decreased considerably, as had American butternut, with the latter species becoming virtually nonexistent as a result of an introduced fungus. Sugar maple and pignut hickory were more dominant in the overstory at the beginning of the twenty-first century, as were eastern hemlock and red maple. In 1920, however, those four species had comprised less than 2 percent of the total area, making them clear winners in the game of postblight forest succession.[45]

A similar observation was made near Highlands, North Carolina, where biologist Catherine Keever found northern red oak, chestnut oak, and pignut hickory the most dominant chestnut-replacement species. In 1953, as a result of her findings, she predicted forests previously classified as "oak-chestnut" would eventually be labeled "oak-hickory."[46] By the late 1970s, her prediction came true in parts of the Blue Ridge, especially in areas where chestnut had been absent for at least a half-century. Indeed, a study completed by biologists J. Frank McCormick and Robert B. Platt in southwest Virginia found pignut hickory to be the most common tree species replacing chestnut, followed by northern red oak, chestnut oak, white oak, and red maple.[47]

Not all chestnut-dominant stands yielded to oaks or hickories, especially those on northern-facing slopes or in deep, isolated hollows. In New Jersey, such locations saw an increase in tuliptree, red maple, and American beech, and in western North Carolina, hemlock, tuliptree, and red maple rose in number.[48] Thus, when region, soil type, and directional aspect are taken into account, broad generalizations can be made about which species replaced chestnuts in the forest canopy. In the most comprehensive study of the postblight forest—which investigated seventy-nine chestnut stands of every aspect and elevation in the southern Appalachians—the replacement species were, in descending order: chestnut oak (17% of all trees), northern red oak (16%), red maple (13%), eastern hemlock (6%), Carolina silverbell (5%), sourwood (4%), black locust (4%), scarlet oak (4%), tuliptree (4%), sweet birch (3%), white oak (2%), American beech (2%), flowering dogwood (2%), pitch pine (2%), and black oak (2%). Another thirty-six species were found in those same forests, although representing no more than 1 percent of all trees.[49]

The above study, conducted by biologist Frank W. Woods in 1956, is remarkable in that it provided the first comprehensive survey of chestnut-dominant forests after the arrival of the blight.[50] His findings, which included the plants and shrubs in the forest understory, provide ample evidence that chestnuts impacted not only woodland composition, but entire forest ecosystems. As Woods noted, on upland cove sites the removal of the American chestnut resulted in "more mesic forest types," whereas on the driest slopes and ridges chestnut was "being replaced by species more xeric than itself" (meaning dry locations became drier and less biologically diverse).[51] The same observation was made by the noted biologist E. Lucy Braun, who found that on southern-facing slopes where chestnut had formerly been the dominant species, understory growth became thinner or even nonexistent after the blight.[52]

Later studies confirm the trees aided in maintaining moisture levels in the soil, as well as promoted the recycling of essential nutrients, including carbon and nitrogen. In 2007, U.S. Forest Service biochemist Charles C. Rhoades discovered American chestnut leaves contained higher concentrations of nutrients than other deciduous trees, including oaks and hickories. Atop silty soils, the leaves also retained higher amounts of nitrogen, phosphorous, potassium, and magnesium, and beneath the leaf litter the soil possessed more carbon and nitrogen.[53] Researchers in Connecticut found higher amounts of nitrogen in chestnut leaves, but discovered the leaves decayed more quickly than other species, including American beech and northern red oak. The authors of the study concluded the faster decomposition meant more available energy for other plants and microbes, which improved overall nutrient recycling.[54] Such findings suggest that chestnut leaf litter promoted not only a greater abundance of nitrogen-loving organisms in the soil, but also an overall healthier ecosystem.

Chestnut leaves were also beneficial to numerous aquatic insects, including caddisflies, stoneflies, and craneflies. In 1988, two Virginia Commonwealth University biologists discovered that when stonefly larvae were fed decaying chestnut leaves, they had "significantly faster specific growth rates and [larger] adult body mass than individuals reared on oak." They also found adult female stoneflies reared more offspring as a result of daily consumption of chestnut leaves.[55] Obviously, freshwater fish also benefited from the leaf litter, as caddisflies and stoneflies are among their most preferred foods.

The American chestnut improved stream quality in yet another way. When large limbs or logs of the tree were submerged in mountain streams, they decayed very slowly—perhaps more so than all other tree species. As a result, more organic matter was captured in the water course, which, over time, created higher concentrations of nutrients beneficial to both macro-invertebrates and vertebrates.[56] The deeper pools and eddies caused by the woody debris also reduced soil erosion, minimized flooding, and even lowered water temperatures, thus benefitting cold-water fish species like native trout.[57] Remarkably, a study conducted in the southern Appalachians during the mid-1990s found that woody chestnut debris was still having a measurable impact on riparian ecosystems.[58] In another study, also conducted in the southern Appalachians, researchers found that 24 percent of the woody debris in a single mountain stream was still comprised of chestnut, more than sixty years after the blight first struck the area.[59]

Not all living things profited from the presence of chestnuts, however. The leaves of the tree are also high in tannin and possess flavonoids that have a debilitating or "allelopathic" effect on some plants.[60] Eastern hemlock, rosebay rhododendron, and black and yellow birch are among the species thought to be negatively impacted by the trees. In laboratory settings, chestnut leaf extract prevented them from germinating as well as slowed seedling growth when germination did occur.[61] Beyond those species, it appears the American chestnut has no more deleterious impact on understory vegetation than other allelopathic trees, including northern red oak, black locust, sassafras, and American elm.[62] In some locales, chestnut even increased the germination and growth of some trees, such as white pine.[63]

Archival photographs provide ample evidence that vegetation commonly grew underneath chestnuts, including ferns, mosses, and a number of herbaceous plants. In Catherine Keever's study of blight in western North Carolina—where living chestnut trees remained abundant "on north-facing slopes"—thirty-eight plant species were documented growing near or directly beneath the trees, including New York fern, hay-scented fern, Virginia galax, white snakeroot, false Solomon's seal, and Catesby's trillium.[64] In eastern Kentucky, E. Lucy Braun found those and other plants growing beneath chestnuts, with mountain laurel, Cumberland azalea, wild huckleberry, and eastern teaberry the most prevalent.[65]

Keever and Braun's observations suggest that if native chestnuts did affect understory vegetation, it was most likely as the result of their acidic leaves or the deep shadows cast by large old-growth trees. The high tannin content of chestnut leaves certainly lowered the pH of the soil, discouraging the growth of some plants and organisms. However, at least one study found that because tannins leached out of their leaves relatively quickly, the trees were probably no more detrimental to surrounding ecosystems than common oaks.[66] Moreover, more than a dozen mushroom species are known to have a symbiotic relationship with the tree and ninety additional fungi have been documented growing on or near the species, including the ink-stain bolete, waxy laccaria, and hairy fairy cap mushroom.[67] Perhaps the most common species associated with the tree is the shelf fungus known as "chicken of the woods" (*Laetiporus sulphureus*) and the similar but more reddish-colored beefsteak fungus (*Fistulina hepatica*). In 1909, William A. Murrill reported that individuals in New York City were harvesting beefsteak fungi from the trunks and stumps of decaying chestnuts, a practice that became more common as the blight swept across Manhattan and the surrounding boroughs.[68] Even before the blight, beefsteak fungus was so closely associated with the tree that it was commonly referred to as "chestnut tongue."[69]

a banquet table for wildlife

Perhaps the most significant impact of the blight on the landscape was the elimination of chestnut mast from the forest floor. Although oak and hickory trees eventually lessened the shortfall, in areas where the American chestnut represented nearly half of all nut-producing species, overall mast production declined by as much as 34 percent.[70] However, a more recent study found the species produced higher amounts of mast than even northern red oaks—"the next highest nut-producing trees"—and may have accounted for as much as "80% of the hard mast in any given year."[71] Computer simulation models project a precipitate loss in mammal populations as a result of chestnut blight, with white-tailed deer, gray squirrel, eastern chipmunk, and the white-footed mouse all declining measurably in numbers.[72] There is also considerable evidence the endangered Allegheny woodrat was heavily dependent on chestnuts, as the mammals cached literally hundreds in their winter larders.[73]

These findings are also corroborated in oral histories, further evidence the trees played a vital role in forest health. In southern Appalachia, the relationship between wildlife and chestnut mast was so well known it often became the subject of community folklore. Walter Cole, who grew up in the Sugarlands community of the Great Smoky Mountains, recalled in the 1960s, "the worst thing that ever happened in this country [was] when the chestnut trees died. Turkeys disappeared and the squirrels were not one-tenth as many as they was before. . . . Bears got fat on chestnuts, coons got fat on chestnuts . . . most all game ate chestnut."[74] Will Effler, a neighbor of Cole's, remembered shooting a wild turkey near their homes that contained "ninety-two chestnuts, still in the hulls and undigested" in its swollen craw.[75] Earl R. Cady, a forester trained at the University of Michigan and one of the first naturalists at the Great Smoky Mountains National Park, even referred to the annual chestnut crop as "a banquet table for wildlife." Cady believed the annual bounty was so significant that it allowed mammals to store additional "layers of fat in their bodies" as well as "nourish larger and healthier litters of young."[76] Former Cades Cove resident Maynard Ledbetter echoed similar sentiments when he jocularly exclaimed, "back when they was chestnuts, bear got so fat they couldn't run fast, now the poor bear run like a fox."[77]

Predator species also suffered as a result of chestnut blight, as they consumed birds and mammals dependent on chestnut mast.[78] In 1992, James M. Hill, a former Randolph-Macon College biologist, ascribed the decline of goshawk, Cooper's hawk, eastern cougar, and bobcat in the southern Appalachians to the loss of the American chestnut.[79] Although Hill's evidence was largely anecdotal, wildlife managers were certainly aware of the relationship between mammal and game-bird populations and the availability of chestnuts. A report published by the North Carolina Wildlife Resources Commission in 1957, for example, stated "the fruit was a staple in the diets of squirrels, turkeys, bear, and deer. The loss of the chestnut as a wildlife food is immeasurable."[80]

Nongame species also relied on the trees, including several moths that only ate chestnut leaves as their primary food source. In 1978, Paul A. Opler of the U.S. Fish and Wildlife Service estimated seven moth species became extinct as a result of chestnut blight, including the American chestnut moth, the chestnut ermine moth, the phleophagan chestnut moth, the chestnut clearwing, the chestnut casebearer, the chestnut yponomeutid moth, and the Confederate microbagworm.[81] Although two of the above species were later identified in the wild (chestnut clearwing and Confederate microbagworm), the other five represent a significant portion of all known invertebrate extinctions since the last ice age. As University of Connecticut entomologist David L. Wagner stated it, "American chestnut extinction correlates to the greatest invertebrate extinctions on earth. . . . There are only 61 invertebrate extinctions in the modern era . . . 41 in North America[,] and of those, 5 are directly related to loss of chestnut."[82] The loss of the trees impacted other insect populations as well, including native bees, wasps, and butterflies.[83] According to Douglas W. Tallamy, an entomologist specializing in the propagation of native plants and wildflowers, chestnut leaves provided larval food for as many as 125 different *Lepidoptera* species.[84]

Thus, in hindsight, the loss of the tree was a national calamity, although not only for the sentimental reasons most often associated with the species. Yes, the tree provided holiday treats to millions and gave the young and old alike an enjoyable autumn pastime. It inspired seasonal desserts, music, and poetry, and directly influenced the development of American material culture. It helped build America's transportation and communication networks and was an economic engine providing employment for tens of thousands of individuals. Yet, at the same time, wildlife also greatly depended on the tree; so much so, that numerous animal species suffered as a direct result of its disappearance. The trees also provided important ecosystem services, including the retention of moisture in forest soils and essential habitat for mammals, birds, insects, and fungi. For those reasons and more, the functional extinction of the American chestnut was not only a human loss, but an ecological one as well.

PART FOUR

Chestnut Revival

CHAPTER 12

genes for blight resistance

By the mid-1940s, most Americans had given up hope that a blight-free native chestnut tree would ever be returned to the forest. This did not stop scientists from working on restoration efforts, including Russell B. Clapper of the USDA research station in Glen Dale, Maryland. At Glen Dale, Clapper crossed American chestnuts with Chinese/American parents, producing a tree that became known as the Clapper hybrid.[1] In 1949, the hybrid showed so much promise that numerous seedlings were planted at the Crab Orchard National Wildlife Refuge in southern Illinois.[2] A decade later, when USDA senior plant pathologists R. Kent Beattie and Jesse Diller shared their views in an article entitled "Fifty Years of Chestnut Blight in America," the pair suggested that the Clapper hybrid might serve as the future seed source for the return of chestnuts to the eastern United States.[3]

However, before the Clapper hybrid could be introduced into the wild, researchers would need to observe large, mature trees in order to record their height, form, and growth rates. In 1964, one Clapper tree—labeled B26–3146—showed so much promise the U.S. Forest Service issued a press release about its growth and progress. "The largest, blight-free, forest-type hybrid, B26:#3146 USDA, occurs in the test plot near Carterville, Illinois," announced the release. "It is an American x Chinese backcrossed on to an American; the cross was made by R. B. Clapper in 1946. After 17 growing seasons, this tree measured 7.3 inches d.b.h. and 45 feet in height. . . . It apparently has a high degree of blight resistance, as chestnut blight is present in the plot."[4] Others, too, saw great promise in the Clapper hybrid, including Richard A. Jaynes of the Connecticut Agricultural Experiment Station. In the mid-1960s, Jaynes collected scions from the tree and later grafted them onto nursery stock at the Sleeping Giant Chestnut Plantation. Jaynes had spent many hours at the plantation, which was owned by Arthur H. Graves of the New York Botanical Garden. In 1962, Graves bequeathed the property to the experiment station, in effect creating the largest hybrid chestnut orchard in the United States.[5] After the closure of the U.S.

chestnut breeding program in 1960, Jaynes's work was critical to restoration efforts and remained so for several decades.

In 1963, federal authorities asked Jaynes, Clapper, and Diller to evaluate all surviving chestnut trees in the former government test orchards, including the hybrid trees at Carterville. After doing so, they concluded the vigorous trees—"those averaging approximately 2 feet height growth per year"—were too small in number to warrant large-scale forest plantings.[6] The news was disappointing for chestnut enthusiasts as it meant the Clapper hybrid could not compete with faster-growing tree species and thus could not reproduce in most natural forest settings. As a result, the B26–3146 tree was largely forgotten and only occasionally visited by chestnut researchers or forest technicians.[7] Nevertheless, by 1968 the tree was sixty feet tall and eleven inches in diameter.[8] It eventually showed signs of the blight, however, leading chestnut breeders to conclude that "tall timber-like" trees were unlikely to survive the blight fungus.[9] The tree was declared dead in 1976.

By the mid-1970s, with little national interest and no federal funding to solve the blight problem, American chestnut restoration efforts were left to either state-funded experiment stations or private individuals. Perhaps the most notable individual taking up the chestnut cause was Charles R. Burnham, the geneticist who worked with Nobel laureate Barbara McClintock in the late 1920s and early 1930s (see fig. 12.1).[10] At Cornell, McClintock and Burnham were engaged in the study of cytogenetics, which combined careful field observation with the use of microscopes in laboratory settings (the discovery of DNA was two decades away). Under McClintock's tutelage, Burnham had the ability to observe specific chromosomal traits in different corn varieties, thus allowing him to predict the color and size of their maize kernels.[11]

After arriving at the University of Minnesota in 1938, Burnham had already established himself as one of America's leading plant cytologists. Much of his research was focused on the movement of genes across chromosomes as well as the importance of chromosome pairing in developing cereal hybrids. Incredibly prolific, Burnham produced more than sixty publications during his career, including the textbook *Discussions in Cytogenetics*.[12] After his retirement from the University of Minnesota in 1972, Burnham began applying his knowledge of plant genetics to the problem of breeding a blight-resistant American chestnut. Although it would take him nearly a decade to fully formulate his thoughts on the

FIG. 12.1 Charles Burnham (far left) and Barbara McClintock (far right) at Cornell University, 1929. Courtesy of the Cold Spring Harbor Laboratory Archives, Cold Spring Harbor, New York.

subject, Burnham believed that if the principles of genetics were strictly followed, chestnut breeders could, in fact, produce a blight-resistant tree. In order to do so, however, they would need to backcross American trees with Chinese/American hybrids over many successive generations. Only after years of careful selective breeding and culling, he maintained, would the resulting offspring possess the necessary blight-resistant genes.[13]

Burnham formally introduced his breeding program in 1981, in a letter submitted to *Plant Disease*, an academic journal circulated among horticulturalists and plant pathologists. Burnham, or perhaps the journal's editor, entitled the letter "Blight-Resistant American Chestnut: There's Hope."[14] What made Burnham's breeding program different from previous ones was his insistence that "the recurrent parent must be the variety being improved, the American chestnut."[15] In other words, all progeny of Chinese/American crosses would again need to be pollinated with American chestnuts to ensure that each successive generation was more genetically similar to the native tree. At the same time, any offspring identified as susceptible to the blight would be removed from the breeding program. Advancing in this way, predicted Burnham, trees from the third and later backcrosses would closely resemble the American chestnut, as well as carry the "genes for blight resistance."[16] Burnham concluded the letter by asking individuals to locate pollen sources for his effort and even published his office address and phone number in the plea. Initially, he sought pollen not from American chestnuts, but from surviving trees in former government breeding programs, including the hybrid trees produced by Clapper and Graves.[17] The pollen would be used to pollinate American chestnut trees growing at the University of Minnesota's arboretum as well as trees growing in the wild. Progeny from those crossings would then be planted in open-pollinated seed orchards containing the most blight-resistant trees.

Burnham was so confident that his breeding program would be a success that he asked Phillip Rutter, a researcher from Canton, Minnesota, and Harold Pellett, a colleague from the University of Minnesota horticultural faculty, to join the effort. In 1983, Burnham, Rutter, Pellett, and lawyer Donald C. Willeke founded the American Chestnut Foundation (TACF), a 501(c)(3) nonprofit incorporated in Washington, D.C. According to Rutter, this was done to support the breeding program financially, as well as to "insulate it against the changing priorities so often encountered at public institutions."[18] By the end of 1985, TACF was holding annual meetings, publishing a journal, and monitoring first-generation backcrosses in Minnesota, the Great Smoky Mountains National Park, and Blacksburg, Virginia, among other locations. In December 1987, Rutter announced in the *Journal of the American Chestnut Foundation* that the board of directors had also approved the establishment of a research center, a facility to be located "within the central part of the old range of the American chestnut."[19] According to Rutter, the center would include a building for research purposes, a library of published and archival materials, and orchards for preserving American chestnut germplasm.[20]

In 1989, the center became a reality when TACF acquired a lifetime lease on a parcel of land known as the Wagner Farm in Meadowview, Virginia.[21] The Meadowview property offered a longer growing season as well as additional acreage if the facility needed expanding. The purchase of the Wagner Farm included the hiring of full-time staff member Fred Hebard, a Virginia

Tech graduate whose doctoral dissertation focused specifically on chestnut blight. Hebard's postdoctoral work on chestnut hypovirulence at the University of Kentucky gave him additional insight into the life cycle and behavior of the *Cryphonectria* fungus, as well as the ability to reproduce it in the laboratory.[22] After the formal dedication of the Wagner Farm, Hebard and volunteers planted numerous backcross hybrids on TACF grounds, including trees that were, in theory, three-fourths American chestnut. In 1993, Hebard tested the trees for resistance, applying the deadly fungus to their outer bark. After finding many of the trees resistant to the blight, he relayed the news to Burnham and Rutter. The results were evidence the trees possessed the necessary Chinese chestnut genes to weaken the fungus, if not subdue it entirely. The breeding program continued as planned.[23]

By 2000, the Meadowview research center had grown to forty-five acres and was home to nearly fifteen thousand chestnut trees of various types and sizes, including 4,800 trees considered third-generation backcrosses, making them 15/16th American.[24] Not all trees showed the same level of blight resistance, however, and generally only one in seven remained in the breeding program after being inoculated with the fungus. Nor did all blight-resistant trees demonstrate uniform growth, as some were shorter in stature or more rounded in form.[25] In addition, some of the backcrosses formed large cankers or possessed leaves and twigs more closely resembling those of the Chinese chestnut. A significant number of trees opened their leaf buds too early—before May 1—making them susceptible to late frosts.[26] Despite these shortcomings, a team of scientists evaluating the program gave it high marks, including their official endorsement. The report did offer several recommendations, including increasing the number of American parents in the program and using additional sources of blight resistance, such as the inclusion of more Chinese chestnut cultivars in the backcross breeding stock.[27]

Although these recommendations were perceived by TACF as minor issues, the report exposed several inherent shortcomings in the backcross breeding program. If American chestnut trees were to be fully blight resistant, they would need to possess, in perpetuity, Chinese chestnut genes. As a result, TACF trees would likely be viewed as hybrids by the general public and not the actual native tree.[28] There was also concern the American/Chinese trees lacked genetic purity and might not breed true-to-form or would produce infertile nuts. Some insisted the hybrids would pass unwanted Chinese chestnut characteristics to native trees surviving in the wild.[29] By the early 1980s, many Americans viewed the Chinese chestnut as an invasive species, which made the hybrids less desirable to the general public. The shift in perception had occurred during the 1970s, largely as the result of Charles Elton's book *The Ecology of Invasions by Animals and Plants*. Widely popular in the United States, the book focused considerable attention on the impact of invasive species on native ecosystems and even used chestnut blight as a case study.[30]

At least one of the above concerns had already been proven in the field: the notion that crosses between Chinese and American chestnuts produce infertile offspring. Richard A. Jaynes observed this phenomenon as early as 1964 and chestnut growers were also keenly aware of the problem. When Fred Hebard and colleagues addressed the issue in 2000, they conceded that "male sterility occurs in progeny when a Chinese chestnut male is crossed to an American chestnut female."[31] In other words, when male pollen from a Chinese chestnut is applied to the female stigma of an American tree, the resulting offspring produces no fertile pollen. However, when pollen from an American chestnut is gathered from a male catkin and placed on the female stigma of a Chinese chestnut, sterility does not occur. But even those trees might produce infertile offspring, as Hebard and TACF research scientist Yan Shi discovered in 1996.[32] Indeed, when second-generation Chinese/American hybrids were crossed with female American chestnut trees, male sterility ranged from 12 percent to 67 percent, depending on which Chinese/American parents were the source of the male pollen. More importantly, Hebard and Shi observed the frequency of male sterility did not necessarily decrease with each backcross generation, as scientists and chestnut breeders had previously predicted.[33]

Hebard's and Shi's findings are not surprising given "Haldane's Rule of Sterility," a premise postulated by British geneticist John B. S. Haldane in 1922. Haldane's Rule states that when the offspring of a cross between two animal species is sterile, it is generally a male.[34] In tree breeding, Haldane's Rule does not strictly apply, although the same mechanism producing sterility in animals is also present in plants. For this reason, the phenomenon is sometimes used by evolutionary biologists to document species divergence, or what is more commonly known as speciation.[35] Regarding the American and Chinese chestnut, the two species do, in fact, have different chromosomal structures as a result of divergent evolutionary histories. So when genes are exchanged during recombination, several chromosomes do not properly align. As Zoë Hoyle, a science writer for the U.S. Forest

Service explained it, "in the Chinese/American chestnut hybrid, four chromosomes come together into a cross shape instead of the normal linear shape."[36] This "translocation," as geneticists call it, prevents the successful exchange of genes between the two species, causing some pollen to be aborted. It also means, says Hoyle, that a larger portion of the Chinese chestnut genetic code is carried over when the two chromosomes combine.[37] Obviously, translocation is a long-term obstacle for TACF's breeding program, especially if important American chestnut traits are found within the abnormal chromosomes. It also means a fully blight-resistant American chestnut–type tree might be impossible to breed, since the most advanced hybrids will always exhibit some Chinese chestnut characteristics.

Despite setbacks, Hebard and his colleagues remained committed to the TACF backcross breeding program and by 2008 had created blight-resistant trees that were, in essence, 94 percent American chestnut and 6 percent Chinese chestnut.[38] They were also more morphologically similar to the American chestnut, meaning their twigs, bark, leaves, and form more closely resembled the native parents.[39] Representing no fewer than six generations of progeny, the trees were labeled B_3F_3s and were considered, at the time, the gold standard of the TACF breeding program. The trees also possessed significant blight resistance, with initial tests ranking them near, but not equal to, Chinese chestnuts.[40] In fact, the U.S. Forest Service was so impressed with the "Restoration Chestnut 1.0" that it began planting them on public lands to test their long-term blight resistance and adaptability to local conditions. By 2013, the federal agency had invested nearly $800,000 on the B_3F_3 trees and even established seed orchards for selecting future progeny.[41] Although the survival rate of the hybrids was not as high as initially hoped, their growth performance was nearly on par with the native tree. As a result, in 2014 the U.S. Forest Service claimed the return of the American chestnut was "theoretically possible" and offered additional institutional support for achieving that goal.[42]

As TACF and its staff worked toward improving American/Chinese hybrids, others adopted a different strategy. Gary Griffin, a Virginia Tech professor emeritus and lifelong chestnut researcher, believed large surviving American chestnut trees, although extremely rare, possessed a common genetic proclivity for resisting the blight.[43] For Griffin, the strategy was simple: intercross all surviving trees in a breeding program that captured blight resistance in lines of pure-American chestnuts.[44] In 1985, he and several colleagues formed the American Chestnut Cooperators' Foundation

(ACCF) in order to restore the American chestnut "to its former place in our Eastern hardwood forests."[45] Although as late as 2005 fewer than several hundred large, fully mature, blight-resistant native trees were growing in the wild, Griffin put the plan into motion with the aid of graduate students and hundreds of volunteers. By 2007, ACCF had distributed 160,000 nuts and seedlings across the former range of the species and inoculated hundreds of trees with hypoviruses designed to debilitate the *Cryphonectria parasitica* fungus.[46]

In 2009, several ACCF orchards were demonstrating significant levels of blight resistance in as much as 10 percent of their progeny, suggesting the breeding program was moving toward its stated goal.[47] In 2012, cooperators gathered some 3,725 nuts from select trees, distributing them to twenty-eight states and the province of Quebec.[48] Over the next several years, ACCF (as well as TACF) experienced numerous setbacks, including tree mortality due to deer, rodents, *Phytophthora*, and ambrosia beetles. Gypsy moths and cicadas also took their toll on ACCF test plots and plantings, especially in Tennessee and Virginia.[49] However, ACCF Executive Director Lucille Griffin (wife of Gary Griffin) was able to graft many of the infected trees before their demise, thus preserving their germplasm and reported blight resistance. In 2017, Griffin could report that ACCF cooperators had grown 3,103 trees in some fifteen states, with 22 percent of those plantings under her direct care in southwest Virginia.[50]

Although few individuals—including Fred Hebard and Charles Burnham—were publicly critical of ACCF's initial breeding program, they were certainly aware of its shortcomings. Firstly, not all native trees that appear to be blight resistant actually are, as some temporarily avoid the effects of the fungus due to geographic isolation or other anomalies, such as the presence of colder temperatures, which limits the growth and spread of the fungus.[51] Secondly, reported levels of resistance in ACCF breeding progeny are not always reliable, since data is sometimes gathered by untrained volunteers in uncontrolled settings.[52] Thirdly, a few of the so-called "survivors" may already possess some Chinese chestnut genes, since thousands of those trees, as noted in previous chapters, were planted in the former range of the American chestnut. And finally, because the genes for blight resistance appear to be more randomly transferred to each generation of trees, so that creating a fully blight-resistant tree may take many decades, if not a full century. What members of ACCF seem to possess, however, is patience. As promising as TACF's breeding program has been in the short term, many

ACCF growers question the approach, placing genetic purity over program expediency. For most ACCF members, the goal of species preservation is only truly solved "*when American chestnuts have blight resistance.*"[53]

a massive and irreversible experiment

A more controversial attempt at returning the American chestnut to the eastern deciduous forest involves twenty-first-century biotechnology. Although several individuals have produced genetically engineered (GE) chestnut trees, the most notable is the late William Powell, a former professor at the State University of New York's College of Environmental Science and Forestry (SUNY-ESF). Powell, who also directed the American Chestnut Research and Restoration Center at SUNY-ESF, was recognized as "forest biotechnologist of the year" in 2013. Much of Powell's initial research was done in collaboration with Charles Maynard, and both are credited for developing a GE American chestnut possessing "enhanced blight resistance."[54] The breakthrough came in 2005, after the pair successfully inserted a wheat gene into the DNA of the American chestnut. According to Powell and Maynard, the wheat gene causes the tree to produce large amounts of oxalate oxidase (OxO), an enzyme that detoxifies the harmful oxalic acid created by the blight fungus. Although initial tests of the GE trees found levels of resistance to be between "susceptible American chestnut and resistant Chinese chestnut," later attempts proved more successful.[55] By 2013, Powell and Maynard had produced a tree demonstrating resistance beyond even Chinese chestnuts, causing Timothy Tschaplinski, a scientist at Oak Ridge National Laboratory, to proclaim, "the team has accomplished a major goal, the generation of a blight-resistant American chestnut tree."[56]

As promising as the accomplishment is to chestnut enthusiasts and the general public, it was not the panacea many had hoped—for reasons even the researchers were reluctant to admit. Presently, the trees are officially classified as genetically modified organisms (GMOs) by the USDA, the Food and Drug Administration, and the Environmental Protection Agency. The trial plantings are also closely monitored by the USDA's Animal and Plant Health Inspection Service (APHIS), the federal entity that has authority over their release into the wild. As a result, the regulatory process will be a lengthy one, involving numerous government protocols and oversight procedures. However, Powell believed if all government procedures were strictly followed, and no problems or issues were raised, approval of the GE tree could happen as early as 2023.[57]

However, some think the approval process could take much longer, including Paul Sisco, a former staff geneticist with TACF. Sisco, who holds a PhD in plant genetics from Cornell University, is concerned that Powell and Maynard's GE chestnuts could impact the surrounding environment in yet unknown ways. In 2014, in an email written to friends and colleagues, Sisco stated that the SUNY-ESF trees needed "to be tested for a long time—probably fifty years or more." He added, "in the case of the chestnut, we are dealing with a long-lived organism. Powell and Maynard have only tested young engineered trees for blight resistance and they obviously do not know whether these genes will have unintended side effects on the trees or the organisms that interact with the trees as the trees mature."[58]

Others shared Sisco's concern, including Martha Crouch, a former Indiana University biologist intimately familiar with the science of GE trees. Crouch, who obtained her PhD in developmental biology from Yale University, not only believes there is need for longer-term testing of GE chestnuts, but thinks the methods used by Powell and Maynard will likely prove ineffective. In a November 14, 2014, letter published at Syracuse.com, the website of the *Syracuse Post-Standard*, she noted, "as a biologist familiar with GE tree research, I think release [of GE chestnuts] is premature. Their wheat gene is unproven outside of the lab, with no real-world track record even in crops, much less in a tree that can live hundreds of years. And the blight is likely to quickly defeat single gene resistance. These prototype trees should be confined to test plots, not let loose."[59]

Undeterred by such comments, Powell and Maynard remained committed to their restoration efforts, and by 2015 had raised an additional $100,000 in order to plant ten thousand GE chestnut trees on private and public lands.[60] They also obtained a $500,000 USDA Biotechnology Risk Assessment Grant to help facilitate the regulatory process, making it more likely they will obtain APHIS approval.[61] In 2017, Powell (Maynard retired in 2016) tested additional GE chestnuts for blight resistance, as well as the effects of the OxO enzyme on other organisms. The tree with the most promise, he concluded, was "Darling 58," which had been grown in tissue culture using embryos extracted from American chestnut seed. By 2019, Powell was confident the Darling 58 chestnut was not only fully blight-tolerant, but had complied with all USDA and APHIS safety protocols. As a result, on January 17, 2020, he filed a petition with APHIS, asking them to deregulate the Darling 58 tree. After a formal review and public comment period, the response to the petition could come as early as spring 2025. If APHIS does not find the GE trees a significant plant-pest risk, it would be the first major step toward their release in the wild.[62]

Without question, the American Chestnut Research and Restoration Center has been highly successful in swaying public opinion about the safety of GE chestnuts. However, not everyone is impressed with the work of SUNY-ESF, including Anne Petermann of the New York–based Global Justice Ecology Project. In 2015, in a widely circulated op-ed, Petermann noted that a quarter of a million people had signed a petition to stop the release of genetically engineered trees.[63] A number of other nonprofit organizations oppose their introduction into the wild, including the Center for Food Safety, a Washington, D.C., based nonprofit at the forefront of the GMO debate. In 2016, the Center for Food Safety actually filed several Freedom of Information Act (FOIA) petitions with APHIS in order to learn more about the SUNY-ESF trial plantings. In one instance, as a direct result of their FOIA request, the staff learned that six mature burrs, all created with transgenic pollen, had gone missing.[64] The chestnuts had apparently been eaten or buried by squirrels as a result of improper bagging before harvest. Oddly, no remedial action was taken by APHIS, since, in their words, there was "insufficient evidence and inadequate scientific information to determine the risks associated with this incident."[65]

Prior to final government approval, both the Center for Food Safety and Global Justice Ecology Project plan to challenge the release of the GE American chestnut via coordinated legal action and/or direct-action protest. "Once the field trials are complete, we will no doubt use every means at our disposal to prevent the introduction of GE chestnuts into the Eastern forest ecosystem," stated Andrew Kimbrell, executive director of the Center for Food Safety.[66] "Based on current regulatory protocols, the release of GE chestnuts is not only premature but irresponsible. The trees will undoubtedly contaminate the germplasm of native American chestnuts and cannot be recalled once they are released into the wild."[67] The same two groups spearheaded a similar campaign in 2017 challenging the regulatory status of a freeze-tolerant GE eucalyptus tree for commercial use. In response to the draft environmental impact statement, which announced preliminary government approval for releasing the GE trees, more than 284,000 individuals signed a petition opposing the decision, as did five hundred nonprofit organizations, including the Indigenous Environmental Network, Dogwood Alliance, Biofuelwatch, and the Sierra Club.[68]

The Global Justice Ecology Project anticipates a similar response to the GE American chestnut regulatory petition and will very likely coordinate public opposition if the government recommends release. In 2019, their Campaign to Stop GE Trees published a forty-seven-page whitepaper on transgenic chestnuts, authored by Rachel Smolker and Anne Petermann. Although the whitepaper focuses largely on Powell's GE chestnuts, it also addresses concerns with all genetically modified trees.[69] At issue is the unpredictability of GE trees, including their potential negative impact on native forest ecosystems. As Petermann and Smolker stated it, "To assess how the GE AC [American chestnut] will affect other trees, understory plants, insects, soils, fungi, and wildlife over time, we would need to have a far better understanding of both American chestnut and overall forest ecology."[70] Although Powell has done preliminary testing anticipating such concerns, Petermann and Smolker believe his studies are limited in both duration and scope.[71] In many respects, their concerns echo those of others wanting to see more rigorous, long-term studies before the trees are fully released into the wild.

Petermann and Smolker's anti-GMO stance, it should be noted, is influenced by the noted author and geneticist David Suzuki, who produced the 2006 documentary *A Silent Forest: The Growing Threat of Genetically Engineered Trees*.[72] Suzuki believes GE trees have serious containment problems, as it is impossible to control the movement of insects, birds, mammals, wind, and rain—all agents of pollen dispersal.[73] As a result, GE trees are capable of transferring pollen over many miles, including the genes for insect resistance, herbicide tolerance, and even sterility. For the American chestnut, the outcome of such an experiment could be devastating, especially if unwanted traits are transferred to native trees.[74] If, for example, the amount of OxO enzyme becomes higher in large adult trees, the leaves, after falling to the forest floor, could have a measurable impact on soil ecology. And if those same trees exhibit high levels of blight resistance and survive many decades or more, they will pollinate surviving American chestnut trees and/or hybrids, spreading the undesired trait over large areas. For this reason, say Petermann and Smolker, the release of GE chestnut trees in the wild would be "a massive and irreversible experiment."[75]

From the perspective of Petermann, Smolker, Suzuki, Kimbrell, and others, GE chestnuts have little redeeming value. From the point of view of the American Chestnut Foundation—which continues to endorse the introduction of transgenic trees—the return of the American chestnut is not only possible but likely inevitable as a result of GE advances.[76] TACF is certainly benefitting from a renewed interest in chestnuts by the American public, who are consuming imported and domestically harvested varieties in ever-greater numbers. U.S. chestnut growers currently produce more than two million pounds of chestnuts annually and another eight million pounds are imported each year from Italy, China, Turkey, and South Korea.[77] In Michigan alone, one hundred thousand pounds of chestnuts are yearly harvested, with some growers predicting a fivefold increase by 2024.[78] With the growing demand for chestnuts and chestnut products, it is little wonder that discussions regarding the return of the American chestnut are met with unbridled enthusiasm.

In 2023, TACF alone had more than five thousand members, twenty-two full-time paid staff, sixteen state chapters, and a $3.4 million dollar budget.[79] Presently, there are tens of thousands of trees in various stages of development growing at the Meadowview research farm, including both advanced hybrids and genetically modified trees.[80] Many TACF state chapters also maintain their own seed and breeding orchards, including the Pennsylvania chapter, which oversees 150 orchards and no fewer than thirty thousand trees across the state.[81] Even the smaller Maine chapter has many breeding orchards and spends no fewer than five thousand volunteer hours annually maintaining them.[82] The Maine chapter also monitors the health of one of the tallest living American chestnuts growing in the wild, a 115-foot-tall specimen located forty miles north of Portland.[83] In Georgia, the TACF chapter maintains at least fifty orchard sites, including plantings at both the governor's mansion and the Carter Center in Atlanta.[84] Many TACF members also maintain their own private orchards, growing not only American trees and Chinese/American hybrids, but also European/American hybrids. Many such trees go undocumented in TACF inventories, so the overall impact of their chestnut plantings on the North American landscape is even greater than the published data reflects.

the law of unintended consequences

Tracking the location and provenance of chestnut trees—American, hybrid, GE, or otherwise—will become increasingly difficult as more trees are planted in or near native forests. The plantings will certainly increase the likelihood that cross-pollination among the varieties will occur, making it nearly impossible to determine the genetic provenance of future offspring—at least not without considerable expense and meticulous record keeping. The best way to illustrate the conundrum is to highlight the work of the Appalachian Regional Reforestation Initiative (ARRI), a coalition of federal and state agencies formed in 2003 to replant forests on former strip-mine sites across the eastern United States.[85] By 2005, the ARRI partnership included the U.S. Office of Surface Mining and several nonprofits, including TACF and the National Wild Turkey Federation. According to founding documents, ARRI's initial goal was to plant tens of thousands of hardwood trees on 1.2 million acres of abandoned mine lands.[86] The American chestnut was selected as a preferred species at these sites even though its chances of survival, due to blight, were relatively small. It should be noted that no hardwood trees could survive into maturity on strip-mined lands without the development of a site-preparation technique called the Forestry Reclamation Method, which makes compacted soils atop surface mines more compatible with tree growth.[87]

By 2008, with the help of local community groups and volunteer labor, ARRI had planted forty-five hundred American chestnut and hybrid seedlings across no fewer than six states. After gaining national recognition for its efforts, ARRI created a nonprofit—Green Forest Works—to increase the plantings and to solicit additional funds from donors and charitable foundations.[88] Between 2009 and 2011, Green Forest Works planted 644,000 hardwood seedlings on nearly one thousand acres of abandoned mine lands, including thousands of trees donated by the American Chestnut Foundation.[89] In 2011, a $541,000 grant by the USDA Natural Resources Conservation Service resulted in even more chestnut plantings—and additional publicity for both ARRI and TACF.[90] Among the planted trees were Chinese chestnuts, which served as experimental controls as well as an additional food source for wildlife. By 2015, after the USDA grant cycle had ended, ARRI and TACF had planted nearly three hundred thousand American and hybrid chestnuts on mined lands in Pennsylvania, Ohio, Kentucky, West Virginia, and Virginia. Of those, thirteen thousand were Restoration Chestnut 1.0 trees possessing high levels of blight resistance.[91]

Although the majority of the American chestnut trees planted by ARRI will die before reaching maturity, some will bear nuts and produce seedlings, even after being infected with the blight. They will also provide important ecosystem services, including habitat for birds and insects, erosion control, and carbon sequestration. A few native trees might even prove to be blight resistant, although the chances are infinitely small, as the lessons of environmental history have shown. A significant number of the thirteen thousand Restoration 1.0 chestnuts planted on mine sites will likely survive into maturity, however, producing thousands of additional nuts and seedlings.[92] In fact, a 2018 study at two West Virginia mine sites found that eight years after planting, 33 percent of the American trees and 30 percent of the advanced hybrid backcrosses were still alive.[93] Curiously, in terms of growth and vigor, the Chinese chestnuts outperformed all others, achieving significantly taller heights during that period.[94]

The success of hybrid chestnuts at these locations will certainly make documentation of restoration efforts more problematic, particularly as the trees and nuts spread over the surrounding landscape. As noted in chapter 1, jays and crows can carry the nuts several miles before burying them, and undoubtedly will do so in the future. Moreover, only trained experts can differentiate a Restoration Chestnut 1.0 from a native tree, so the identification, tracking, and on-site improvement of the hybrids will be extremely difficult.[95] Some Restoration 1.0 chestnuts will, in the near future, also cross-pollinate with American chestnuts, since the latter trees outnumber the former by 24 to 1. Indeed, the genetic "introgression" of the two trees was an initial goal of the plantings, based on the assumption the advanced hybrids possessed no unwanted characteristics.[96] However, the long-term mortality and growth rates of the Restoration Chestnut 1.0 are still unknown, which means it could still be rejected as the ideal type.[97]

In fact, recent DNA analyses have determined the trees will need to possess more genomic material from Chinese chestnuts in order to be fully blight resistant. In a May 2019 memorandum to collaborators and partners, TACF president and CEO Lisa Thomson shared the news: "Through partnerships with HudsonAlpha, Virginia Tech, and other collaborators . . . TACF now has a clearer understanding of how many genes are involved in blight tolerance and the proportion of backcross trees' genomes inherited from American chestnut. The genomics research has revealed that there are likely at least nine regions of the genome involved in blight tolerance, rather than just two or three as originally hypothesized in the Burnham Plan." Thus these findings suggest Restoration 1.0 Chestnuts possess more Chinese chestnut DNA than previously believed. As Thomson put it, "our very best-looking trees are not as 'American' as we had hoped. The problem we optimistically set out to solve in 1983 has now become more challenging and complex."[98]

Simply put, the TACF trees most likely to survive the blight are those with the most Chinese chestnut genes. If that is indeed the case, they will be smaller than American trees, have different blooming and bud-break times, and produce larger, but fewer, nuts. The restoration hybrids could even have a negative impact on forest ecology, something TACF researchers have, to date, been reluctant to admit. In fact, a study published in 2018 evaluating the health and vigor of Chinese and American/Chinese hybrids at mine reclamation sites warned of that very possibility. "The introduction of Chinese chestnut and chestnut hybrids into Eastern North American forests," stated the authors, "may alter tree species competition and forest ecosystem dynamics, shift the potential for exotic insects and pathogens to affect native species, and invade into areas outside the original native range of chestnuts."[99] In fairness, the same authors admit there are potential benefits in introducing the hybrids into the wild, especially if they provide the same "favorable features of American chestnut to forest ecosystems."[100]

Of course, having Chinese, advanced hybrid, or transgenic chestnuts sharing the same landscape would be less problematic if there were no surviving American chestnut trees in the forest. Although the American species is considered functionally extinct, it is far from rare, so there is always the likelihood that cross-pollination, or what is sometimes called outcrossing, will occur. Indeed, according to a government database known as the U.S. Forest Inventory, there may be as many as 430 million American chestnuts surviving in the eastern woodlands, with some 8.5 million trees three inches or more in diameter.[101] A smaller number are blooming specimens of considerable height and stature, as the trees require sunlight and an open canopy for successful reproduction. However, in 2015, Sara Fitzsimmons, a regional science coordinator for the American Chestnut Foundation, estimated there may be as many as 2.5 million blooming-size chestnuts in the tree's native range.[102]

Curiously, Fitzsimmons does not think that number necessarily high, as she believes it reflects a low density of annually blooming trees—about 0.02 trees per acre. And even though outcrossing to wild American chestnut populations will likely occur as a result of the restoration plantings, she believes it will happen with such low frequency that it will have little impact on native trees.[103] However, figures gleaned from the U.S. Forest Inventory found the highest number of wild chestnut survivors are in the Kentucky and West Virginia coalfields, with population densities ranging from 4 to 26 trees per acre.[104] Similarly, the MEGA-Transect Chestnut Project, a biological survey conducted along the two-thousand-mile Appalachian Trail, found an even greater density of living trees, particularly in the southern portion of the tree's former range. In states like Georgia, there are no fewer than seventy-two living trees for every linear mile of the trail, although that figure only represents specimens within fifteen feet of either side of the trail, the parameter used by the MEGA-Transect reporters.[105] Of course, not all chestnut survivors in the coalfields or along the Appalachian Trail are of blooming size, but they are certainly more common than the numbers proposed by Fitzsimmons. But even if the conservative estimate of 0.02 blooming trees per acre is accurate, it still does not provide an adequate buffer to prevent outcrossing, as that figure represents no fewer than twelve trees per square mile.[106] Because viable wind-blown chestnut pollen can travel a mile or more, population densities would need to be substantially lower to prevent cross-pollination between any two trees.

This is precisely the problem facing the staff at the Tervuren Arboretum near Brussels, Belgium, a 250-acre forest containing 460 different species of trees. In 1902, several American chestnut seedlings were planted on the arboretum grounds under the supervision of its first curator, botanist Charles Bommer.[107] The trees are growing in the "Appalachian Highland" section of the arboretum and share the landscape with dozens of tree species from the southern mountains, including butternut, shagbark hickory, cucumber tree, and northern red oak (see plate 5). Several of the American chestnuts on the property are extremely large, including one specimen that is forty-five inches in diameter and more than 130 feet tall. According to Kevin Knevels, the forest manager of the arboretum, many of the seedlings produced by the American trees may be the result of intercrosses with European chestnuts blooming nearby, so he is unable to propagate them without additional DNA testing. To date, such testing has proven too expensive, so the seedlings must be destroyed, removed, or kept strictly segregated. As a result, Knevels is currently growing several hundred seedlings in an isolated nursery at the arboretum, but cannot return them to the grounds proper without first confirming their genetic provenance.[108]

For some observers, promoting the widespread reintroduction of advanced hybrid and GE chestnuts is curious, given there are so many American chestnut trees growing in the wild. The number of native survivors perhaps suggests that the term "functionally extinct" may no longer even apply to the American chestnut, since thousands of trees continue to annually blossom and, in some instances, produce fertile, viable nuts. Although chestnut blight made the trees largely insignificant in the woodland ecosystem after 1950, the high number of survivors does raise the question of why there is such an accelerated push to resurrect the species, given the dangers from doing so, should the restoration experiments go awry. The short-term benefits of introducing near facsimiles of *Castanea dentata* into the wild should not outweigh the risk of permanently altering the germplasm of one of our nation's most iconic tree species. Certainly, there is no better example of "the law of unintended consequences" than the introduction of chestnut blight into North America.[109] This principle should certainly cause those involved in the chestnut restoration arms race to consider the consequences of prematurely releasing GE trees into the wild.

Of the various organizations involved in the restoration efforts, only the American Chestnut Cooperators' Foundation has, to date, avoided altering the genetic blueprint of this foundational tree species.[110] It is possible the work of ACCF will prove most beneficial to the American chestnut's ultimate return, however plodding their efforts may seem to the general public. It is also possible their breeding program will become irrelevant, particularly if ACCF trees come into direct contact with GE trees and TACF's advanced hybrids.[111] In the age of the Anthropocene—in which no natural landscape remains isolated from human influence—the desideratum of genetic purity may be an impossibility as an increasing number of experimental trees are planted on private and public lands.[112]

It is also possible that even the most well-intentioned humans will be unable to restore the American chestnut to its former position in the woodland ecosystem—whatever the outcome of the various breeding programs. When and if American (or American-like) chestnut trees are established in the eastern deciduous forest, they will have to contend with old adversaries like *Phytophthora*, weevils, periodical cicadas, and chestnut timber worms, as well as newer diseases and pests, including the Asiatic oak weevil, Asian chestnut gall wasp, and Asian ambrosia beetle.[113] These obstacles will obviously reduce the number of healthy living trees in the forest, making the successful reintroduction of the species less likely.

Future attempts to restore the American chestnut will also need to better illuminate the tree's evolutionary history. Evolutionary history, in this sense, is not evolutionary biology, but a subfield of environmental history that sees nature-human relationships as ongoing, reciprocal relationships.

Proponents of evolutionary history, such as Edmund Russell of Carnegie-Mellon, argue that when plants and animals evolve with humans, they become altered by that relationship, including in their genomic structure.[114] According to Russell, evolutionary history allows one to marry biology to history in unique and important ways, offering a perspective not found in either discipline alone. A good example of the phenomenon would be any domesticated plant, such as New World cotton, which possesses longer fibers as the result of long-term human selection and breeding.[115] Although *Castanea dentata* evolved for millions of years without the presence of humans, the trees have, over the last several millennia, been directly influenced by anthropogenic forces. Twenty-first-century breeding efforts have also altered the DNA structure of the American chestnut, although the jury is still out regarding what this ultimately means for the future of the species.

If the U.S. government does recommend the release of GE chestnuts, tens of thousands of trees will likely be planted over the next several decades across the eastern deciduous forest. Initially, the young saplings will probably possess tolerance to the blight, but as the trees grow older and larger there is also the possibility they will become less resistant to the fungus. If the OxO enzyme proves ineffective in large older trees, the consequences are hardly inconsequential. If that happens, the GE chestnuts would serve as hosts and vectors for the *Cryphonectria* fungus, making it more likely the disease will spread to blight-free native trees in the surrounding forest. As the law of unintended consequences advises, we must always anticipate the unexpected, especially when dealing with living organisms. In India, when Monsanto introduced a GE cotton containing genes from *Bacillus thuringiensis* (Bt), a bacterium commonly used as a pesticide, it made the problem worse. The transgenic cotton was designed to prevent infestations of pink bollworm, a small gray moth found over much of the growing area. After it was planted, the larvae of the bollworm eventually developed immunity to the bacterium, and it now causes even more damage to commercial crops.[116] Although it is too early to know how the fungus will ultimately respond to mature GE chestnuts, having even more blight spores in the forest is not an impossibility. This also means the outcome of the GE chestnut experiment may be many years away, even if the federal government approves, in 2025, their release into the wild.

CONCLUSION

the giving tree

Although few individuals were able to observe a mature living American chestnut after 1950, the trees did not disappear from the North American landscape. As noted in previous chapters, some trees continued to bear nuts after they were attacked by the fungus and did so for numerous years afterward. Many of those survivors were also exceptionally large, which promoted the notion they were the "Mighty Giants" of the eastern deciduous forest—or as some would later state it, "the Redwoods of the East."[1] A few such trees even received attention in the national press, further increasing their renown. The species certainly captured the public imagination in ways none had done before, except possibly the giant sequoia or California redwood.

One such specimen was located at Porters Flats in the Great Smoky Mountains, near Gatlinburg, Tennessee. When photographer Albert Roth made images of the tree in 1933, it measured "twenty-eight and a half feet at four feet from the ground" (nine feet in diameter).[2] The tree also appeared in the official publication of the Southern Appalachian Botanical Club, in the issue debuting the journal's new title, *Castanea*, in May 1937. Its opening pages were penned by West Virginia forester Alonzo B. Brooks, who was appropriately asked to summarize the ecological importance of the American chestnut to the southern Appalachian region.[3] In his appraisal of the Porters Flat tree, Brooks referred to it as a "magnificent specimen," although he noted it had suffered severe blight damage in 1936.[4]

In 1942, the American Forestry Association brought additional attention to the Porters Flat tree, designating it their national champion in the native chestnut category. Stanley A. Cain, a University of Tennessee botanist, nominated the tree after discovering several large living specimens in the same location.[5] When the association introduced the tree in the November 1942 edition of *American Forests*, it was given the title "King Chestnut" and labeled "the largest American chestnut in the world" (see fig. C.1).[6] Curiously, the association omitted the fact that its crown was dead or dying, perhaps anticipating criticism from readers had they not done so.[7]

Predictably, the tree did not survive beyond the end of the decade, as was the case with other large surviving American chestnuts. To see future national champions, one would have to travel to Wisconsin, Michigan, or as far away as the state of Washington.

In Washington, the trees survived in small isolated stands after being planted by newly arriving settlers in the nineteenth century.[8] Jesse Ferguson brought chestnuts to the town of Tumwater, Washington, in 1845, for example, after traveling along the Oregon Trail. Although it is not known exactly where Ferguson obtained the nuts or seedlings, he was accompanied on the journey by George Washington Bush, a nurseryman and free person of color. Bush had considerable knowledge of plant propagation and himself transported various scions, seeds, and seedlings to Tumwater.[9] Without the assistance of Bush, it is doubtful the nuts or seedlings would have survived the trip, as they required constant attention during the sixteen-month-long journey.[10]

By the mid-twentieth century, two of the Tumwater trees had grown to exceptionally large size, becoming notable landmarks in the town's Olympic Memorial Gardens cemetery. The largest currently stands ninety-four feet tall and is more than six feet in diameter. In 2018, it was designated national champion by American Forests (formerly the American Forestry Association), even though Washington State tree enthusiast Ron Brightman, who inspected the tree in 1993, thought it to be an American/European hybrid and not a native chestnut.[11] The Tumwater chestnuts, like others in the Pacific Northwest, produce few seedlings, as the annual mast crop is generally consumed by squirrels or humans. And if seedlings do occasionally sprout from the fallen nuts, they are eaten by rabbits or deer or removed by cemetery groundskeepers. Presently, the Tumwater trees remain blight-free, as the *Cryphonectria* fungus remains relatively uncommon west of the Rocky Mountains.

In Wisconsin and western Michigan, the trees spread more widely over the landscape, forming, in some places, large naturalized stands. Native chestnuts were growing in Wisconsin as early as 1867, although one source traces their accession to 1851, when Jacob Lowe is believed to have planted a single seedling on his Columbia County homestead.[12] Bestwick Beardsley grew chestnuts on his Trempealeau County farm as early as the 1880s, after acquiring several seedlings from a New York acquaintance. The trees quickly multiplied, becoming a small grove by the early twentieth century.[13] In the early 1960s, a decade after the Beardsley property was sold to Einar

FIG. C.1. University of Tennessee botanist Harry M. Jennison (left) and an unknown individual measuring the Porters Flat Chestnut, November 19, 1933. A. G. "Dutch" and Margaret Ann Roth Papers, Betsey B. Creekmore Special Collections and University Archives, University of Tennessee Libraries, Knoxville. Reproduced with permission from Charlie Roth.

Lunde, twenty-six large trees survived in the woodlot behind the main home, along with several hundred seedlings, saplings, and young trees.[14] The largest tree eventually measured 3.5 feet in diameter, and in 1970 was designated "the state-record American chestnut."[15]

The most notable chestnut stand in Wisconsin, however, is located in West Salem, about twelve miles east of La Crosse. The trees became established there in 1885, after Martin Hicks brought nine saplings from Pennsylvania and planted them along a fencerow on his small farm.[16] By the 1950s, they were scattered over sixty acres and numbered in the several thousands. In 1985, the trees represented what many considered the last blight-free American chestnut stand in the United States, a distinction that came to an end two years later.[17]

In the decade that followed, nearly every imaginable treatment was used to save the trees, including the application of hypovirulent strains of the fungus on the infected bark. Hypovirulent strains of the fungus possess a debilitating virus, which neutralizes the blight and prevents it from killing the host tree.[18] However, because the hypovirulent fungus passes the virus to the virulent fungus by anastomosis, the two organisms must come into direct contact with each other in order for the process to successfully occur.[19] In the end, the hypovirulent strains spread over only a small portion of the West Salem stand and did not stop the disease from killing trees. By 2000, many of the survivors were still visibly weakened but continued producing viable nuts, albeit in a more limited fashion. Obviously, the application of the hypovirulent fungus onto infected trees gave some hope to those wanting a more immediate solution to the blight problem. At the time, however, the method proved costly, impractical to administer over large areas, and not very effective on saplings or small trees.[20]

A similar story unfolded in Michigan, albeit one with a more promising ending. In the late nineteenth century, farmers planted hundreds of American chestnuts along the western edge of the Lower Peninsula, where climate and soil conditions favored their growth and reproduction.[21] A 1910 planting near Crystal Lake in Benzie County, for example, numbered in the several thousands by the 1960s. In nearby Missaukee County, a survey done in the late 1970s found 144 large trees in a single wooded tract and 4,450 smaller trees and saplings.[22] Although chestnut blight entered the state as early as the 1930s, trees in the interior exhibited few signs of the disease until the 1970s, with a few areas escaping the fungus well into the 1980s. Remarkably, in several locations the blight developed its own hypovirulent strains of the fungus, which allowed the trees to survive without help from humans. The phenomenon was first observed in 1976 when cultures of a hypovirulent fungus were retrieved from chestnut cankers near Rockford.[23] By the late 1990s, chestnut growers in Michigan were optimistic about their ability to control the blight and predicted a statewide resurgence in native chestnut production.[24]

In the eastern United States, very few large trees survived beyond the mid-1950s, and almost none were able to subdue the fungus—with a few notable exceptions.[25] However, in many parts of the Appalachians only a small portion of the trees died suddenly, so they were still usable for acidwood, sawtimber, and firewood. Nevertheless, many trees were harvested immediately upon infection, which provided economic benefits to rural communities and timber companies alike. In parts of western North Carolina, North Georgia, and East Tennessee, salvaged chestnut timber paid substantial dividends well into the early 1950s. In an article entitled "We Still Have Chestnut," Paul H. Russell, assistant U.S. forest supervisor for the Nantahala National Forest in western North Carolina, claimed the value of "extract-wood" in the seven-county area in 1949 was $486,520 ($6.4 million in 2024 currency). The trees contributed an additional two million board feet to Nantahala timber sales, with lumber worth $90,850 and another $32,270 in stumpage fees—monies paid directly to the federal government.[26] As those figures demonstrate, the harvesting of chestnut continued to have an important impact on the Appalachian economy more than two decades after the blight first entered the mountain region.

With the adoption of synthetic tannins in leather manufacturing and the closure of the extract division of Champion Paper and Fibre, very little acidwood was harvested in the southern mountains after 1955. In fact, by 1950, two-thirds of all tannin used in leather making in the United States was imported from other countries.[27] Chestnut sawtimber also became scarce during the mid-1950s after the remaining infected trees were removed from the forest. In North Carolina, a few isolated stands above four thousand feet in elevation survived into the late 1950s, as the higher altitudes protected them from the wind-blown spores.[28] After 1960, virtually no large healthy trees could be found anywhere within the original range of the American chestnut, marking an end to its tenure as an integral forest species. Although *Castanea dentata* had occupied North American soils for more than forty million years, it had in a half-century been eliminated from four hundred thousand square miles of territory.

In many areas, all that remained of the tree were a few spindly coppice sprouts that grew from diseased and dying stumps. And although many of these sprouts developed into small saplings, most perished from the blight before reaching nut-bearing age.

With the tree effectively removed from its former range and government breeding programs shuttered, the fate of the American chestnut was virtually sealed. But the tree hardly faded from the American psyche or disappeared from everyday personal experience. As late as 1964, a reporter for the North Carolina *Lexington Dispatch* noted that even though the trees had been "dead and gone these 30 years but their skeletons still remain bare, white and ghostlike by the millions in the forests of the western North Carolina mountains."[29] Even dead, the trunks and limbs were gathered by local residents, who used them for stovewood, fencing, and construction material. Some even crafted the stumps and largest roots into commercial products. As late as the 1970s, Lloyd Fish of Madison County, North Carolina, dug and hauled dozens of chestnut stumps out of the woods to make hand-split shingles and decorative boards.[30] Some coppice sprouts produced eatable nuts, but then perished as their "mother" trees had done decades earlier. In 2021, reclaimed chestnut lumber is still in high demand, with prices exceeding ten dollars per board foot.[31]

In retrospect, for those who shared the same geographic space with the American chestnut, its value was irreducible to a single utility. Indeed, it was the *totality* of its uses that made the tree so important to humans. In this and other respects, the American chestnut was not unlike the sylvan protagonist of Shel Silverstein's best-selling *The Giving Tree.*[32] In the book, a young boy asks a tree to provide him with everything he desires—from leaves to make a crown to apples for selling at the market. The tree grants the boy's every wish, including branches to construct a house and its very trunk to build a boat. Much later, when the grown man asks for a place to sit, all that remains of the tree is a large stump. The tree again obliges, even though the man seems totally unaware that he himself has caused its demise.

For eleven thousand years, the American chestnut offered up its nuts, limbs, trunks, and stumps for human consumption, and in return was brought closer to death, perhaps beyond resuscitation. Barring a miraculous recovery due to unforeseen events, the tree will likely need human assistance if it is to survive into the next century. The species remains vulnerable due to the deadly fungus, as well as new pests and pathogens, a changing climate, and suburban development. And even though the chance of an American chestnut rebirth is perhaps greater now than it has been in decades, the future of the species is far from certain. While many believe twenty-first-century genomics will ultimately solve the blight problem, others are not so sure. Public opposition to the genetic manipulation of the species will likely continue, and possibly intensify if GE chestnuts are approved for release in the wild. In 2019, two board members of the Massachusetts/Rhode Island chapter of TACF even resigned over the GMO issue.[33]

Some in the biotech community would no doubt use the term "anti-scientific" to describe individuals opposing the GE approach to chestnut restoration. But such language would be inaccurate and based on a very narrow understanding of the scientific method. Trees are complex organisms that interact with other living things over many growing seasons, if not centuries. For this reason, more research is needed if we are to fully understand the impact of GE chestnuts on larger forest ecosystems. Since *Castanea dentata* emerged on the planet forty million years ago, humans have interacted with the trees for eleven millennia but only studied the *Cryphonectria* fungus for a single century. The supposition that genetically modified trees (or the attacking fungus) will behave in a specific and predictable way, based only on a decade of research, is premature if not bad science. Indeed, studies have shown that the genomic structure of transgenic plants can mutate as a result of gene insertion events and exhibit unexpected traits after reproducing.[34] It is also possible that GE chestnuts, as they grow older and larger, will not be unable to repel the blight, particularly if the OxO enzyme becomes less prevalent in mature trees.[35]

Scientists must be able to predict the future outcomes of their experiments and cannot reliably do so in the case of GE chestnuts.[36] Only by observing GE chestnuts over a considerable length of time will they be able to make such claims. In some respects, this same desideratum applies to those embracing the backcross breeding method, as the advanced hybrids have not been observed as large, fully mature trees. If the BC_3F_3 and later-generation hybrids do not exhibit the requisite characteristics of the American chestnut—including exceptional height and stature—then they should be segregated from native trees and possibly even removed from the wild.[37] Unfortunately, by the time such decisions are made, both GE chestnuts and TACF advanced hybrids may have permanently altered the germplasm of surviving American chestnuts, making future restoration efforts more difficult.

An equally important concern is the potential impact of GE chestnuts on Native American communities. Although the Cherokee Nation or Iroquois Confederacy have not taken official positions on transgenic chestnuts, members of both groups have publicly denounced their release. Lisa Montelongo of the Eastern Band of Cherokees, for example, believes the trees could have a detrimental impact on the production and marketing of tribal crafts. At a Campaign to Stop GE Trees event in western North Carolina, she stated, "I'm very concerned that GE trees would impact our future generations and their traditional uses of trees. Our basket makers, people that use wood for the natural colors of our clay work—there would be no natural life, no cycle of life in GE tree plantations."[38] In 2015, Montelongo, along with tribal member Mary "Missy" Crowe, were responsible for passing a council resolution rejecting the "genetic engineering [of the] natural world," which would, in principle, prohibit the future planting of GE chestnuts on Cherokee lands.[39] Sid Hill, a spiritual leader of the Haudenosaunee or Iroquois Confederacy, also worries about the ecological consequences of releasing GE chestnuts into the wild. "You don't know what it's going to do to the whole food chain," he told a *Los Angeles Times* reporter in 2019. "I am afraid of the domino effect."[40]

Whatever the path forward for chestnut restoration, the lessons of environmental history are important ones and have implications for how the trees are both perceived and managed. In 2013, when the U.S. Forest Service published *The Silvics of Castanea dentata*, which outlined policy guidelines for restoring and managing the species, there was minimal discussion about the negative impact of fire on the species, despite considerable historical evidence that it not only killed trees but accelerated the spread of the blight fungus.[41] Although the authors admitted the effects of fire on the American chestnut are not well understood, eyewitness accounts spanning more than a century suggest the trees were severely injured by fire, especially younger saplings and seedlings. Periodic fires were certainly beneficial to chestnuts by opening woodland canopies and eliminating competitor species, but frequent fires, particularly those in warmer months, greatly reduced their numbers.[42]

The most important lesson found within the pages of this book is, unfortunately, a reoccurring one. With respect to the cultivation and management of chestnut trees, humans have made repeatedly poor decisions. George Washington spent considerable energy growing European chestnuts at his Mount Vernon estate, despite having large and viable American chestnuts on the property. His importation and celebration of chestnuts from Spain not only fueled the perception that native chestnuts were inferior but increased the risk of introducing *Phytophthora* into the United States. A century later, nurserymen ignored protocols regarding the importation and distribution of trees from Japan, even after San José scale had caused millions of dollars of damage to the U.S. fruit industry. When the *Cryphonectria* fungus was linked to Japanese chestnuts during the early twentieth century, chestnut growers insisted a nationwide quarantine involving a halt to the interstate shipment of nursery stock was unnecessary. Then, as the blight spread southward, many assumed that cutting and burning all chestnut trees would somehow eradicate the fungus, despite warnings that such salvage methods might do more harm than good. And most recently, geneticists are choosing to alter the DNA code of the species, even in the face of growing public opinion against GMOs and the decision by hundreds of food manufacturers to stop selling GE products.[43]

However cliché it might be to state that those who do not learn the past are doomed to repeat it, in the case of the American chestnut this could very well be true. Chestnut restoration is a noble undertaking, but the process should be done carefully as possible, without harming the genomic heritage of this iconic tree. A wiser approach would be to adopt what the United Nations refers to as the precautionary principle, which restricts actions that permanently harm a species or ecosystem, especially if there is no absolute certainty about their safety.[44] Although the widespread cultivation and planting of American/Chinese hybrids would probably not be prohibited under the guidelines of the principle, transgenic chestnuts do not, at present, meet that threshold. Obviously, the best option moving forward would be to have *Castanea dentata* thriving again in the eastern deciduous forest, as it was *that* tree, and not others, that shaped the natural and human communities of North America. If the only benefit of the American chestnut was giving character to the landscape, as noted in the introduction, it would be worthy of our attention. That the tree fed, housed, and employed us makes its future, indeed our own future, no trivial pursuit.

acknowledgments

Although this book was written almost entirely in solitude over a period of five years, it was hardly a solitary effort. A number of individuals shaped its content, including several librarians who provided assistance in locating original documents and archival materials.

I am especially indebted to Margaret Dittemore, a reference librarian at the John Wesley Powell Library of Anthropology of the Smithsonian National Museum of Natural History. Ms. Dittemore provided carte-blanche access to the library for several hours, including their open stacks. The discovery that chestnuts were visually depicted in several Iroquois folktales was a serendipitous outcome of that visit.

Eric Frazier deserves special mention for his assistance in the Rare Book and Special Collections Division of the Library of Congress. Mr. Frazier retrieved many of the volumes cited in chapter 2, particularly the first published accounts of the American chestnut in its native habitat. Several of those texts were not written in the English language or translated by individuals with expertise in botany, environmental history, or forest ecology. As a result, I often provided my own translations or consulted experts for assistance. I also owe special thanks to Dr. Michael J. Ferreira of Georgetown University, who translated the words of the Portuguese soldier the "Gentleman of Elvas" into modern English.

A visit to the Mertz Library at New York Botanical Garden resulted in the discovery of additional archival materials, including a detailed map of the garden grounds, c. 1900. Senior Reference Librarian Stephen Sinon provided invaluable assistance during my visit, as he retrieved several documents unavailable in their digitized collection, including an annotated map of the adjoining Bronx Zoo. Archival documents were also retrieved from digital libraries, namely, HathiTrust.org, JSTOR.org, and Google Books. These online databases served as important finding aids, as well as assisted in vetting documents containing possibly redundant or irrelevant material. The Library of Congress's digital newspaper archives were also invaluable in documenting the role that chestnuts played in American life.

As a recipient of an Alfred D. Bell Jr. Travel Grant, I also had the fortunate opportunity to work with Cheryl Oakes at the Forest History Society archives in Durham, North Carolina. Ms. Oakes's enthusiasm for this project is reflected in the book's endnotes, several of which appear as the direct result of her assistance. She was also responsible for locating an early copy of the black-and-white photograph taken by Sidney Streator in 1909. Although details surrounding the Streator photograph were largely unknown, University of Wisconsin forest ecologist Craig Lorimer had identified the site in the mid-1970s. Armed with Lorimer's recollections and details from his field notes, Owen McConnell, Hugh Irwin, and I were able to locate the remains of the trees in the Joyce Kilmer Memorial Forest. A debt of gratitude also goes to Hugh Erwin and Brent Martin for spending, at my request, a day in the national forest supervisor's office archives in Asheville, North Carolina. Their visit uncovered a detailed timber inventory of the Joyce Kilmer Memorial Forest prior to its official dedication, a document providing additional evidence the trees in the Streator photograph were still alive in 1936.

Lynn Stanko, archivist at the USDA National Agricultural Library in Beltsville, Maryland, also provided important assistance. On short notice, Ms. Stanko made available the Frank N. Meyer collection, housed in the National Agricultural Library's Special Collections. Ms. Stanko also retrieved files from the Isabel S. Cunningham collection, as well as several Japanese nursery catalogues dating from the late nineteenth century.

Sincere thanks are also owed Fred Hebard, scientist emeritus of the American Chestnut Foundation. Hebard invited me into his home in Meadowview, Virginia, and, although he is now retired, responded graciously to numerous questions in a lengthy recorded interview. Some of his views influenced the book's narrative, especially several paragraphs in chapter 12. Few individuals in the United States understand the pitfalls and platitudes of breeding American chestnuts as does Dr. Hebard.

I am also indebted to John McNeill and his graduate students enrolled in his environmental history seminar (Hist-704) at Georgetown University. They provided input and suggestions regarding the organization and content of chapter 6, which they read as a course assignment. Those individuals are Sydney Browning, Daniel Graham, Anna-Sophia Haub, Robert Jones, Douglas McRae, Susan Peavy, Jackson Perry, Jeff Reger, Alan Roe, and Fábio Sanson. Denisa Paun, a graduate student in the cultural innovation master's program at the Transylvania University of Braşov (Romania), also deserves special mention. Ms. Paun proofread and reformatted the endnotes to chapters 6 and 7 in remarkably quick fashion.

Miroslav Mirovic, a freelance digital designer, was responsible for reformatting and cropping many of the archival images that appear in the book. Jon Davies, staff editor at the University of Georgia Press, also provided assistance in this regard, as well as offering helpful suggestions concerning the layout and placement of the book's many illustrations.

I owe additional gratitude to the many individuals who read chapters of the working manuscript—or portions of chapters—and provided feedback verbally or in writing. Those persons are, in alphabetical order, Jean Thomson Black, Marti Crouch, Fenny Dane, Paul E. Hoffman, Andrew C. Isenberg, Marc LaFountain, Owen McConnell, Neil Pederson, the late Mary Belle Price, Frederick J. Rich, and several anonymous reviewers. Mick Gusine-Duffy, executive editor of scholarly and digital projects at the University of Georgia Press, also deserves many thanks. His dedication to the project has been steadfast from beginning to end, and is very much appreciated.

Finally, I wish to thank all who have offered support and encouragement either in person or via correspondence over the last decade. Those individuals include Paul Arnold, Rodney Bailey, Chris Baker, Janet Biehl, George Brosi, Lee Ann Cline, Tom Colkett, Jelena Davis, Mike Edmondson, Ronald Eller, Yvonne Federowicz, William Heyen, Hugh Irwin, Doug Gillis, Rosann Kent, Dave Kimbrough, Brenda Lewis, Regina Markey, Brent Martin, Lou Martin, Brian Miller, Kathy Newfont, John Preston, Anita Puckett, Brian Smith, Dianne Smith, Jerry Smith, Mark Stoakes, Jim Veteto, and Barry Whittemore. Although the above individuals may not agree with my assessment of the tree's past, present, or future, all are friends of the American chestnut. The arguments and opinions regarding the appropriate path for restoring the species are strictly my own, however, and do not reflect the point of view of any one institution or entity.

notes

PREFACE

1. Bernd Heinrich, "Revitalizing Our Forests," *New York Times*, December 20, 2013, http://www.nytimes.com/2013/12/21/opinion/revitalizing-our-forests.html.

2. Bernd Heinrich, *The Homing Instinct: Meaning and Mystery in Animal Migration* (New York: Houghton Mifflin, 2014), 233.

3. Jessica C. Barnes and Jason A. Delbourne, "Rethinking Restoration Targets for American Chestnut Using Species Distribution Modeling," *Biodiversity and Conservation* 28, no. 12 (October 2019): 3199–3220.

INTRODUCTION. Giving Character to the Landscape

1. Yi-Fu Tuan, *Space and Place: The Perspective of Experience* (Minneapolis: University of Minnesota Press, 2001), 27–36, 114–17. For the role forests played in American history, see Michael Williams, *Americans and Their Forests: A Historical Geography* (Cambridge: Cambridge University Press, 1992), 3–20; Gordon G. Whitney, *From Coastal Wilderness to Fruited Plain: A History of Environmental Change in Temperate North American from 1500 to the Present* (Cambridge: Cambridge University Press, 1994), 39–52, 131–71; and Eric Rutkow, *American Canopy: Trees, Forests, and the Making of a Nation* (New York: Scribner, 2012), 11–39, 211–18.

2. "The Costly Blight of the Chestnut Canker," *New York Times Magazine*, May 31, 1908, 9; Irvin C. Williams, "A History of the Early Pennsylvania Effort to Combat the Chestnut Bark Disease," in Pennsylvania Chestnut Tree Blight Commission, *Final Report of the Pennsylvania Chestnut Tree Blight Commission, January 1 to December 15, 1913* (Harrisburg, Pa.: William Stanley Ray, 1914), 19; G. F. Gravatt and L. S. Gill, *Chestnut Blight*, USDA Farmers' Bulletin no. 1641 (Washington, D.C.: GPO, 1930), 1, 16–17; E. George Kuhlman, "The Devastation of American Chestnut by Blight," in *Proceedings of the American Chestnut Symposium, College of Agriculture and Forestry, West Virginia University, Morgantown, West Virginia, January 4–5, 1978*, ed. William L. McDonald et al. (Morgantown: West Virginia University College of Agriculture and Forestry, 1978), 1–3.

3. Donald Edward Davis, *Where There Are Mountains: An Environmental History of the Southern Appalachians* (Athens: University of Georgia Press, 2000), 192–98;

Cory Joe Stewart, "Chestnut Blight," in *Encyclopedia of Appalachia*, ed. Rudy Abramson and Jean Haskell (Knoxville: University of Tennessee Press, 2006), 112; K. L. Burke, "History of Chestnut Survival in the Appalachians (Prehistory to Present)," *Journal of the American Chestnut Foundation* 21, no. 2 (Fall 2007): 18–25; Scott Osborne, "The Influence of the American Chestnut in Appalachian History and Culture, Part I," *Journal of the American Chestnut Foundation* 24, no. 2 (July 2010): 12–15; Scott Osborne, "The Influence of the American Chestnut in Appalachian History and Culture, Part II," *Journal of the American Chestnut Foundation* 24, no. 3 (September 2010): 12–15.

4. A June 2021 search of the United States Geological Survey's Geographic Names Information System (GNIS), the official database for U.S. geographic nomenclature, revealed 1,179 chestnut place names in the tree's native range. The database does not list roads or boulevards in the official repository, however. A June 2021 Google search of "Chestnut Street" yielded 6,740,000 results. See also Songlin Fei, "The Geography of American Tree Species and Associated Place Names," *Journal of Forestry* 105, no. 2 (March 2007): 89–90.

5. Ramseur quoted in Bethany N. Baxter, "An Oral History of the American Chestnut in Appalachia" (master's thesis, University of Tennessee at Chattanooga, 2009), 15.

6. William S. Stryker, *The Affair at Egg Harbor, New Jersey, October 15, 1778* (Trenton, N.J.: Naar, Day & Naar, 1894), 8–9, 25–27; Leah Blackman, *Old Times: Country Life in Little Egg Harbor Fifty Years Ago* ([Tuckerton, N.J.]: Tuckerton Historical Society, 2000), 185–90. Other examples are Chestnut Flat, Alabama, elev. 883 (Marshall County); Chestnut Gap, Kentucky, elev. 922 (Owsley County); Chestnut Grove, Maryland, elev. 692 (Washington County); Castanea, Pennsylvania, elev. 599 (Clinton County); Chestnut Ridge, New Jersey, elev. 427 (Bergen County); Chestnut Ridge, New York, elev. 413 (Rockland County); Chestnut Mound, Tennessee, elev. 971 (Smith County); Chestnut Grove, Virginia, elev. 673 (Buckingham County); and Chestnut Grove, West Virginia, elev. 835 (Doddridge County).

7. "Cheap and Lasting Monuments," *New York Times*, July 9, 1859, 4.

8. "Roland Park" (advertisement), *Baltimore Morning Herald*, May 3, 1893, 4.

9. "Gives Chestnut Hunt," *Baltimore Sun*, October 19, 1902, 7.

10. Farmer quoted in *Lewiston (Maine) Evening Journal*, February 28, 1893, 4.

11. Susan Freinkel, *American Chestnut: The Life, Death, and Rebirth of a Perfect Tree* (Berkeley: University of California Press, 2007), 4, 6, 97; Ellen Mason Exum, "Tree in a Coma," *American Forests* 98, no. 11–12, (November–December 1992): 21–59; Chad McGrath, "The Perfect Tree," sidebar article in Exum, "Tree in a Coma," 59.

12. Philip L. Buttrick, "Commercial Uses of Chestnut," *American Forestry* 21, no. 262 (October 1915): 961.

13. Ibid., 961–62.

14. *Boston Daily Globe*, October 23, 1904, 41; *Lawrence Daily Journal*, November 25, 1904, 7; *Milwaukee Sentinel*, December 13, 1912, 11. See also Scott Enebak, "The Holiday Nut," *The World and I* (December 1990), 333–34.

15. *Boston Evening Transcript*, November 15, 1898, 4.

16. Frederick Law Olmsted, *A Journey in the Back Country* (New York: Mason Brothers, 1860), 224; Karen Cecil Smith, *Orlean Puckett: The Life of a Mountain Midwife, 1844–1939* (Boone, N.C.: Parkway, 2003), 58–59; Ralph H. Lutts, "Like Manna from God: The American Chestnut Trade in Southwestern Virginia," *Environmental History* 9, no. 3 (July 2004): 501–3.

17. Marie Washburn quoted in Baxter, "Oral History," 22.

18. Henry David Thoreau, *The Writings of Henry David Thoreau*, vol. 19, *Journal XIII: December 1, 1859–July 31, 1860*, ed. Bradford Torrey (Boston: Houghton Mifflin, 1906), 400.

19. Roland D. Sawyer, "New Hampshire's Most Friendly Trees," *Granite Monthly: New Hampshire's State Magazine* 52, no. 8 (August 1920): 334. Sawyer found the aroma "delicate," but noted the blossoms had "an odor that mosquitoes do not like," which made the area beneath the trees "a fine place to camp."

20. Wilbur G. Zeigler and Ben S. Grosscup, *The Heart of the Alleghanies, or Western North Carolina* (Raleigh, N.C.: Alfred Williams, 1883), 194.

21. United States Geological Survey, Geographic Names Information System (GNIS), accessed May 1, 2020, http://geonames.usgs.gov/apex/f?p=136:1:192913189548. The town of Castanea, Pennsylvania, was named from the Latin word for chestnut (*castanea*), as it was a major transfer point for chestnuts shipped to the East Coast during the nineteenth and early twentieth centuries.

22. A. S. Fuller, "Nuts and Nut-Bearing Trees," *Popular Monthly* 9, no. 5 (May 1880): 551; "More About Chestnuts," *American Gardening: An Illustrated Journal of Horticulture* 15, no. 10 (March 10, 1894): 174; J. W. Kerr, "The Plain Truth about Chestnut Culture," *Garden Magazine* 6 (November 1904): 194–95.

23. George H. Powell, *The European and Japanese Chestnuts in the Eastern United States*, Delaware College Agricultural Experiment Station Bulletin no. 42 (Newark, Del.: Delaware College Agricultural Experiment Station, 1898), 2–3; Granville Lowther and William Worthington, eds., *The Encyclopedia of Practical Horticulture*, vol. 2 (North Yakima, Wash.: Encyclopedia of Horiculture Corporation, 1914), 800.

24. Henry Stephens Randall, *The Life of Thomas Jefferson*, vol. 1 (New York: Derby and Jackson, 1858), 71; James Parton, *Life of Thomas Jefferson: Third President of the United States* (Boston: James R. Osgood, 1874), 43.

25. Charles S. Sargent, *The Silva of North America*, vol. 9, *Cupuliferae-Salicaceae* (Boston: Houghton Mifflin, 1896), 9; G. Harold Powell, "The Historical Status of Commercial Chestnut Culture in the United States," *American Gardening* 20, no. 220 (March 11, 1899): 178; Lowther and Worthington, *Encyclopedia of Practical Horticulture*, 800.

26. Daniel Zohary, and Maria Hopf, *Domestication of Plants in the Old World*, 3rd ed. (New York: Oxford University Press, 2000), 189–90; Patrick Krebs et al., "Quaternary Refugia of the Sweet Chestnut (*Castanea sativa* Mill.): An Extended Palynological Approach," *Vegetation History and Archaeobotany* 13, no. 3 (August 2004): 145–60; Claudia Mattioni et al., "Microsatellite Markers Reveal a Strong Geographical Structure in European Populations of *Castanea sativa* (Fagaceae): Evidence for Multiple Glacial Refugia," *American Journal of Botany* 100, no. 5 (May 2013): 951–61.

27. Paolo Squatriti, *Landscape and Change in Early Medieval Italy: Chestnuts, Economy, Culture* (Cambridge: Cambridge University Press, 2013), 62–63. See also Della Hooke, *Trees in Anglo-Saxon England* (Woodbridge, UK: Boydell, 2010), 275; Paolo Squatriti, "Trees, Nuts, and Woods at the End of the First Millennium: A Case from the Amalfi Coast," in *Ecologies and Economies in Medieval and Early Modern Europe*, ed. Scott G. Bruce (Boston: Brill 2010), 25–44.

28. Powell, *European and Japanese Chestnuts*, 4–5; Liberty Hyde Bailey, *The Standard Cyclopedia of Horticulture*, vol. 2 (New York: Macmillan, 1914), 745–46.

29. William P. Corsa, comp., *Nut Culture in the United States, Embracing Native and Introduced Species* (Washington, D.C.: GPO, 1896), 87–88; John R. Parry, ed., *Nuts for Profit: A Treatise on the Propagation and Cultivation of Nut-Bearing Trees* (Camden, N.J.: Sinnickson Chew, 1897), 12–13; George H. Powell, "The Chestnut Industry," in *Transactions of the Peninsula Horticultural Society, Eleventh Annual Session* (Dover, Del.: Press of the Delawarean, 1898), 66–67.

30. Ernest A. Sterling, *Chestnut Culture in the Northeastern United States: Reprinted from Seventh Report of the Forest, Fish and Game Commission, State of New York* (Albany, N.Y.: J. B. Lyon, 1903), 98.

31. Ibid., 100.

32. Powell, *European and Japanese Chestnuts*, 5; Haven Metcalf, *The Immunity of the Japanese Chestnut to the Bark Disease*, USDA Bureau of Plant Industry Bulletin no. 121, part 6 (Washington, D.C.: GPO, 1908), 3–4; Philip J. Pauly, *Fruits and Plains: The Horticultural Transformation of America* (Cambridge, Mass.: Harvard University Press, 2008), 94, 125, 153.

33. James Hill Craddock, quoted in the PBS documentary *Appalachia: A History of Mountains and People*, episode four, "Power and Place," directed by Ross Spears (Riverdale Park, Md., Agee Films, 2009).

34. This estimate is extrapolated from observations made by forest professionals who counted the number of trees per acre in four different regions of the United States. See, for example, Ralph C. Hawley and Austin F. Hawes, *Forestry in New England: A Handbook of Eastern Forest Management* (New York: John Wiley and Sons, 1912), 467. The biomass estimate assumes chestnuts comprised, on average, 15 percent of all forested lands in their preblight range. See also Hazel R. Delcourt, Darrell C. West, and Paul A. Delcourt, "Forests of the Southeastern United

States: Quantitative Maps for Aboveground Woody Biomass, Carbon, and Dominance of Major Tree Taxa, *Ecology* 62, no. 4 (August 1981): 879–87; USDA, *All Live Tree Biomass Per Acre of Timberland, 2002*, USDA Forest Service map, Northern Research Station, 2002, accessed July 28, 2019, http://nrs.fs.fed.us/fia/maps/nfr/descr/xlivebio.asp; Douglass F. Jacobs, Marcus F. Selig and Larry R. Severeid, "Aboveground Carbon Biomass of Plantation-Grown American Chestnut (*Castanea dentata*) in Absence of Blight," *Forest Ecology and Management* 258, no. 3 (June 2009): 288–94.

35. Fred V. Hebard, "The Backcross Breeding Program of the American Chestnut Foundation," in *Restoration of American Chestnut to Forest Lands: Proceedings of a Conference and Workshop Held at the North Carolina Arboretum, Asheville, N.C., U.S.A., May 4–6, 2004*, Natural Resources Report NPS/NCR/CUE/NRR—2006/01, ed. Kim C. Steiner and John E. Carlson (Washington, D.C.: U.S. Department of the Interior, National Park Service, 2006), 61–77; Douglass F. Jacobs, "Toward Development of Silvical Strategies for Forest Restoration of American Chestnut (*Castanea dentata*) Using Blight-Resistant Hybrids," *Biological Conservation* 137, no. 4 (July 2007): 497–506.

36. See, for example, Donald Worster, *Nature's Economy: The Roots of Ecology* (San Francisco: Sierra Club Books, 1977), 341; J. Donald Hughes, *What is Environmental History?* (Cambridge: Polity, 2006), 40–42. Other important texts include Alfred W. Crosby Jr., *The Columbian Exchange: Biological and Cultural Consequences of 1492* (Westport, Conn.: Greenwood, 1972); Carolyn Merchant, *The Death of Nature: Women, Ecology, and the Scientific Revolution* (San Francisco: Harper & Row, 1980); Keith Thomas, *Man and the Natural World: A History of the Modern Sensibility* (New York: Pantheon, 1983); and William Cronon, *Changes in the Land: Indians, Colonists, and the Ecology of New England* (New York: Hill and Wang, 1983).

37. Donald Worster, "Transformations of the Earth: Toward an Agroecological Perspective in History," *Journal of American History* 76, no. 4 (March 1990): 1089.

38. Crosby, *Columbian Exchange*, 1–34. See also Charles C. Mann, *1493: Uncovering the New World Columbus Created* (New York: Alfred A. Knopf, 2011), 6–7, 9–12.

39. Frederica Bowcutt, *The Tanoak Tree: An Environmental History of a Pacific Coast Hardwood* (Seattle: University of Washington Press, 2015), xiv–xv, 5, 111–21.

40. William Thomas Okie, *The Georgia Peach: Culture, Agriculture, and Environment in the American South* (Cambridge: Cambridge University Press, 2017).

41. Jared Farmer, *Trees in Paradise: A California History* (New York: W. W. Norton, 2013).

42. Gary Paul Nabhan, *Mesquite: An Arboreal Love Affair* (White River Junction, Vt.: Chelsea Green, 2018).

43. See, for example, Donald Worster, *The Wealth of Nature: Environmental History and the Ecological Imagination* (New York: Oxford University Press, 1993), 66–69, 84–86; Timothy Silver, *Mount Mitchell and the Black Mountains: An Environmental History of the Highest Peaks in Eastern America* (Chapel Hill: University of North Carolina Press, 2003), 209–56; Martin V. Melosi and Joseph A. Pratt, "Introduction," in *Energy Metropolis: An Environmental History of Houston and the Gulf Coast*, ed. Martin V. Melosi and Joseph A. Pratt (Pittsburgh, Pa.: University of Pittsburgh Press, 2007), 1–14; and Zachary J. S. Falck, *Weeds: An Environmental History of Metropolitan America* (Pittsburgh, Pa.: University of Pittsburgh Press, 2010), 5–14, 174–84. When surveying the important works in environmental history, Paul S. Sutter found the studies he most admired were those "reaching for innovative ways to reengage and improve environmental advocacy." Sutter, "The World with Us: The State of Environmental History," *Journal of American History* 100, no. 1 (June 2013): 119.

44. Frederick L. Paillet, "Chestnut and Wildlife," in Steiner and Carlson, *Restoration of American Chestnut*, 41–49; William Lord, "A Noticeable Decrease in the Game," in *Mighty Giants: An American Chestnut Anthology*, ed. Chris Bolgiano and Glenn Novak (Bennington, Vt.: American Chestnut Foundation, 2007), 62–65; Meghan Jordan, "The American Chestnut: A Legacy to Come," *Compass* 11 (June 2008): 3–5 (*Compass* is published by the Science Delivery Group of the Southern Research Station, U.S. Forest Service, Asheville, North Carolina).

CHAPTER 1. The Evolutionary History of the Species

1. William L. Crepet and Kevin C. Nixon, "The Earliest Megafossil Evidence of *Fagaceae*: Phylogenetic and Biogeographic Implications," *American Journal of Botany* 76, no. 6 (June 1989): 842–55; Alan Graham, *Late Cretaceous and Cenozoic History of North American Vegetation* (Oxford: Oxford University Press, 1999), 217–18; K. J. Willis and J. C. McElwain, *The Evolution of Plants* (Oxford: Oxford University Press, 2002), 168, 190, 251.

2. William Crepet, "History and Implications of the Early North American Fossil Record of Fagaceae," in *Evolution, Systematics, and Fossil History of the Hamamelidae*, vol. 2, *"Higher" Hamamelidae*, ed. Peter R. Crane and Stephen Blackmore (Oxford: Clarendon, 1989), 45–66; Fenny Dane and Ping Lang, "Biodiversity and Evolution of *Castanea*," in *Plant Genome: Biodiversity and Evolution*, vol. 1, part E, ed. A. K. Sharma and Archana Sharma (Enfield, N.H.: Science Publishers, 2008), 88.

3. Paul S. Manos and Alice M. Stanford, "The Historical Biogeography of Fagaceae: Tracking the Tertiary History of Temperate and Subtropical Forests of the Northern Hemisphere," *International Journal of Plant Sciences* 162, no. s6 (November 2001): s77–s93; Ping Lang et al., "Molecular Evidence for an Asian

Origin and a Unique Westward Migration of Species in the Genus *Castanea* via Europe to North America," *Molecular Phylogenetics and Evolution* 43, no. 1 (April 2007): 57.

4. Ping Lang et al., "Molecular Evidence for an Asian Origin," 56. Paleoecologists speculate some *Castanea* species existed in North America sixty million years ago and fossil evidence supports such claims. However, no trees would have survived the K-T meteor event, which, as scientist Tim Flannery has pointed out, "carbonized" nearly all of North America. See Flannery, *The Eternal Frontier: An Ecological History of North America and Its Peoples* (New York: Grove, 2001), 17.

5. Jun Wen, "Evolution of Eastern Asian and Eastern North American Disjunct Distributions in Flowering Plants," *Annual Review of Ecology and Systematics* 30 (1999): 421–55; Fenny Dane, "Comparative Phylogeography of *Castanea* Species," *Acta Horticulturae* 844 (2009): 211–22.

6. Patrik Krebs et al., "Quaternary Refugia of the Sweet Chestnut (*Castanea sativa* Mill.): An Extended Palynological Approach," *Vegetation History and Archaeobotany* 13, no. 3 (August 2004): 146.

7. Lang et al., "Molecular Evidence for an Asian Origin," 56; Dane and Lang, "Biodiversity and Evolution of *Castanea*," 88.

8. The nine species are American chestnut (*Castanea dentata*), European chestnut (*Castanea sativa*), Japanese chestnut (*Castanea crenata*), Chinese chestnut (*Castanea mollissima*), Dode's chestnut (*Castanea davidii*), Chinese chinquapin (*Castanea henryi*), Seguin's chestnut (*Castanea seguinii*), Allegheny chinquapin (*Castanea pumila*), and the Ozark chinquapin (*Castanea ozarkensis*). Some biologists recognize eleven species and consider the Florida chinquapin (*Castanea alnifolia*) a separate species. The most authoritative summary of the classification problem associated with the *Castanea* genus is George P. Johnson, "Revision of *Castanea* Sect Balanocastanon (*Fagaceae*)," *Journal of the Arnold Arboretum* 69, no. 1 (January 1988): 25–49.

9. Lang et al., "Molecular Evidence for an Asian Origin," 56; Dane and Lang, "Biodiversity and Evolution of *Castanea*," 88; Dane, "Comparative Phylogeography of Castanea Species," 214.

10. Krebs et al., "Quaternary Refugia," 146.

11. William C. Rember, "The Clarkia Flora of Northern Idaho," Tertiary Research Center, University of Idaho, accessed July 24, 2019, http://www.webpages.uidaho.edu/tertiary/images/dirfour/Castanea2.jpg.

12. Krebs et al., "Quaternary Refugia," 146.

13. The BP (before present) abbreviation refers to calendar years, not radiocarbon years. Studies published prior to 1990 often rely on antiquated radiocarbon dating methods, causing the pollen grains to register younger than they actually are. All radiocarbon dates were converted to calendar years using the

CalPal radiocarbon age calibration curve program: http://www.calpal-online.de (accessed September 1, 2020).

14. CLIMAP, *Seasonal Reconstructions of the Earth's Surface at the Last Glacial Maximum*, Map and Chart Series Technical Report MC-36 (Boulder, Colo.: Geological Society of America, 1981), 1–18; Neil Roberts, *The Holocene: An Environmental History* (Oxford: Blackwell, 2002), 101–103; George H. Denton et al., "The Last Glacial Termination," *Science*, n.s., 328, no. 5986 (June 25, 2010): 1652–56.

15. Burkhard Frenzel et al., *Atlas of Paleoclimates and Paleoenvironments of the Northern Hemisphere* (Budapest: Geographical Research Institute, 1992), 28–45; Stephen T. Jackson et al., "Vegetation and Environment in Eastern North America during the Last Glacial Maximum," *Quaternary Science Reviews* 19, no. 6 (February 2000): 489–508; Douglas E. Soltis et al., "Comparative Phylogeography of Unglaciated Eastern North America," *Molecular Ecology* 15, no. 14 (December 2006): 4261–63, 4280–81; Jesse Bellemare and David A. Moeller, "Climate Change and Forest Herbs of Temperate Deciduous Forests," in *The Herbaceous Layer in Forests of Eastern North America*, 2nd ed., ed. Frank S. Gilliam (New York: Oxford University Press, 2014), 460–79.

16. W. A. Watts, "Postglacial and Interglacial Vegetation History of Southern Georgia and Central Florida," *Ecology* 52, no. 4 (July 1971): 676–90; Paul A. Delcourt and Hazel R. Delcourt, *Long-Term Forest Dynamics of the Temperate Zone: A Case Study of Late-Quaternary Forests in Eastern North America* (New York: Springer-Verlag, 1987), 94–96; Robert K. Booth, Fredrick J. Rich, and Gale A. Bishop, "Palynology and Depositional History of Late Pleistocene and Holocene Coastal Sediments from St. Catherines Island, Georgia, USA," *Palynology* 23, no. 1 (1999): 67–86.

17. Most palynologists refer to themselves as paleoecologists and work in a variety of disciplines, including evolutionary biology and forest ecology. An excellent summary of the history and methods of palynological studies is found in Michael Hesse et al., *Pollen Terminology: An Illustrated Handbook* (New York: SpringerWienNewYork, 2009).

18. Generally, no fewer than three hundred pollen grains are counted in each sediment cross-section, a practice validated by statistical sampling methods. See Margaret Kneller, "Pollen Analysis," in *Encyclopedia of Paleoclimatology and Ancient Environments*, ed. Vivien M. Gornitz (Dordrecht, Netherlands: Springer, 2009), 815–23.

19. David R. Foster et al., "The Environment and Human History of New England," in *Forests in Time: The Environmental Consequences of 1,000 Years of Change in New England*, ed. David R. Foster and John D. Aber (New Haven, Conn.: Yale University Press, 2004), 45–46.

20. Paul A. Delcourt, "Goshen Springs: Late Quaternary Vegetation Record for Southern Alabama," *Ecology* 61, no. 2 (April 1980): 372.

21. Hazel R. Delcourt, *Forests in Peril: Tracking Deciduous Trees from Ice-Age Refuges into the Greenhouse World* (Blacksburg, Va: McDonald & Woodward, 2002), 143–50.

22. Paul A. Delcourt, "Goshen Springs," 376–77.

23. Richard H. W. Bradshaw, "Prof. Dr. William A. Watts (1930–2010)," *Review of Palaeobotany and Palynology* 162, no. 2 (September 2010): 119–21; Brian Huntley, "William A. Watts," *Quaternary Newsletter* 122 (October 2010): 38–45; Fraser J. G. Mitchell, "William A. Watts (1930–2010)," *Biology and Environment: Proceedings of the Royal Irish Academy* 110B, no. 2 (2010): iii–iv.

24. William A. Watts, Barbara C. S. Hansen, and Eric C. Grimm, "Camel Lake: A 40,000-YR Record of Vegetational and Forest History from Northwest Florida," *Ecology* 73, no. 3 (June 1992): 1058.

25. Goshen Springs was more northerly in latitude than Camel Lake and closer to a major water course—the Conecuh River. See Delcourt and Delcourt, *Long-Term Forest Dynamics*, 14–16; David S. Leigh, Pradeep Srivastava, and George A. Brook, "Late Pleistocene Braided Rivers of the Atlantic Coastal Plain, USA," *Quaternary Science Reviews* 23, nos. 1–2 (January 2004): 65–84.

26. John W. Williams, Bryan N. Shuman, and Thompson Webb III, "Dissimilarity Analyses of Late-Quaternary Vegetation and Climate in Eastern North America," *Ecology* 82, no. 12 (December 2001): 3358. See also Neil Roberts, *Holocene*, 61, 62.

27. William A. Watts and Minze Stuiver, "Late Wisconsin Climate of Northern Florida and the Origin of Species-Rich Deciduous Forest," *Science*, n.s., 210, no. 4467 (October 17, 1980): 325–27; William A. Watts, "The Late Quaternary Vegetation History of the Southeastern United States," *Annual Review of Ecology and Systematics* 11 (November 1980): 387–409; Eric C. Grimm et al., "Evidence for Warm Wet Heinrich Events in Florida," *Quaternary Science Reviews* 25, nos. 17–18 (September 2006): 2197–211.

28. Kendra Gurney and Paul Schaberg, "Life in the Cold: Another Challenge to American Chestnut Restoration?," *Journal of the American Chestnut Foundation* 24, no. 3 (November 2010): 13. See also Thomas M. Saielli et al., "Nut Cold Hardiness as a Factor Influencing the Restoration of American Chestnut in Northern Latitudes and High Elevations," *Canadian Journal of Forest Research* 42, no. 5 (2012): 849–57.

29. William A. Watts and Barbara C. S. Hansen, "Pre-Holocene and Holocene Pollen Records of Vegetation History from the Florida Peninsula and Their Climatic Implications," *Palaeogeography, Palaeoclimatology, Palaeoecology* 109, nos. 2–4 (June 1994): 163–76; Paul L. Koch, Noah S. Diffenbaugh, and Kathryn A. Hoppe, "The Effects of Late Quaternary Climate and $_{p}CO_2$ Change on C_4 Plant Abundance in the South-Central United States," *Palaeogeography, Palaeoclimatology, Palaeoecology* 207, nos. 3–4 (May 2004): 331–57.

30. Richard A. Davis, *Sea-Level Change in the Gulf of Mexico* (College Station: Texas A&M University Press, 2011), 47; James J. Miller, *An Environmental History of Northeast Florida* (Gainesville: University Press of Florida, 1998), 43.

31. Eric C. Grimm et al., "A 50,000-Year Record of Climate Oscillations from Florida and Its Temporal Correlation with Heinrich Events," *Science* 261, no. 5118 (July 9, 1993): 198–201; Melissa Trend-Staid and Warren L. Prell, "Sea Surface Temperature at the Last Glacial Maximum: A Reconstruction Using the Modern Analog Technique," *Paleoceanography and Paleoclimatology* 17, no. 4 (December 2002): 1065; Timothy Herbert et al., "Tropical Ocean Temperatures over the Past 3.5 Million Years," *Science* 328, no. 5985 (June 18, 2010): 1530–34.

32. Dale A. Russell et al., "A Warm Thermal Enclave in the Late Pleistocene of the South-Eastern United States," *Biological Review* 84, no. 2 (May 2009): 183–84.

33. Jackson et al., "Vegetation and Environment," 501; John W. Williams et al., "Late-Quaternary Vegetation Dynamics in North America: Scaling from Taxa to Biomes," *Ecological Monographs* 74, no. 2 (February 2004): 312.

34. Johan J. Groot and Catharina R. Groot, "Marine Palynology: Possibilities, Limitations, Problems," *Marine Geology* 4, no. 6 (December 1966): 387–95; Watts and Stuiver, "Late Wisconsin Climate of Northern Florida," 325; Barbara C. S. Hansen, "Setting the Stage: Fossil Pollen, Stomata, and Charcoal," in *First Floridians and Last Mastodons: The Page-Ladson Site in the Aucilla River*, ed. S. David Webb (Dordrecht, Netherlands: Springer, 2006), 159–80.

35. For an informed discussion about interpreting chestnut pollen frequencies in paleoenvironmental contexts, see Frederick L. Paillet, Marjorie G. Winkler, and Patricia R. Sanford, "Relationship between Pollen Frequency in Moss Polsters and Forest Composition in a Naturalized Stand of American Chestnut: Implications for Paleoenvironmental Interpretation," *Bulletin of the Torrey Botanical Club* 118, no. 4 (Fall 1991): 432–43.

36. Fredrick J. Rich, personal communication, August 30, 2011. Several Late Pleistocene sites in the Ohoopee River drainage area of the Georgia coastal plain yielded *Castanea* pollen in quantities ranging from 1 to 12 percent. See also Fredrick J. Rich, Fredric L. Pirkle, and Eric Arenberg, "Palynology and Paleoecology of Strata Associated with the Ohoopee River Dune Field, Emanuel County, Georgia," *Palynology* 26 (2002): 239–56.

37. Fredrick J. Rich, "*Castanea* in the American Southeast," unpublished manuscript, Statesboro, Georgia, Georgia Southern University, 2011, 9–10.

38. Russell et al., "Warm Thermal Enclave," 183.

39. Booth, Rich, and Bishop, "Palynology and Depositional History," 80; Robert K. Booth and Fredrick J. Rich, "Identification and Paleoecological Implications of a Late Pleistocene Pteridophyte-Dominated Assemblage Preserved in Brown Peat from St. Catherines Island, Georgia," *Castanea* 64, no. 2 (June 1999): 125.

40. Fredrick J. Rich, personal communication, August 30, 2011. See also Fredrick J. Rich, R. Kelly Vance, and Clara R. Rucker, "The Palynology of Upper Pleistocene and Holocene Sediments from the Eastern Shoreline and Central Depression of St. Catherines Island, Georgia, USA," *Palynology* 39, no. 2 (2015): 234–47.

41. Watts, "Postglacial and Interglacial Vegetation History," 679; Mary B. Davis, "Quaternary History of Deciduous Forests of Eastern North American and Europe," *Annals of the Missouri Botanical Garden* 70, no. 3 (Fall 1983): 553–54; Phil Leduc, "Pollen Viewer 3.2," National Climatic Data Center, NOAA Paleoclimatology, Asheville, N.C., accessed October 9, 2018, http://www.ncdc.noaa.gov.paleo/paelo/html.

42. Hazel R. Delcourt and Paul A. Delcourt, "Presettlement Magnolia-Beech Climax of the Gulf Coastal Plain: Quantitative Evidence from the Apalachicola River Bluffs, North-Central Florida," *Ecology* 58, no. 5 (September 1977): 1085–93; Annisa Karim and Martin B. Main, "Tropical Hardwood Hammocks in Florida," USDA Cooperative Extension Service, University of Florida, Florida A & M University, 2004 (updated December 2015), accessed July 12, 2018, http://edis.ifas.ufl.edu/uw206.

43. Watts and Stuiver, "Late Wisconsin Climate of Northern Florida," 326; Watts, Hansen, and Grimm, "Camel Lake: A 40,000-YR Record," 1059; Booth, Rich, and Bishop, "Palynology and Depositional History," 80–83.

44. See Stephen T. Jackson and Jonathan T. Overpeck, "Responses of Plant Populations and Communities to Environmental Changes of the Late Quaternary," *Paleobiology* 26, no. 4 (Autumn 2000): 194–220; Stephen T. Jackson and John W. Williams, "Modern Analogs in Quaternary Paleoecology: Here Today, Gone Yesterday, Gone Tomorrow?," *Annual Review of Earth and Planetary Sciences* 32 (2004): 495–537.

45. Fredrick J. Rich and Robert K. Booth, "Quaternary Vegetation and Depositional History of St. Catherines Island," in *Geoarchaeology of St. Catherines Island, Georgia: Proceedings of the Fourth Caldwell Conference, St. Catherines Island, Georgia, March 27–29, 2009*, ed. Gale A. Bishop, Harold B. Rollins, and David H. Thomas (New York: American Museum of Natural History, 2011), 133–35, 137.

46. See Donald R. Whitehead and Kiat W. Tan, "Modern Vegetation and Pollen Rain in Bladen County, North Carolina," *Ecology* 50, no. 2 (March 1969): 244; Rachel A. Jones, "Transition in the Ozarks: A Paleoecological Record of the Last Deglaciation at Cupola Pond, Missouri, U.S.A." (master's thesis, University of Wyoming, 2010), 47–50.

47. Davis, "Quaternary History of Deciduous Forests," 554–59; Stephen T. and Chengyu Weng, "Late Quaternary Extinction of a Tree Species in Eastern North America," *Proceedings of the National Academy of Sciences U.S.A.* 96, no. 24 (November 23, 1999): 13847–52; David N. Wear and John G. Greis, *Southern*

Forest Resource Assessment: Summary Report (Asheville, N.C.: U.S. Forest Service, Southern Research Station, 2002), 588–89.

48. Meagan A. Binkley, "The Phylogeography of North American Chestnuts and Chinquapins" (master's thesis, University of Tennessee at Chattanooga, 2008), 33–40; Xiaowei Li, "DNA Fingerprinting of *Castanea* Species in the U.S.A." (master's thesis, Auburn University, 2011), 56–61; Xiaowei Li and Fenny Dane, "Comparative Chloroplast and Nuclear DNA Analysis of *Castanea* Species in the Southern Region of the USA," *Tree Genetics and Genomes* 9, no. 1 (2013): 107–16.

49. Dane, "Comparative Phylogeography of *Castanea* Species," 215–16.

50. Soltis et al., "Comparative Phylogeography of Unglaciated Eastern North America," 4261–93. See also Nathan G. Swenson and Daniel J. Howard, "Clustering of Contact Zones, Hybrid Zones, and Phylogeographic Breaks in North America," *American Naturalist* 166, no. 5 (November 2005): 581–91.

51. Binkley, "Phylogeography of North American Chestnuts," 33–34; Li, "DNA Fingerprinting of *Castanea* Species," 56–57; Sarah E. Kennedy, "Chloroplast DNA Analysis of Putative Chestnut Chinquapin Hybrids" (honors thesis, University of Tennessee at Chattanooga, 2008), 10–11; Li and Dane, "Comparative Chloroplast and Nuclear DNA Analysis," 109–12.

52. Davis, "Quaternary History of Deciduous Forests," 558; James S. Clark et al., "Reid's Paradox of Rapid Plant Migration: Dispersal Theory and Interpretation of Paleoecological Records," *Bioscience* 48, no. 1 (January 1998): 14; Stephen B. Vander Wall, "The Evolutionary Ecology of Nut Dispersal," *Botanical Review* 67, no. 1 (January 2001): 103–4; Jason S. McLachlan, James S. Clark, and Paul S. Manos, "Molecular Indicators of Tree Migration Capacity under Rapid Climate Change," *Ecology* 86, no. 8 (August 2005): 2088–98.

53. W. Carter Johnson and Thompson Webb III, "The Role of Blue Jays (*Cyanocitta cristata* L.) in the Postglacial Dispersal of Fagaceous Trees in Eastern North America," *Journal of Biogeography* 16, no. 6 (November 1989): 564; Jeffrey E. Moore and Robert K. Swihart, "Nut Selection by Captive Blue Jays: Importance of Availability and Implications for Seed Dispersal," *Condor* 108, no. 2 (May 2006): 377–88.

54. Johnson and Webb, "Role of Blue Jays," 563.

55. John Burroughs, *Riverby* (Boston: Houghton Mifflin, 1894), 92.

56. Raphael Zon, *Chestnut in Southern Maryland*, USDA Bureau of Forestry Bulletin no. 53 (Washington, D.C.: GPO, 1904), 19.

57. Quoted in Edwin R. Kalmbach, *The Crow and Its Relation to Man*, USDA Bulletin no. 621 (Washington, D.C.: GPO, 1918), 52. See also Walter B. Barrows and E. A. Schwarz, *The Common Crow of the United States*, USDA Division of Ornithology and Mammalogy Bulletin no. 6 (Washington, D.C.: GPO, 1895), 76.

58. Daniel A. Cristol, "Walnut-Caching Behavior of American Crows," *Journal of Field Ornithology* 76, no. 1 (Winter 2005): 27.

59. Ibid.

60. Joshua W. Ellsworth and Brenda C. McComb, "Potential Effects of Passenger Pigeon Flocks on the Structure and Composition of Presettlement Forests of Eastern North America," *Conservation Biology* 17, no. 6 (December 2003): 1548–58.

61. William B. Mershon, *The Passenger Pigeon* (New York: Outing Publishing, 1907), 132; Arlie W. Schorger, *The Passenger Pigeon: Its Natural History and Extinction* (Madison: University of Wisconsin Press, 1955), 35; Christopher Cokinos, *Hope Is a Thing with Feathers: A Personal Chronicle of Vanished Birds* (New York: Warner Books, 2001), 201; Joel Greenberg, *A Feathered River across the Sky: The Passenger Pigeon's Flight to Extinction* (New York: Bloomsbury, 2014), 8–9.

62. John James Audubon, *Ornithological Biography, or An Account of the Habits of the Birds of the United States of America* (Edinburgh: Adam Black, 1831), 324.

63. Sara L. Webb, "Potential Role of Passenger Pigeons and Other Vertebrates in the Rapid Holocene Migrations of Nut Trees," *Quaternary Research* 26, no. 3 (November 1986): 372. See also Vander Wall, "Evolutionary Ecology of Nut Dispersal," 103–4; Greenberg, *Feathered River*, 4–5.

64. Webb, "Potential Role of Passenger Pigeons," 372.

65. Watts, Hansen, and Grimm, "Camel Lake: A 40,000-YR Record," 1060.

66. Barbara E. Taylor et al., "Late Pleistocene and Holocene Vegetation Changes in the Sandhills, Fort Jackson, South Carolina," *Southeastern Geology* 48, no. 3 (October 2011): 152, Table 2; Rich, Pirkle, and Arenberg, "Palynology and Paleoecology of Strata," 239–56; Hazel R. Delcourt, Paul A. Delcourt, and Elliott C. Spiker, "A 12,000-Year Record of Forest History from Cahaba Pond, St. Clair County, Alabama," *Ecology* 64, no. 4 (August 1983): 881; Stephen T. Jackson and Donald R. Whitehead, "Pollen and Macrofossils from Wisconsinan Interstadial Sediments in Northeastern Georgia," *Quaternary Research* 39, no. 1 (January 1993): 99–106.

67. National Climatic Data Center, NOAA Paleoclimatology, Asheville, North Carolina, accessed January 11, 2018, http://www.ncdc.noaa.gov.paleo/paelo/html. See also William A. Watts, "The Full-Glacial Vegetation of Northwestern Georgia," *Ecology* 5, no. 1 (January 1970): 28.

68. William A. Watts, "The Vegetation Record of a Mid-Wisconsin Interstadial in Northwest Georgia," *Quaternary Research* 3, no. 2 (August 1973): 261; Hazel R. Delcourt, "Late Quaternary Vegetation History of the Eastern Highland Rim and Adjacent Cumberland Plateau of Tennessee," *Ecological Monographs* 49, no. 3 (September 1979): 255–80; Margaret Kneller and Dorothy Peteet, "Late-Glacial to Early Holocene Climate Changes from a Central Appalachian Pollen and Macrofossil Record," *Quaternary Research* 51, no. 2 (March 1999): 134, 138.

69. Kandace Hollenbach, *Foraging in the Tennessee River Valley, 12,500 to 8,000 Years Ago* (Tuscaloosa: University of Alabama Press, 2009), 210.

70. Paul A. Delcourt and Hazel R. Delcourt, *Prehistoric Native Americans and Ecological Change: Human Ecosystems in Eastern North America since the Pleistocene* (Cambridge: Cambridge University Press, 2004), 53, 69–73; Samuel T. Turvey, "Holocene Mammal Extinctions," in *Holocene Extinctions*, ed. Samuel T. Turvey (Oxford: Oxford University Press, 2010), 41–62.

CHAPTER 2. The Seasonal Bounty of Nuts and Acorns

1. Paul A. Delcourt et al., "Holocene Ethnobotanical and Paleoecological Record of Human Impact on Vegetation in the Little Tennessee River Valley, Tennessee," *Quaternary Research* 25, no. 3 (May 1986): 336; Richard A. Yarnell and M. Jean Black, "Temporal Trends Indicated by a Survey of Archaic and Woodland Plant Food Remains from Southeastern North America," *Southeastern Archaeology* 4, no. 2 (Winter 1985): 98; C. Margaret Scarry, ed., *Foraging and Farming in the Eastern Woodlands* (Gainesville: University Press of Florida, 1993), 36.

2. Paul A. Delcourt et al., "Prehistoric Human Use of Fire, the Eastern Agricultural Complex, and Appalachian Oak-Chestnut Forests: Paleoecology of Cliff Palace Pond, Kentucky," *American Antiquity* 63, no. 2 (April 1998): 266–67.

3. Delcourt et al., "Holocene Ethnobotanical and Paleoecological Record," 336; Jefferson Chapman and Andrea B. Shea, "The Archaeobotanical Record: Early Archaic Period to Contact in the Lower Little Tennessee River Valley," *Tennessee Anthropologist* 6, no. 1 (Spring 1981): 64–66; Jefferson Chapman et al., "Man-Land Interaction: 10,000 Years of American Indian Impact on Native Ecosystems in the Lower Little Tennessee River Valley, Eastern Tennessee," *Southeastern Archaeology* 1, no. 2 (Winter 1982): 118.

4. Bruce D. Smith, "The Archaeology of the Southeastern United States: From Dalton to de Soto, 10,500–500 B.P.," in *Advances in World Archaeology*, vol. 5, ed. Fred Wendorf and Angela E. Close (New York: Academic Press, 1986), 9–10, 16–17; Lynne P. Sullivan and Susan C. Prezzano, "Introduction: The Concept of Appalachian Archaeology," in *Archaeology of the Appalachian Highlands*, ed. Lynne P. Sullivan and Susan C. Prezzano (Knoxville: University of Tennessee Press, 2001), xxvii; Kandace Hollenbach, *Foraging in the Tennessee River Valley, 12,500 to 8,000 Years Ago* (Tuscaloosa: University of Alabama Press, 2009), 235.

5. Kenneth E. Sassaman, *The Eastern Archaic, Historicized* (Lanham, Md.: AltaMira, 2010), 173.

6. Michael D. Wiant, Kenneth B. Farnsworth, and Edwin R. Hajic, "The Archaic Period in the Lower Illinois River Basin," in *Archaic Societies: Diversity and Complexity across the Midcontinent*, ed. Thomas E. Emerson, Dale L. McElrath, and Andrew C. Fortier (Albany: State University of New York Press, 2009), 246; Juliet E. Morrow, "Earliest Inhabitants of the Northeast," in *Archaeology in America: An Encyclopedia*, vol. 1, *Northeast and Southeast*, ed.

Francis P. McManamon (Westport, Conn.: Greenwood, 2009), 16; UNC Research Laboratories of Archaeology, "The Forest People: The Archaic Period: 8000–1000 B.C.," University of North Carolina, accessed January 24, 2018, http://www.learnnc.org/lp/editions/intrigue/1268.

7. William A. Watts, "Late Quaternary Vegetation of Central Appalachia and the New Jersey Coastal Plain," *Ecological Monographs* 49, no. 4 (December 1979): 452.

8. Margaret Kneller and Dorothy Peteet, "Late-Glacial to Early Holocene Climate Changes from a Central Appalachian Pollen and Macrofossil Record," *Quaternary Research* 51, no.2 (March 1999): 134; Phil Leduc, "Pollen Viewer 3.2," National Climatic Data Center, NOAA Paleoclimatology, Asheville, N.C., accessed January 24, 2018, http://www.ncdc.noaa.gov.paleo/paelo/html.

9. Watts, "Late Quaternary Vegetation," 463; Dean R. Snow, *The Archaeology of New England* (New York: Academic Press, 1980), 176–82; Guy Gibbon, "Eastern Woodlands Middle Archaic," in *Archaeology of Prehistoric Native America: An Encyclopedia*, ed. Guy Gibbon (London: Garland, 1998), 256–57; Lucinda McWeeney and Douglas C. Kellogg, "Early and Middle Holocene Climate Changes and Settlement Patterns along the Eastern Coast of North America," *Archaeology of Eastern North America* 29 (2001): 198–99.

10. Joseph W. Michels and James S. Dutt, eds., *A Preliminary Report of Archaeological Investigations at the Sheep Rock Shelter Site, Huntingdon, Pennsylvania* (University Park: Pennsylvania State University, Department of Anthropology, 1971), 28; James Adovasio, "Meadowcroft Rockshelter and Nearby Paleoindian Sites: Pennsylvania and New Jersey: Paleoindian Sites in the Mid-Atlantic Area," in McManamon, *Archaeology in America*, 43.

11. Watts, "Late Quaternary Vegetation," 458.

12. Margaret Bryan Davis, "Holocene Vegetational History of the Eastern United States," in *Late-Quaternary Environments of the United States*, vol. 2, *The Holocene*, ed. H. E. Wright Jr. (Minneapolis: University of Minnesota Press, 1983), 172–73, 175; McWeeney and Kellogg, "Early and Middle Holocene Climate Changes," 198.

13. Watts, "Late Quaternary Vegetation," 432.

14. Elizabeth Martin, "Holocene Environmental History of Panthertown Valley in the Blue Ridge Mountains of North Carolina" (master's thesis, Western Carolina University, 2014), 21.

15. Watts, "Late Quaternary Vegetation," 462–63; Calvin J. Heusser, "Vegetational History of the Pine Barrens," in *Pine Barrens: Ecosystem and Landscape*, ed. Richard T. T. Forman (Piscataway, N.J.: Rutgers University Press, 1998), 221–22; Eric C. Grimm and George L. Jacobson Jr., "Late-Quaternary Vegetation History of the Eastern United States," in *The Quaternary Period in the United States*, ed. Alan R.Gillespie, Stephen C. Porter, and Brian F. Atwater

(Amsterdam: Elsevier, 2003), 391–92; Paul A. Delcourt and Hazel R. Delcourt, *Prehistoric Native Americans and Ecological Change: Human Ecosystems in Eastern North America since the Pleistocene* (Cambridge: Cambridge University Press, 2004), 68.

16. Ralph A. Ibe, "Postglacial Montane Vegetational History around Balsam Lake, Catskill Mountains, New York," *Bulletin of the Torrey Botanical Club* 112, no. 2 (April–June 1985): 180.

17. Nut-harvesting, as archaeobotanist Mary Simon stated it, involved "intentional movement to predictable spaces, rather than opportunistic foraging," as doing so guaranteed longer periods of sustenance. Mary L. Simon, "A Regional and Chronological Synthesis of Archaic Period Plant Use in the Midcontinent," in Emerson, McElrath, and Fortier, *Archaic Societies*, 82.

18. Gary Wilkins, "Prehistoric Mountaintop Occupations of Southern West Virginia," *Archaeology of Eastern North America* 6 (Summer 1978): 15–19.

19. William A. Ritchie, *The Archaeology of New York State* (Garden City, N.Y.: Natural History Press, 1969), 57–58; Nina M. Versaggi et al., "Adding Complexity to Late Archaic Research in the Northeastern Appalachians," in Sullivan and Prezzano, *Archaeology of the Appalachian Highlands*, 125–26; Suanna C. Selby, "The Geoarchaeology of the Upper Susquehanna River: Anthropological Approaches to Archaic and Woodland Period Landscapes and Settlement Systems in New York State" (PhD diss., New York University, 2007), 80–82, 90.

20. Sassaman, *Eastern Archaic*, 174, 176; Edward V. Curtin, "The Enigmatic Archaic Site at Lamoka Lake, New York," Curtin Archaeological Consulting, Inc., November 23, 2015, accessed August 5, 2018, http://www.curtinarch.com/blog/2015/11/23/archaic-lamoka-lake.

21. Watts, "Quaternary Vegetation of Central Appalachia," 457; Emily Russell et al., "Recent Centuries of Vegetational Change in the Glaciated North-Eastern United States," *Journal of Ecology* 81, no. 4 (December 1993): 654–60; Nina M. Versaggi, "Hunter-Gatherer Settlement Models and the Archaeological Record: A Test Case from the Upper Susquehanna Valley of New York" (PhD diss., State University of New York at Binghamton, 1988), 73.

22. In the nineteenth century, Audubon noted that passenger pigeons killed in New York state with crops full of rice had flown directly from the fields of South Carolina. See John James Audubon, *Ornithological Biography, or An Account of the Habits of the Birds of the United States of America* (Edinburgh: Adam Black, 1831), 319–20.

23. Dean R. Snow, "Lamoka Lake Site," in Gibbon, *Archaeology of Prehistoric Native America*, 442.

24. Thomas R. Whyte, "Proto-Iroquoian Divergence in the Late Archaic–Early Woodland Period Transition of the Appalachian Highlands," *Southeastern Archaeology* 26, no. 1 (Summer 2007): 140. See also Sassaman, *Eastern Archaic*,

93; Andrew MacDougall, "Did Native Americans Influence the Northward Migration of Plants during the Holocene?," *Journal of Biogeography* 30, no. 5 (May 2003): 642.

25. Deborah A. Bolnick, Daniel I. Bolnick, and David G. Smith, "Asymmetric Male and Female Genetic Histories among Native Americans from Eastern North America," *Molecular Biology and Evolution* 23, no. 11 (December 2006): 2161. See also Deborah A. Bolnick and David G. Smith, "Unexpected Patterns of Mitochondrial DNA Variation among Native Americans from the Southeastern United States," *American Journal of Physical Anthropology* 122, no. 4 (December 2003): 336–54; Whyte, "Proto-Iroquoian Divergence," 136, 140.

26. Kristen J. Gremillion, "Plant Husbandry at the Archaic/Woodland Transition: Evidence from the Cold Oak Shelter, Kentucky," *Midcontinental Journal of Archaeology* 18, no. 2 (Fall 1993): 161–89; Gail E. Wagner, "Eastern Woodlands Anthropogenic Ecology," in *People and Plants in Ancient Eastern North America*, ed. Paul E. Minnis (Washington, D.C.: Smithsonian Books, 2003), 126–71; C. Margaret Scarry, "Patterns of Wild Plant Utilization in the Prehistoric Eastern Woodlands," in Minnis, *People and Plants in Ancient Eastern North America*, 50–104.

27. Yarnell and Black, "Temporal Trends," 97–98.

28. Paul S. Gardner, "The Ecological Structure and Behavioral Implications of Mast Exploitation Strategies," in *People, Plants, and Landscapes: Studies in Paleoethnobotany*, ed. Kristen J. Gremillion (Tuscaloosa: University of Alabama Press, 1997), 162.

29. Ibid., 176.

30. Yarnell and Black, "Temporal Trends," 93. See also Richard Yarnell, "Problems of Interpretation of Archaeological Plant Remains of the Eastern Woodlands," *Southeastern Archaeology* 1, no. 1 (Summer 1982): 1–7; Richard Yarnell, "The Importance of Native Crops during the Late Archaic and Woodland Periods," in Scarry, *Foraging and Farming*, 21.

31. Lara K. Homsey, Renee B. Walker, and Kandace D. Hollenbach, "What's For Dinner? Investigating Food-Processing Technologies at Dust Cave, Alabama," *Southeastern Archaeology* 29, no. 1 (Summer 2010): 188.

32. Kristen J. Gremillion, "The Paleoethnobotanical Record," in *Archaeology of the Mid-Holocene Southeast*, ed. Kenneth E. Sassaman and David G. Anderson (Gainesville: University Press of Florida, 1996), 104; Richard Jefferies, *Holocene Hunter-Gatherers of the Lower Ohio River Valley* (Tuscaloosa: University of Alabama Press, 2008), 163; Sassaman, *Eastern Archaic, Historicized*, 173–74.

33. Yarnell and Black, "Temporal Trends," 98. See also Neal H. Lopinot, "Archaeobotanical Formation Processes and Late Middle Archaic Human-Plant Interrelationships in the Midcontinental U.S.A." (PhD diss., Southern Illinois University, 1984), 151–53; Deborah M. Pearsall, *Paleoethnobotany: A Handbook of*

Procedures, 2nd ed. (San Diego, Calif.: Academic Press, 2000), 209–24; Simon, "Regional and Chronological Synthesis," 97–98.

34. Yarnell and Black, "Temporal Trends," 98.

35. Ibid.

36. Gremillion, "Plant Husbandry," 180–81. At another location inside the same rock shelter, from another Late Archaic period of occupation, chestnuts comprised 30 percent of the total consumed nutmeat.

37. Gardner, "Ecological Structure and Behavioral Implications," 162–72.

38. Gardner's data comes from a 1984 USDA handbook that measures the caloric value of a single hickory species, presumably shagbark hickory (*Carya ovata*). However, hickory nutmeats vary in protein, carbohydrate, and lipid levels, with pignut hickory (*Carya glabra*) having the lowest nutritional value. See Marie A. McCarthy and Ruth H. Matthews, *Composition of Foods: Nut and Seed Products: Raw, Processed, Prepared*, USDA Agriculture Handbook 8–12 (Washington, D.C.: GPO, 1984), 79.

39. In Europe as early as the eleventh century, the shelf life of chestnut meal allowed it to be exported across the entire Mediterranean basin. See Paolo Squatriti, "Trees, Nuts, and Woods at the End of the First Millennium: A Case from the Amalfi Coast," in *Ecologies and Economies in Medieval and Early Modern Europe*, ed. Scott G. Bruce (Boston: Brill 2010), 39–41.

40. Seth Diamond et al., "Hard Mast Production before and after the Chestnut Blight," *Southern Journal of Applied Forestry* 24, no. 4 (November 2000): 199. See also Seth Diamond, "Vegetation, Wildlife, and Human Foraging in Prehistoric Western Virginia" (master's thesis, Virginia Polytechnic Institute and State University, 1989), 84–90; John E. Keller, "Prehistoric Subsistence and Hardwood Nut Yields," *Midcontinental Journal of Archaeology* 12, no. 2 (1987): 185–86.

41. Diamond et al., "Hard Mast Production," 199.

42. Knowledge of chestnut mast yields comes largely from oral histories. See, for example, Jake Waldroop in the chapter "Memories of the American Chestnut," in *Foxfire 6*, ed. Eliot Wigginton (Garden City, N.Y.: Anchor Books, 1980), 402–3; Vic Weals, *Last Train to Elkmont: A Look Back at Life on Little River in the Great Smoky Mountains* (Knoxville, Tenn.: Olden Press, 1991), 128; Leland R. Cooper and Mary Lee Cooper, *The Pond Mountain Chronicle: Self-Portrait of a Southern Appalachian Community* (Jefferson, N.C.: McFarland, 1998), 55, 84–86; Richard Ashe, *Chinquapins and Chestnuts* (Mooresville, N.C.: Lone Deer Press, 1998), 4–5.

43. Kandi Hollenbach, "Late Archaic–Early Woodland Transitions at the Townsend Sites," Tennessee Council for Professional Archaeology, September 9, 2015, accessed September 16, 2018, https://tennesseearchaeologycouncil.wordpress.com/2015/09/22/30-days-of-tennessee-archaeology-2015-day-22.

44. Ibid.

45. Ibid..

46. Stephen B. Carmody and Kandace D. Hollenbach, "The Role of Gathering in Middle Archaic Social Complexity in the Mid-South: A Diachronic Perspective," in *Barely Surviving or More than Enough? The Environmental Archaeology of Subsistence, Specialisation and Surplus Food Production*, ed. Maaike Groot, Daphne Lentjes, and Jørn Zeiler (Leiden, Netherlands: Sidestone, 2013), 48.

47. Jeffries, *Holocene Hunter-Gatherers*, 163, 239–40; Ernest Small, *North America Cornucopia: Top 100 Indigenous Food Plants* (Boca Raton, Fla.: CRC, 2014), 359–66.

48. William P. Corsa, comp., *Nut Culture in the United States, Embracing Native and Introduced Species* (Washington, D.C.: GPO, 1896), 68; Thomas S. Elias, *Field Guide to North American Trees* (Danbury, Conn.: Grolier Book Clubs, 1989), 286.

49. Donald E. Davis, *Southern United States: An Environmental History* (Santa Barbara, Calif.: ABC-CLIO, 2006), 38–39; Dale L. McElrath and Thomas E. Emerson, "Concluding Thoughts on the Archaic Occupation of the Eastern Woodlands," in Emerson, McElrath, and Fortier, *Archaic Societies*, 841–56.

50. Gayle J. Fritz, "Levels of Native Biodiversity in Eastern North America," in *Biodiversity and Native America*, ed. Paul E. Minnis and Wayne J. Elisens (Norman: University of Oklahoma Press, 2000), 223–47; Stephen B. Carmody, "From Foraging to Food Production on the Southern Cumberland Plateau of Alabama and Tennessee, USA," (PhD diss., University of Tennessee, 2014), 8–11, 295, 371–88.

51. David H. Thomas, *Exploring Ancient Native America: An Archaeological Guide* (New York: Routledge, 1999), 124; DeeAnne Wymer, "The Origins of Agriculture in the Midwest: The Eastern Agriculture Complex," in McManamon, *Archaeology in America*, 36–39; Dean R. Snow, *Archaeology of Native North America* (New York: Routledge, 2016), 86–89.

52. Gerald McCarthy, "The Chestnut and Its Weevil," in *Annual Report of the North Carolina Agricultural Experiment Station* (Raleigh: North Carolina College of Agriculture and Mechanic Arts, 1895), 267–72; F. H. Chittenden, "The Chestnut Weevils, with Notes on other Nut-Feeding Species," in *Some Miscellaneous Results of the Work of the Division of Entomology*, vol. 7 USDA, Division of Entomology (Washington, D.C.: GPO, 1904), 24–38.

53. For an anthropological, albeit mostly uncritical summary of fire use among Native Americans, see Omer C. Stewart, *Forgotten Fires: Native Americans and the Transient Wilderness* (Norman: University of Oklahoma Press, 2002), 70–112. A more measured assessment is found in Glenn R. Matlack, "Reassessment of the Use of Fire as a Management Tool in Deciduous Forests of Eastern North America," *Conservation Biology* 27, no. 5 (October 2013): 916–26; W. Wyatt Oswald et al., "Conservation Implications of Limited Native American Impacts in Pre-contact New England," *Nature Sustainability* 3, no. 3 (March 2020): 241–46.

54. Some forest ecologists believe intentional burning began in the Late Archaic, which explains the underrepresentation of chestnuts in the archaeological record. If chestnuts were burned on the forest floor before being processed and eaten, and then burned again as refuse, their hulls would have been destroyed beyond recovery. For fire use among Native Americans, particularly as a horticultural technique, see Delcourt and Delcourt, *Prehistoric Native Americans and Ecological Change*, 34, 176–77; Marc D. Abrams and Gregory J. Nowacki, "Native Americans as Active and Passive Promoters of Mast and Fruit Trees in the Eastern USA," *Holocene* 18, no. 7 (November 2008): 1123–26.

55. Wagner, "Eastern Woodlands Anthropogenic Ecology," 135–38; Delcourt and Delcourt, *Prehistoric Native Americans and Ecological Change*, 34, 72; Gregory S. Springer et al., "Micro-Charcoal Abundances in Stream Sediments from Buckeye Creek Cave, West Virginia, USA," *Journal of Cave and Karst Studies* 74, no. 1 (2012): 58–64.

56. Hazel R. Delcourt and Paul A. Delcourt, "Pre-Columbian Native American Use of Fire on Southern Appalachian Landscapes," *Conservation Biology* 11, no. 4 (August 1997): 1011. After European contact, both fires and chestnuts again increase in prevalence. Although chestnut densities peak at 25 percent in 1830, burning levels peak much later, c. 1900.

57. Ibid., 1013. The honey locust (*Gleditsia triacanthos*) also benefitted from human presence and intentionally set fires, as the trees became more common near Cherokee settlements. Robert J. Warren II, "Ghosts of Cultivation Past: Native American Dispersal Legacy Persists in Tree Distribution," *PLoS ONE* 11, no. 3 (March 2016): e0150707, https://doi.org/10.1371/journal.pone.0150707.

58. Springer et al., "Micro-Charcoal Abundances in Stream Sediments," 60–61. See also Cecil Ison, "Fire on the Edge: Prehistoric Fire along the Escarpment Zone of the Cumberland Plateau," in *Proceedings: Workshop on Fire, People, and the Central Hardwoods Landscape, March 12–14, 2000, Richmond, Kentucky*, General Technical Report NE-274, comp. Daniel A. Yaussy (Newtown Square, Pa.: U.S. Forest Service, Northeastern Research Station, 2000), 36–45; Kristen J. Gremillion, "Prehistoric Upland Farming, Fuelwood, and Forest Composition on the Cumberland Plateau, Kentucky, USA," *Journal of Ethnobiology* 35, no. 1 (March 2015): 60–84.

59. Jeffrey M. Kane, J. Morgan Varner, and Michael R. Saunders, "Resurrecting the Lost Flames of American Chestnut," *Ecosystems* 22, no. 5 (August 2019): 995–1006.

60. Marc D. Abrams and Sarah E. Johnson, "The Impacts of Mast Year and Prescribed Fires on Tree Regeneration in Oak Forests at the Mohonk Preserve, Southeastern New York, USA," *Natural Areas Journal* 33, no. 4 (October 2013): 427–34; Cathryn H. Greenberg et al., "Acorn Viability following Prescribed Fire in Upland Hardwood Forests," *Forest Ecology and Management* 275 (July 2012): 79–86.

61. William W. Ashe, *Chestnut in Tennessee*, Tennessee Geological Survey Series Bulletin no. 10–B (Nashville, Tenn.: Baird-Ward, 1912), 10–11, 35; Quentin Bass, "Talking Trees: The Appalachian Forest Ecosystem and the American Chestnut," *Journal of the American Chestnut Foundation* 16, no. 1 (Fall 2002): 46; Gretel E. Hengst and Jeffrey O. Dawson, "Bark Properties and Fire Resistance of Selected Tree Species from the Central Hardwood Region of North America," *Canadian Journal of Forest Research* 24, no. 4 (April 1994): 688–96.

62. See, for example, Jeffrey M. Kane et al., "Reconsidering the Fire Ecology of the Iconic American Chestnut," *Ecosphere* 11, no. 10 (October 2020): e03267, https://doi.org/10.1002/ecs2.3267.

63. William F. Fox, "Extracts from Remarks Appended to the Reports of the Fire Wardens," in *Sixth Annual Report of the Forest, Fish and Game Commission of the State of New York* (Albany, N.Y.: James B. Lyon, State Printer, 1901), 144–55; USDA, *Forest Fire Protection by the States: As Described by Representative Men at the Weeks Law Forest Fire Conference, January 9–10, 1913* (Washington, D.C.: GPO, 1914), 2–14; Laurence C. Walker, *The Southern Forest: A Chronicle* (Austin: University of Texas Press, 1991), 178–84.

64. William W. Ashe, *Forest Fires: Their Destructive Work, Causes and Prevention*, North Carolina Geological Survey Bulletin no. 7 (Raleigh, N.C.: Josephus Daniels, State Printer, 1895), 49.

65. Horace B. Ayers and William W. Ashe, *The Southern Appalachian Forests*, U.S. Geological Survey Professional Paper no. 37 (Washington, D.C.: GPO, 1905), 81–82.

66. Ibid., 212.

67. Ashe, *Chestnut in Tennessee*, 10.

68. Delcourt and Delcourt, "Pre-Columbian Native American Use of Fire," 1013; Delcourt and Delcourt, *Prehistoric Native Americans and Ecological Change*, 34, 176–77; Abrams and Nowacki, "Native Americans as Active and Passive Promoters," 1123; Kurt A. Fesenmyer and Norman L. Christensen Jr., "Reconstructing Holocene Fire History in a Southern Appalachian Forest Using Soil Charcoal," *Ecology* 91, no. 3 (March 2010): 662–70. Some chestnut proliferation likely occurred by removing undesirable trees with stone axes or uprooting unwanted saplings by hand. If large groups were involved in such work, several acres could be cleared in a single day.

69. David S. Shafer, "Flat Laurel Gap Bog, Pisgah Ridge, North Carolina: Late Holocene Development of a High-Elevation Heath Bald," *Castanea* 51, no. 1 (March 1986): 4. Naturally occurring fires may explain some of the chestnut increase atop Pisgah Ridge, although in the modern era such fires burned relatively small areas downslope.

70. David R. Foster, "New England's Forest Primeval," *Wild Earth* 11, no. 1 (Spring 2001): 42. See also David R. Foster and Tad M. Zebryk, "Long-Term

Vegetation Dynamics and Disturbance History of a Tsuga-Dominated Forest in New England," *Ecology* 74, no. 4 (June 1993): 992–93; David R. Foster et al., "Oak, Chestnut and Fire: Climatic and Cultural Controls of Long-Term Forest Dynamics in New England, USA," *Journal of Biogeography* 29, nos. 10–11 (October 2002): 1366–68; W. Wyatt Oswald et al., "Post-Glacial Changes in Spatial Patterns of Vegetation across Southern New England," *Journal of Biogeography* 34, no. 5 (May 2007): 906–10.

71. Stanley W. Bromley, "The Original Forest Types of Southern New England," *Ecological Monographs* 5, no. 1 (January 1935): 61–89; Margaret B. Davis, "Pleistocene Biogeography of Temperate Deciduous Forests," *Geoscience and Man* 13 (March 1976): 22; Watts, "Late Quaternary Vegetation," 447–48; Emily W. B. Russell, *People and the Land through Time: Linking Ecology and History* (New Haven, Conn.: Yale University Press, 1997), 225–26.

72. Sullivan and Prezzano, "Introduction: The Concept of Appalachian Archaeology," xxx; Charles H. Faulkner, "Woodland Cultures of the Elk and Duck River Valleys, Tennessee: Continuity and Change," in *The Woodland Southeast*, ed. David G. Anderson and Robert C. Mainfort Jr. (Tuscaloosa: University of Alabama Press, 2002), 185–203; Kenneth Sassaman and David Anderson, "The Late Holocene Period, 3750 to 650 B.C.," in *Handbook of North American Indians*, vol. 14, *Southeast*, ed. Raymond D. Fogelson (Washington, D.C.: Smithsonian Institution, 2004), 104–13.

73. Davis, *Southern United States*, 46. See also Dean Snow, "Northeast Late Woodland," in *Encyclopedia of Prehistory*, vol. 6, *North America*, ed. Peter N. Peregrine and Melvin Ember (Boston: Springer US, 2001), 339–57; Southeast Archaeological Center, "Late Woodland Period," National Park Service website, accessed March 2, 2018, http://www.nps.gov/seac/outline/04-woodland/index-3.htm.

74. See, for example, Office of Indian Affairs, "Report Concerning Indians in North Carolina," in *Annual Report of the Commissioner of Indian Affairs, for the Year 1902*, part 1 (Washington, D.C.: GPO, 1902), 260.

75. In the precontact Mississippian period, Native American settlements in the southeast and Mid-Atlantic regions seldom exceeded a thousand individuals. In New England, village sizes were smaller until the adoption of maize after 1200 AD. For a general discussion of Native American population estimates prior to European contact, see Sherburne F. Cook, *The Indian Population of New England in the Seventeenth Century* (Berkeley: University of California Press, 1976), 1–12; David E. Stannard, *American Holocaust: The Conquest of the New World* (Oxford: Oxford University Press, 1992), 27–56; Russell Thornton, "The Demography of Colonialism and 'Old' and 'New' Native Americans," in *Studying Native America: Problems and Prospects*, ed. Russell Thornton (Madison: University of Wisconsin Press, 1998), 17–39.

76. Charles C. Jones, *Antiquities of the Southern Indians* (New York: D. Appleton, 1873), 99–100; Horatio B. Cushman, *History of the Choctaw, Chickasaw and Natchez Indians* (Greenville, Tex.: Headlight Printing House, 1899), 537; John R. Swanton, *Indian Tribes of the Lower Mississippi Valley and Adjacent Coast of the Gulf of Mexico*, Smithsonian Institution Bureau of American Ethnology Bulletin no. 43 (Washington: GPO, 1911), 109–10.

CHAPTER 3. Wherever There Are Mountains

1. The figure includes portions of all states and provinces within the historic range of the American chestnut, after subtracting the surface area of lakes, ponds, streams, and rivers. The range boundaries were drawn from fossil pollen records, witness-tree surveys, botanical studies, and historical sources. See, for example, George B. Emerson, *A Report on the Trees and Shrubs Growing Naturally in the Forests of Massachusetts* (Boston: Dutton and Wentworth, 1846), 165; Alvan W. Chapman, *Flora of the Southern United States* (New York: Ivison, Phinney, 1865), 424; Charles S. Sargent, *A Catalogue of the Forest Trees of North America* (Washington, D.C.: GPO, 1880), 54; Henry A. Gleason, "Additional Notes on Southern Illinois Plants," *Torreya* 4, no. 11 (1904): 167–70.

2. Thomas Hariot, *A Briefe and True Report of the New Found Land of Virginia* (Frankfurt-am-Main: Johann Wechel, 1590), 13, 18; James Mooney, "Myths of the Cherokee," in *Nineteenth Annual Report of the Bureau of American Ethnology*, part 1 (Washington, D.C.: GPO, 1900), 53, 179, 317–18, 328, 439; James A. Robertson, ed., *True Relation of the Hardships Suffered by Governor Fernando de Soto . . . during the Discovery of the Province of Florida*, 2 vols. (Deland: Florida State Historical Society, 1932–33), 2:311; Lawrence A. Clayton, Vernon J. Knight Jr., and Edward C. Moore, eds., *The De Soto Chronicles: The Expedition of Hernando de Soto to North America in 1539–1543*, 2 vols. (Tuscaloosa: University of Alabama Press, 1993), 1:170, 263, 292, 308.

3. Paul E. Hoffman, "Lucas Vázquez de Ayllón's Discovery and Colony," in *The Forgotten Centuries: Indians and Europeans in the American South, 1521–1704*, ed. Charles Hudson and Carmen Chaves Tesser (Athens: University of Georgia Press, 1994), 36–38; Paul E. Hoffman, *A New Andalucia and a Way to the Orient: The American Southeast during the Sixteenth Century* (Baton Rouge: Louisiana State University Press, 2004), 8–12. John Lawson reported seeing chinquapins in a native Sewee cabin in 1701, but noted chestnuts were not found "near the Sea or Salt Water, tho' they are frequently in such places in Virginia," Lawson, *A New Voyage to Carolina* (London: J. Knapton, 1709), 17.

4. Douglas T. Peck, "Lucas Vásquez de Ayllón's Doomed Colony of San Miguel de Gualdape," *Georgia Historical Quarterly* 85, no. 2 (Summer 2001): 186–87.

5. Thomas Suárez, *Shedding the Veil: Mapping the European Discovery of America and the World* (River Edge, N.J.: World Scientific, 1992), 92–93.

6. George P. Winship, *Sailors' Narratives of Voyages along the New England Coast, 1524–1624* (Boston: Houghton Mifflin, 1905), 1–23; William H. Hobbs, "Verrazzano's Voyage along the North American Coast in 1524," *Isis* 41, no. 3/4 (December 1950): 268–77; Lawrence C. Wroth, *The Voyages of Giovanni da Verrazzano, 1524–1528* (New Haven, Conn.: Yale University Press, 1970), 133–43.

7. Quoted in Winship, *Sailors' Narratives*, 10. See also Hobbs, "Verrazzano's Voyage," 273–74; Wroth, *Voyages of Giovanni da Verrazzano*, 137–38; Carl O. Sauer, *Sixteenth Century North America: The Land and the People as Seen by the Europeans* (Berkeley: University of California Press, 1975), 52–58. In the Julian calendar, the date of Verrazzano's anchoring in New York harbor was April 17, 1524.

8. Richard Hakluyt, *Divers Voyages Touching the Discoverie of America and the Ilands Adjacent* (London: Thomas Woodcocke, 1582), 67. See also Luca Codignola, "Another Look at Verrazzano's Voyage, 1524," *Acadiensis* 29, no. 1 (Autumn 1999): 29–42.

9. In 1642, Roger Williams observed a Narragansett Indian constructing a chestnut canoe using only a hatchet, fire, and hot stones. Upon completion—the process took ten full days—Williams observed the individual launching the canoe, who then ventured out to fish in the ocean. Williams, *A Key into the Language of America* (London: Gregory Dexter, 1643), 106–7.

10. Gonzalo Fernández de Oviedo y Valdéz, *Historia general y natural de las Indias, islas y tierra-firme del Mar Océano*, vol. 2, part 2 (Madrid: Real Academia de la Historia, 1853), 631. Unless otherwise noted, all Spanish translations were made by the author, with the assistance of Ms. Liliana Silva, a native Spanish speaker.

11. Ibid., 629.

12. A more nuanced analysis of the writings of Oviedo, including his descriptions of North American flora and fauna as well as his views on the Spanish treatment of Native Americans, is found in Kathleen Ann Myers, *Fernández de Oviedo's Chronicle of America: A New History for a New World* (Austin: University of Texas Press, 2007), 63–81.

13. Garcilaso de la Vega's *La Florida del Inca* also relies on eyewitness accounts, but is written as a standard history of the de Soto conquest. It offers the most sympathetic portrayal of Native Americans, as de la Vega was of Incan heritage. See Raquel Chang-Rodriguez, "Introduction," in *Beyond Books and Borders: Garcilaso de la Vega and "La Florida del Inca,"* ed. Raquel Chang-Rodriguez (Lewisburg, Pa.: Bucknell University Press, 2006), 26–33.

14. Fernández de Oviedo, *Historia general y natural*, 1:544–77.

15. David E. Duncan, *Hernando de Soto: A Savage Quest in the Americas* (Norman: University of Oklahoma Press, 1997), xxv; Patricia Galloway, "The Incestuous Soto Narratives," in *The Hernando de Soto Expedition: History, Historiography, and "Discovery" in the Southeast*, ed. Patricia Galloway (Lincoln: University of Nebraska Press, 2005), 12–18; George E. Lankford, "How Historical

Are the De Soto Chronicles?," in *The Search for Mabila: The Decisive Battle between Hernando de Soto and Chief Tascalusa*, ed. Vernon James Knight Jr. (Tuscaloosa: University of Alabama Press, 2009), 33–44.

16. Charles Hudson, *Knights of Spain, Warriors of the Sun: Hernando de Soto and the South's Ancient Chiefdoms* (Athens: University of Georgia Press, 1997), 106; Fernández de Oviedo, *Historia general y natural*, 1:551.

17. Hudson, *Knights of Spain*, 106. Thomas Nuttall described the plant in 1817, noting the shrub might be as small as "12 inches high." Nuttall, *The Genera of North American Plants, and a Catalogue of the Species to the Year 1817*, vol. 2 (Philadelphia: D. Heartt, 1818), 217. It is described as the "Dwarf Chestnut" in Thomas Nuttall, *The North American Sylva*, vol. 4 (Philadelphia: J. Dobson, 1842), 19–20.

18. Fernández de Oviedo, *Historia general y natural*, 1:551.

19. Ibid.

20. Patricia Galloway, "Incestuous Soto Narratives," 18. An original copy of the Elvas volume is found in the John Carter Library at Brown University and has been digitized for public use, accessed December 8, 2019, http://openlibrary.org/books/OL24609510M/Relaçam_verdadeira_dos_trabalhos_q_ue_ho_gouernador_do_m_Ferna_n_do_d_e_Souto_e_certos_fidalgos_port.

21. On the validity of the Elvas narrative, see Hudson, *Knights of Spain*, 443–45; Martin M. Elbl and Ivana Elbl, "The Gentleman of Elvas and His Publisher," in Galloway, *Hernando de Soto Expedition*, 45–96; Lankford, "How Historical Are the De Soto Chonicles?," 33–38. For commentary on North American forests in the de Soto chronicles, see E. Thomas Shields Jr., "Imagining the Forest: Longleaf Pine Ecosystems in Spanish and English Writings of the Southeast, 1542–1709," in *Early Modern Ecostudies: From the Florentine Codex to Shakespeare*, ed. Thomas Hallock, Ivo Kamps, and Karen L. Raber (New York: Palgrave Macmillan, 2008), 251–68.

22. Edward Gaylord Bourne, ed., *Narratives of the Career of Hernando De Soto*, 2 vols. (New York: Allerton, 1904), 1:134. See also Robertson, *True Relation*, 2:189; Clayton, Knight, and Moore, *De Soto Chronicles*, 1:123.

23. Bourne, *Narratives of the Career of Hernando De Soto*, 2:219; Robertson, *True Relation*, 2:310; Clayton, Knight, and Moore, *De Soto Chronicles*, 1:170.

24. Robertson, *True Relation*, 1:311.

25. Richard Hakluyt, ed., *Virginia Richly Valued, by the Description of the Maine Land of Florida, Her Next Neighbor* (London: Felix Kyngston for Matthew Lownes, 1609), 178.

26. Michael J. Ferreira, personal communication, April 11, 2012. Fidalgo d'Elvas, *A True Account of the Travails Experienced by Governor Hernando de Soto and Some Portuguese Gentlemen in the Discovery of the Province of Florida* (Evora, Portugal: André de Burgos, 1557), 179 (folio z, 3r). Very few sixteenth-century books contain

individual page numbers, as their pages were printed on large sheets of paper before being cut and folded. Most books of the period possess folio numbers and/or printer signatures.

27. Michael J. Ferreira, personal communication, April 14, 2012. According to Ferreira, "it was not uncommon at the time to have such typos: between the author and the text were the editor and, at least, one assistant" (personal communication, June 6, 2012).

28. Dr. Ernesto González-Seoane of the Instituto da Lingua Galega, Santiago de Compostela, Spain, also thinks *terras* may have been the intended word choice. González-Seoane to Michael J. Ferreira, personal communication, April 13, 2012.

29. See, for example, Fernández de Oviedo, *Historia general y natural*, 1:568; Bourne, *Narratives of the Career of Hernando de Soto*, 2:123; Clayton, Knight, and Moore, *De Soto Chronicles*, 2:330; Ned D. Jenkins, "The Village of Mabila: Archaeological Expectations," in Knight, ed., *Search for Mabila*, 75.

30. Damiano Avanzato, ed., *Following Chestnut Footprints (Castanea spp.): Cultivation and Culture, Folklore and History, Traditions and Uses*, Scripta Horticulturae 9 (Leuven, Belgium: International Society for Horticultural Science, 2009), 106–11; Rita Costa et al., *Chestnut Varieties in the North and Central Regions of Portugal*, European Union Agricultural Report 448 (Lisbon: Instituto Nacional des Recursos Biologicos, 2008), 39; Santiago Pereira-Lorenzo et al., "Chestnut," in *Handbook of Plant Breeding*, vol. 8, *Fruit Breeding*, ed. Marisa Luisa Badenes and David H. Byrne (New York: Springer, 2012), 741.

31. Michael J. Ferreira, personal communication, June 6, 2012.

32. Mary Ross, "With Pardo and Boyano on the Fringes of the Georgia Land," *Georgia Historical Quarterly* 14, no. 4 (December 1930): 267–85; Chester B. DePratter, Charles M. Hudson, and Marvin T. Smith, "The Route of Juan Pardo's Explorations in the Interior Southeast, 1566–1568," *Florida Historical Quarterly* 62, no. 2 (October 1983): 125–58; Robin Beck, *Chiefdoms, Collapse, and Coalescence in the Early American South* (New York: Cambridge University Press, 2013), 72.

33. Robin A. Beck Jr., "From Joara to Chiaha: Spanish Exploration of the Appalachian Summit Area, 1540–1568," *Southeastern Archaeology* 16, no. 2 (Winter 1997), 165; Robin A. Beck, David G. Moore, and Christopher B. Rodning, "Introduction," in *Fort San Juan and the Limits of Empire: Colonialism and Household Practice at the Berry Site*, ed. Robin A. Beck, Christopher B. Rodning, and David G. Moore (Gainesville: University Press of Florida, 2016), 5–17.

34. Lee Ann Newsom, "Wood Selection and Technology in Structures 1 and 5," in Beck, Rodning, and Moore, *Fort San Juan*, 89–90, 94, 209.

35. DePratter, Hudson, and Smith, "Route of Juan Pardo's Explorations," 135–42; Charles Hudson, *The Juan Pardo Expeditions: Explorations of the Carolinas and Tennessee, 1566–1568* (Tuscaloosa: University of Alabama Press, 2005), 32–35. In the Julian calendar, the departure date was September 1, 1567. Thanks to Neil Pederson

of the Harvard Forest for the suggestion that I provide both Gregorian and Julian calendar dates for relevant historical events. Prior to 1752, individuals in North America used the Julian calendar, which deviates from the modern Gregorian calendar by as much as ten days.

36. Ross, "With Pardo and Boyano," 277. The source is a document copied from an original housed in the Archivo General de Indias in Seville, Spain (AGI 54-5-9, fol. 17). In the early twentieth century, historian Woodbury Lowery deposited his own handwritten copy in the Library of Congress, which I retrieved. See Lowery, "Spanish Settlements in the United States, 1522–1803," Library of Congress, Manuscript Division, container 5, Teresa Martín to Gonzalo Mendez de Canço.

37. Martín Alonso Pedraz, *Enciclopedia del idioma: Diccionario histórico y moderno de la lengua española (siglos XII al XX)*, vol. 1 (Madrid: Aguilar, 1958), 420. According to Pedraz, *apilado* refers to chestnuts that were *secada al humo*, or smoke dried. I am indebted to Paul E. Hoffman for sharing the source and for his expert advice in translating a portion of the Mendez de Canço letter.

38. On the history and process of drying chestnuts in Europe, see "The Chestnut Harvest and Manufacture of Chestnut Meal," *Forest Leaves* (Pennsylvania Forestry Association) 9, no. 5 (October 1903): 73–75; Paolo Squatriti, "Trees, Nuts, and Woods at the End of the First Millennium: A Case from the Amalfi Coast," in *Ecologies and Economies in Medieval and Early Modern Europe*, ed. Scott G. Bruce (Boston: Brill 2010), 39–42; Antoinette Fauve-Chamoux, "Chestnuts," in *The Cambridge World History of Food*, ed. Kenneth F. Kiple and Kriemhild C. Ornelas (Cambridge: Cambridge University Press, 2000), 359–63.

39. Woodbury Lowery, "Spanish Settlements in the United States, 1522–1803," Library of Congress, Manuscript Division, container 5, Juan de Ribas Solvado to Gonzalo Mendez de Canço; Teresa Martín to Gonzalo Mendez de Canço; David Glavin Irlandes to Gonzalo Mendez de Canço; Luisa Mendez to Gonzalo Mendez de Canço.

40. Paul E. Hoffman, trans. and ed., "The 'Short' Bandera Relation," in Hudson, *Juan Pardo Expeditions*, 302. See also John E. Worth, "Recollections of the Juan Pardo Expeditions: The 1584 Domingo de León Account," in Beck, Moore, and Rodning, *Fort San Juan*, 70.

41. Hudson, *Juan Pardo Expeditions*, 104. See also Karen M. Booker, Charles M. Hudson, and Robert L. Rankin, "Place Name Identification and Multilingualism in the Sixteenth-Century Southeast," *Ethnohistory* 39, no. 4 (Autumn 1992): 429.

42. In 1859, an attorney measured a chestnut "thirty-three feet in circumference at four feet from the ground" (10.5 feet in diameter) near the headwaters of the Little Pigeon River. It was "a noble living specimen," he wrote, "apparently sound, and of nearly a uniform diameter upward, for forty or fifty feet," S. B. Buckley,

"The Mountain Regions of North Carolina and Tennessee," *De Bow's Review* 26, no. 6 (June 1859): 704.

43. Hudson, *Juan Pardo Expeditions*, 104.

44. Richard Hakluyt, *The Voyages, Navigations, Traffiques, and Discoveries of the English Nation*, vol. 3 (London: George Bishop, Ralfe Newberie, and Robert Barker, 1600), 243. A firsthand description of the Roanoke Island environs is found in Arthur Barlowe, *The First Voyage to Roanoke, 1584* (Boston: Directors of the Old South Works, 1898).

45. Original documents confer that "Master Hariot" was among those who "remained one whole yeere in Virginia, under the Governement of Master Ralfe Lane." Richard Hakluyt, *The Principall Navigations, Voiages and Discoveries of the English Nation* (London: George Bishop and Ralph Newberie, 1589), 736.

46. Hakluyt, *Principall Navigations*, 733–47. See also David B. Quinn, *Set Fair for Roanoke: Voyages and Colonies, 1584–1606* (Chapel Hill: University of North Carolina Press, 1985), 87–120, 267–68.

47. Hariot, *Briefe and True Report*, 13 (folio b, 3r).

48. Ibid, 18.

49. See Wayne Gisslen, *Professional Baking* (Hoboken, N.J.: John Wiley & Sons, 2013), 61, 154, 167.

50. John Smith, *The Works of Captain John Smith, 1608–1631*, ed. Edward Arber (Birmingham, UK: English Scholar's Library, 1884), 58.

51. Fernández de Oviedo, *Historia general y natural*, 1:568.

52. Ibid.; William Bartram, *The Travels of William Bartram*, ed. Mark Van Doren (New York: Dover, 1955), 321.

53. Linda G. Page and Eliot Wigginton, eds., *The Foxfire Book of Appalachian Cookery* (Chapel Hill, N.C.: University of North Carolina Press, 1992), 179. On the use of chestnut bread among Cherokees, see Mary Ulmer and Samuel E. Beck, eds., *Cherokee Cooklore: To Make My Bread* (Asheville, N.C.: Stephens Press, 1951); Mary Ulmer Chiltoskey, "Cherokee Indian Foods," in *Gastronomy: the Anthropology of Food and Food Habits*, ed. Margaret L. Arnott (The Hague, Netherlands: Mouton, 1975), 235–44; Rhonda D. Terry and Mary A. Bass, "Food Practices of Families in an Eastern Cherokee Township," *Ecology of Food and Nutrition* 14, no. 1 (1984): 63–70.

54. Smith, *Works of Captain John Smith*, 57. See also William R. Gerard, "Virginia's Indian Contributions to English," *American Anthropologist*, n.s., 9, no. 1 (January–March, 1907): 88–90. Gerard believes the word chinquapin derives from the Powhatan word for rattle-nut (*tshitshi-kwe-men*), as the nuts were used in squash-shell rattles. According to Gerard, the change of the suffix *-men* (or *-min*) to *-pin* "occurred at the beginning of the last quarter of the 17th century" (89).

55. Smith, *Works of Captain John Smith*, 56.

56. Ibid., 57.

57. Edward Johnson, *Johnson's Wonder-Working Providence, 1628–1651* (New York: Charles Scribner's Sons, 1910), 162. On the sweetness of American chestnuts, see F. André Michaux, *The North American Sylva*, vol. 3, trans. Thomas Nuttall (Philadelphia: D. Rice & A. N. Hart, 1859), 13; G. Harold Powell, "The Historical Status of Commercial Chestnut Culture in the United States," *American Gardening* 20, no. 220 (March 11, 1899): 178–79.

58. John Josselyn, *New-Englands Rarities Discovered: In Birds, Beasts, Fishes, Serpents, and Plants of that Country* (London: Giles Widdowes, 1672), 51. Twelve pence is the equivalent of eight U.S. dollars if one compares Plymouth Colony probate inventories for the years 1670–1673. A bushel of milled corn was worth twice as much as raw chestnuts. See Plymouth Colony Archive Project, "Analysis of Selected Probate Inventories," accessed January 6, 2018, http://www.histarch.uiuc.edu/plymouth/probates.html.

59. Gertrude Selwyn Kimball, *Providence in Colonial Times* (Boston: Houghton Mifflin, 1912), 28. See also Zachariah A. Mudge, *Foot-Prints of Roger Williams: A Biography* (New York: Carlton and Lanahan, 1871), 100.

60. Adriaen van der Donck, "Description of the New Netherlands," in *Collections of the New York Historical Society, Second Series*, vol. 1 (New York: New York Historical Society, 1841), 151.

61. Josselyn, *New-Englands Rarities*, 51.

62. Thomas Morton, *New English Canaan, or New Canaan* (1637; Washington, D.C.: Peter Force, 1838), 44.

63. See Frederick Pursh, *Journal of a Botanical Excursion in the Northeastern Parts of the States of Pennsylvania and New York, during the Year 1807* (Philadelphia: Brinckloe and Marot, 1869), 9; Ralph C. Hawley and Austin F. Hawes, *Forestry in New England: A Handbook of Eastern Forest Management* (New York: John Wiley & Sons, 1912), 60–62; Edward K. Faison and David R. Foster, "Did American Chestnut Really Dominate the Eastern Forest?," *Arnoldia* 72, no. 2 (October 2014), 18–32.

64. Eric W. Sanderson, *Mannahatta: A Natural History of New York City* (New York: Harry N. Abrams, 2009), 58. See also J. H. Mather and L. P. Brockett, *A Geographical History of the State of New York* (Utica, N.Y.: John W. Fuller, 1853), 140; Andrew M. Greller, "Observations on the Forests of Northern Queens County, Long Island, from Colonial Times to the Present," *Bulletin of the Torrey Botanical Club* 99, no. 4 (July/August 1972): 202–6; Russell Shorto, *The Island at the Center of the World: The Epic Story of Dutch Manhattan and the Forgotten Colony that Shaped America* (New York: Doubleday, 2004), 42, 47, 60.

65. Some naturalists claim lower chestnut distribution estimates for Manhattan, including Alexander Brash, former president of the Connecticut Audubon Society. According to Brash, only 10 percent of Manhattan's "primeval forests" contained chestnut trees. Brash, "New York City's Primeval Forest:

A Review of Characterizing the 'Type Ecosystem,'" in *Natural History of New York City's Parks and Great Gull Island*, Transactions of the Linnaean Society of New York 10 (New York: Linnaean Society of New York, 2007), 65–66.

66. Samuel de Champlain, *Voyages du Sieur de Champlain*, vol. 1 (Paris: [Government of France], 1830), 194. An English version published in the late nineteenth century also translates the phrase as "many chestnut trees." Samuel de Champlain, *The Works of Samuel de Champlain, 1608–1613*, vol. 2, ed. Henry Percival Bigger (Toronto: Champlain Society, 1925), 91.

67. De Champlain, *Voyages du Sieur de Champlain*, 194.

68. William H. H. Murray, *Lake Champlain and Its Shores* (Boston: De Wolfe, Fiske, 1890), 216.

69. Clinton H. Merriam, *The Vertebrates of the Adirondack Region, Northeastern New York* (New York: L. S. Foster, 1884), 119.

70. Samuel de Champlain, "Carte geographique de la Nouvelle Franse faictte par le sieur de Champlain . . . 1612," in *Les voyages du sieur de Champlain xaintongeois, capitaine ordinaire pour le roy, en la marine* (Paris: Jean Berjon, 1613), n.p. The map can be viewed on the website of the Osher Map Library, Smith Center for Cartographic Education, University of Southern Maine, Portland, Maine, accessed June 16, 2018, https://oshermaps.org/browse-maps?id=44809.

71. Victoria Dickenson, "Cartier, Champlain, and the Fruits of the New World: Botanical Exchange in the 16th and 17th Centuries," *Scientia Canadensis* 31, no. 1/2 (2008): 38.

72. René Goulaine de Laudonnière, *L'histoire notable de la Floride* (Paris: Guillaume Auvray, 1586), 3r.

73. Morton, *New English Canaan*, 183. See also William Hancock, "Houses in Early Virginia Indian Society," *Encyclopedia Virginia*, Virginia Humanities, May 30, 2014, accessed December 12, 2018, https://www.encyclopediavirginia.org /Houses_in_Early_Virginia_Indian_Society.

74. Peter Nabokov and Robert Easton, *Native American Architecture* (Oxford: Oxford University Press, 1989), 56–62. Chestnut bark structures were also found among the Virginia Algonquin. See Helen C. Rountree, *The Powhatan Indians of Virginia: Their Traditional Culture* (Norman: University of Oklahoma Press, 1989), 61.

75. Jaspar Dankers [Jasper Danckaerts] and Peter Sluyter, *Journal of a Voyage to New York and a Tour in Several of the American Colonies, 1679–1680*, trans. and ed. Henry C. Murphy (Brooklyn: Long Island Historical Society, 1867), 124. In the Julian calendar, which the pair still observed, the date of the encounter was September 30, 1679.

76. Albert C. Myers, ed., *William Penn's Own Account of the Lenni Lenape or Delaware Indians* (Moorestown, N.J.: Middle Atlantic Press), 27 and 27n1. The passage is from Penn's handwritten letter, which was consulted by Myers when

preparing the reprint edition, leading to Myers's footnote observing that the words "mostly chestnut" had been crossed out in Penn's original draft. The volume was originally published as Albert C. Myers, ed., *William Penn: His Own Account of the Lenni Lenape or Delaware Indians, 1683* (Moylan, Pa.: Albert Cook Myers, 1937).

77. Nabokov and Easton, *Native American Architecture*, 59.

78. William Bartram, *Travels through North and South Carolina, Georgia, East and West Florida* (Philadelphia: James and Johnson, 1791), 365. See also David I. Bushnell Jr., *Native Villages and Village Sites East of the Mississippi*, Bureau of American Ethnology Bulletin 69 (Washington, D.C.: GPO, 1919), 47–50, 52–60.

79. Quoted in Woodbury Lowery, "Spanish Settlements in the United States, 1522–1803," Library of Congress, Manuscript Division, container 5, Luisa Mendez to Gonzalo Mendez de Canço.

80. Van der Donck, "Description of the New Netherlands," 151.

81. Hugh Morrison, *Early American Architecture: From the First Colonial Settlements to the National Period* (New York: Oxford University Press, 1952), 22, 32, 138.

82. Justine McKnight, "*Flotation-Recovered and Hand-Recovered Botanical Remains from the Towne Neck Site (18AN944), Anne Arundel County, Maryland,*" (report submitted to KCI Technologies, Inc., 2000), 4; available at https://apps.jefpat.maryland.gov/archeobotany/ReportPages/18AN944ReportA.pdf. The "upper-middle class" home is identified as being of earthfast construction on the web page providing access to McKnight's archaeobotanical report: "Town [*sic*] Neck: 18AN944," Maryland Archeobotany, https://apps.jefpat.maryland.gov/archeobotany/Sites/18AN944.aspx.

83. Mark R. Edwards, "Work in Progress: Dating Historic Buildings in Lower Southern Maryland with Dendrochronology," *Perspectives in Vernacular Architecture* 1 (1982): 155. See also Cary Carson et. al, "Impermanent Architecture in the Southern American Colonies," *Winterthur Portfolio* 16, no. 2/3 (Summer/Autumn 1981): 158–59; Randy Leffingwell, *The American Barn* (St. Paul, Minn.: Motorbooks International, 2003), 37, 42–43, 78.

84. On deforestation and wood shortages along the Atlantic coast during the seventeenth and eighteenth centuries, see Carville V. Earle, *The Evolution of a Tidewater Settlement System: All Hallow's Parish, Maryland, 1650–1783* (Chicago: University of Chicago, Department of Geography, 1975), 30–34; Thomas R. Cox et al., *This Well-Wooded Land: Americans and Their Forests from Colonial Times to the Present* (Lincoln: University of Nebraska Press, 1985), 10–28; Michael Williams, *Americans and Their Forests: A Historical Geography* (Cambridge: Cambridge University Press, 1992), 60–74.

85. William Waller Hening, ed., *The Statutes at Large: Being a Collection of All the Laws of Virginia, from the First Session of the Legislature, in the Year 1619,*

vol. 1 (New York: R. & W. & G. Bartow, 1823), 126. See also Vanessa Patrick, "Partitioning the Landscape: Fences in Colonial Virginia," *Magazine Antiques* 154, no. 7 (July 1998): 97.

86. John Bowne Account Book, 1649–1703, New York Public Library, Manuscript and Archives Division, Bowne Family Papers, box 6.

87. Lewis M. Norton, *Connecticut Towns: Goshen in 1812*, ed. Thompson R. Harlow (Hartford: Acorn Club of Connecticut, 1949), 12. See also Eugene Cotton Mather and John Fraser Hart, "Fences and Farms," *Geographical Review* 44, no. 2 (April 1954): 201–23; John R. Stilgoe, *Common Landscape of America, 1580–1845* (New Haven, Conn.: Yale University Press, 1982), 189, 191; Terry G. Jordan and Matti Kaups, *The American Backwoods Frontier: An Ethnic and Ecological Interpretation* (Baltimore: Johns Hopkins University Press, 1989), 106–7.

88. Larry Jones, "Uncommon Pickets," *Old House Journal* 12, no. 4 (May 1984): 86.

89. Ted Steinberg, *Gotham Unbound: The Ecological History of Greater New York* (New York: Simon and Schuster, 2014), 36, 405n28.

90. George Washington, *Writings* (New York: Library of America, 1997), 440. See also David M. Ludlum, *Early American Winters, 1604–1820*, vol. 1 (Boston: American Meteorological Society, 1966), 115; Steinberg, *Gotham Unbound*, 36.

91. William Cronon, *Changes in the Land: Indians, Colonists, and the Ecology of New England* (New York: Hill and Wang, 1983), 120.

92. R. V. R. Reynolds and Albert H. Pierson, *Fuel Wood Used in the United States, 1630–1930*, USDA Circular 641 (Washington, D.C.: GPO, 1942), 9, table 2.

93. Ibid., 7. For the heat values of air-dried and green chestnut fuel wood across several U.S. regions, see Henry S. Graves, *The Use of Wood for Fuel*, U.S. Forest Service Office of Forest Investigations, comp., USDA Bulletin no. 753 (Washington, D.C.: GPO, 1919), 29.

94. Reynolds and Pierson, *Fuel Wood Used in the United States*, 9, table 2. According to the same document, the "snapping of chestnut" made it unpopular for general fireplace use (6).

95. David E. Wilkins, *Documents of Native American Political Development: 1500s to 1933* (Oxford: Oxford University Press, 2009), 16. See also Gerald Murphy, "About the Iroquois Constitution," Modern History Sourcebook: The Constitution of the Iroquois Confederacy, Fordham University Internet History Sourcebooks Project, accessed January 8, 2018, http://www.fordham.edu/halsall/mod/iroquois.asp.

96. Although chestnut was felled for the illegal New Jersey timber trade as early as the 1630s, there are few records of those harvests. See Emily W. B. Russell, "Vegetation of Northern New Jersey before European Settlement," *American Midland Naturalist* 105, no. 1 (January 1981): 6. Pollen studies document an increase in chestnut during the colonial period, particularly those areas impacted by seventeenth- and eighteenth- century settlement. See David R.

Foster and Tad M. Zebryk, "Long-Term Vegetation Dynamics and Disturbance History of a Tsuga-Dominated Forest in New England," *Ecology* 74, no. 4 (June 1993): 990, 993; Emily W. B. Russell, *People and the Land through Time: Linking Ecology and History* (New Haven, Conn.: Yale University Press, 1997), 121, 206, 225; W. Wyatt Oswald et al., "Post-Glacial Changes in Spatial Patterns of Vegetation across Southern New England," *Journal of Biogeography* 34, no. 5 (May 2007): 906.

97. David W. Stahle et al., "The Lost Colony and Jamestown Droughts," *Science*, n.s., 280, no. 5363 (April 24, 1998): 564–67; Karen Ordahl Kupperman, *The Jamestown Project* (Cambridge, Mass.: Belknap Press of Harvard University Press, 2008), 169–74; R. Stockton Maxwell et al., "A Multicentury Reconstruction of May Precipitation for the Mid-Atlantic Region Using *Juniperus virginiana* Tree Rings," *Journal of Climate* 25, no. 3 (February 2012): 1045–56.

98. On Native American population losses, see Sherburne F. Cook, *The Indian Population of New England in the Seventeenth Century* (Berkeley: University of California Press, 1976), 13–14, 31–37, 46–47; Henry F. Dobyns, *Their Number Become Thinned: Native American Population Dynamics in Eastern North America* (Knoxville: University of Tennessee Press, 1983), 8–32; Marvin T. Smith, *Archaeology of Aboriginal Culture Change in the Interior Southeast: Depopulation during the Early Historic Period* (Gainesville: University Press of Florida, 1987), 55–60; Peter H. Wood, "The Changing Population of the Colonial South: An Overview by Race and Region, 1685–1790," in *Powhatan's Mantle: Indians in the Colonial Southeast*, rev. and exp. ed., ed. Gregory A. Waselkov, Peter H. Wood, and Tom Hatley (Lincoln: University of Nebraska Press, 2006), 57–132.

99. Wilbur R. Mattoon, "The Origin and Early Development of Chestnut Sprouts," *Forestry Quarterly* 7, no. 1 (1909): 34–47. According to Mattoon, who measured growth rates in both full light and partial shade, chestnut saplings receiving open sun grew, over a period of three years, 2.5 feet, whereas shaded trees grew only 0.9 feet (41).

100. Alice Morse Earle, *Home Life in Colonial Days* (New York: Macmillan, 1898), 159, 306; Edwin Tunis, *Colonial Craftsmen and the Beginnings of American Industry* (Baltimore: Johns Hopkins University Press, 1999), 89; Jane Brox, *Clearing Land: Legacies of the American Farm* (New York: North Point, 2004), 39.

101. The destruction of forests and croplands by livestock is well documented for the colonial and frontier periods. See, for example, Earle, *Evolution of a Tidewater Settlement System*, 30–31; Cronon, *Changes in the Land*, 141–51; Timothy Silver, *A New Face on the Countryside: Indians, Colonists, and Slaves in South Atlantic Forests, 1500–1800* (Cambridge: Cambridge University Press, 1990), 173–80; Brad Alan Bays, "The Historical Geography of Cattle Herding among the Cherokee Indians, 1761–1861" (master's thesis, University of Tennessee, Knoxville, 1991), 81–86; Virginia D. Anderson, "Animals into the Wilderness: The

Development of Livestock Husbandry in the Seventeenth-Century Chesapeake," in *Environmental History and the American South: A Reader*, ed. Paul S. Sutter and Christopher J. Manganiello (Athens: University of Georgia Press, 2009), 36–37, 43–47.

102. Williams, *Americans and Their Forests*, 53–54. See also Gordon G. Whitney, *From Coastal Wilderness to Fruited Plain: A History of Environmental Change in Temperate North America from 1500 to the Present* (Cambridge: Cambridge University Press, 1996), 93–160.

103. Grace S. Brush, "Forests before and after the Colonial Encounter," in *Discovering the Chesapeake: History of an Ecosystem*, ed. Philip D. Curtin, Grace S. Brush, and George W. Fisher (Baltimore: Johns Hopkins University Press, 2001), 47, 51–52. Gordon G. Whitney maintains that forest clearance was extremely slow along the East Coast, as a full century was needed "to clear 50% of the land" (*From Coastal Wilderness*, 153).

CHAPTER 4. The Most Celebrated Hunting Grounds

1. William Byrd, *The Westover Manuscripts: Containing the History of the Dividing Line betwixt Virginia and North Carolina; A Journey to the Land of Eden, A.D. 1733; and a Progress to the Mines* (Petersburg, Va.: Edmund and Julian C. Ruffin, 1841), 1–102.

2. William Byrd, William Byrd's Histories of the Dividing Line *betwixt Virginia and North Carolina* (Raleigh: North Carolina Historical Commission, 1929), 13–267.

3. William Byrd, Richard Fitzwilliam, and William Dandridge, "The 2nd Part of the Journal of the Dividing Line between Virginia & North Carolina Begun the 19th Sept. 1728," in William L. Saunders, ed., *The Colonial Records of North Carolina*, vol. 2, *1713–1728* (Raleigh, N.C.: P. M. Hale, 1886), 782–98.

4. William Byrd, Richard Fitzwilliam, and William Dandridge, "A Journal of the Proceedings of the Commissioners for Settling the Bounds betwixt Virginia and Carolina 1727/8," in Saunders, *Colonial Records of North Carolina*, 2:757. In the Julian calendar (which Byrd observed), the date was March 5.

5. Stephen C. Ausband, *Byrd's Line: A Natural History* (Charlottesville: University of Virginia Press, 2002), 16–63. In the Julian calendar, the date was April 4.

6. Numerous studies have documented the importance of hardwood mast on wildlife populations. See Owen T. Gorman and Roland R. Roth, "Consequences of a Temporally and Spatially Variable Food Supply for an Unexploited Gray Squirrel (*Sciurus carolinensis*) Population," *American Midland Naturalist* 121, no. 1 (January 1989): 41–60; William J. McShea and Georg Schwede, "Variable Acorn Crops: Responses of White-Tailed Deer and Other Mast Consumers,"

Journal of Mammalogy 74, no. 4 (November 1993): 999–1006; William G. Minser et al., "Feeding Response of Wild Turkeys to Chestnuts and Other Hard Mast," *Proceedings of the Annual Conference of the Southeastern Association of Fish and Wildlife Agencies* 49 (1995): 488–97; Michael R. Vaughan, "Oak Trees, Acorns, and Bears," in *Oak Forest Ecosystems: Ecology and Management for Wildlife*, ed. William J. McShea and William M. Healy (Baltimore: Johns Hopkins University Press, 2002), 224–40; Christopher W. Ryan et al., "Influence of Mast Production on Black Bear Non-hunting Mortalities in West Virginia," *Ursus* 18, no. 1 (April 2007): 46–53.

7. Byrd, *William Byrd's Histories*, 195. In the Julian calendar, the date was October 12.

8. Byrd, *Westover Manuscripts*, 49; Ausband, *Byrd's Line*, 98, 101.

9. Colonial land grant to Isaac Cloud, 1748, recorded in Old Survey Book 1, 1746–1782, Pittsylvania County, Virginia, Pittsylvania County Courthouse, Chatham, Virginia, n.p.; N. G. Clement, "Natural Conditions in Pittsylvania County as Shown by Early Land Grants from An Old Surveyor's Book," *Virginia Magazine of History and Biography* 25, no. 3 (July 1917): 299.

10. Maud Carter Clement, *The History of Pittsylvania County, Virginia* (Lynchburg, Va.: J. P. Bell, 1929), 30. The chestnut was located near Callands, Virginia.

11. Byrd, *William Byrd's Histories*, 197.

12. Byrd, *Westover Manuscripts*, 51.

13. William Byrd, *History of the Dividing Line and Other Tracts*, vol. 1, *History of the Dividing Line* (Richmond, Va.: Thomas H. Wynne, 1866), 125.

14. Ausband, *Byrd's Line*, 8–15, 97–128. Bear and deer hunting techniques of the period required individuals to approach their quarry at a slow, measured pace, downwind from the animals. Once the animals were shot and dressed, the meat and carcasses also had be transported back to camp. On their eastbound return trip, over roughly the same terrain, an additional nineteen deer, nine bears, eight turkeys, and one woodland bison were taken by the survey crew (Byrd, *William Byrd's Histories*, 244–89).

15. Byrd, *Westover Manuscripts*, 57. In the Julian calendar, the date was October 19.

16. Ibid.

17. The ornithologist Alexander Wilson estimated a flock of passenger pigeons consumed "seventeen millions, four hundred and twenty-four thousand bushels" of beechnuts, acorns, and chestnuts daily. Wilson, *American Ornithology, or the Natural History of the Birds of the United States*, vol. 3 (Harrison Hall, Pa.: Collins, 1829), 7. John James Audubon—using different assumptions in his calculation—believed passenger pigeons daily consumed "eight millions, seven hundred and twelve bushels" of mast. See Audubon, *Ornithological Biography, or An Account of the Habits of the Birds of the United States of America* (Edinburgh: Adam Black, 1831), 323.

18. Byrd, *William Byrd's Histories*, 216.

19. Peter Bane, "Keystones and Cops: An Eco-mystery Thriller," accessed January 23, 2018, https://permacultureactivist.net/2020/01/02/keystones-and-cops-an-eco-mystery-thriller-by-peter-bane. Studies find the growth and vigor of American chestnuts planted on abandoned mine sites is more influenced by the microbial content and pH of the soil than specific metals. However, the presence of zinc is often an indication the soil is acidic, a condition that improves the growth and vigor of chestnuts. See Jenise M. Bauman, Carolyn H. Keiffer, and Shiv Hiremath, "Facilitation of American Chestnut (*Castanea dentata*) Seedling Establishment by *Pinus virginiana* in Mine Restoration," *International Journal of Ecology* (2012): e257326, https://doi.org/10.1155/2012/257326; Christopher R. Miller, Jennifer A. Franklin, and David S. Buckley, "Effects of Soil Amendment Treatments on American Chestnut Performance and Physiology on an East Tennessee Surface Mine," in *Proceedings of the 2011 National Meeting of the American Society of Mining and Reclamation, Bismarck, ND, June 11–16, 2011*, ed. Richard Barnhisel (Lexington, Ky.: American Society of Mining and Reclamation, 2012), 419–37.

20. Byrd, *Westover Manuscripts*, 112. In the Julian calendar, the date was September 30.

21. Ibid.

22. Timothy Flint, *The History and Geography of the Mississippi Valley*, vol. 1 (Boston: Carter, Hendee, 1833), 351; Annette Kolodny, *The Land before Her: Fantasy and Experience of the American Frontiers, 1630–1860* (Chapel Hill: University of North Carolina Press, 1984), 86–88; Lyman C. Draper, *The Life of Daniel Boone* (Mechanicsburg, Pa.: Stackpole Books, 1998), 101, 187, 135; Malcolm J. Rohrbough, *Trans-Appalachian Frontier: People, Societies, and Institutions, 1775–1850* (Bloomington: Indiana University Press, 2008), 24, 46.

23. John P. Hale, *Trans-Allegheny Pioneers: Historical Sketches of the First White Settlements West of the Alleghenies 1748 and After* (Cincinnati: Graphic Press, 1886), 14–16, 24, 109; J. Stoddard Johnston, *First Explorations of Kentucky* (Louisville: John P. Morton, 1898), 1–19; Donald Jackson, *Thomas Jefferson and the Stony Mountains: Exploring the West from Monticello* (Norman: University of Oklahoma Press, 1993), 7; Harriette Simpson Arnow, *Seedtime on the Cumberland* (Lincoln: University of Nebraska Press, 1995), 20, 145, 227.

24. Johnston, *First Explorations of Kentucky*, 75. In the Julian calendar, which Walker observed, the journal entry date was July 12, 1750. Another Kentucky frontiersman remarked that "such a collection of game was not to be seen in any part of the known World," Stephen Aaron, *How the West was Lost: The Transformation of Kentucky from Daniel Boone to Henry Clay* (Baltimore: Johns Hopkins University Press, 1996), 7.

25. Unknown author, *History of Bedford, Somerset, and Fulton Counties, Pennsylvania* (Chicago: Waterman, Watkins, 1884), 76, 199, 275–76, 394.

26. Samuel P. Hildreth, *Original Contributions to the American Pioneer* (Cincinnati: John S. Williams, 1844), 126/119.

27. George W. Featherstonhaugh, *Excursion through the Slave States, from Washington on the Potomac to the Frontier of Mexico* (New York: Harper & Brothers, 1844), 21.

28. Conway H. Smith, *The Land that is Pulaski County* (Pulaski, Va.: Pulaski County Library Board, 1981), 32; Thomas L. Purvis, *Colonial America to 1763* (New York: Facts on File, 1999), 210; Henry P. Scalf, *Kentucky's Last Frontier* (Johnson City, Tenn.: Overmountain Press, 2000), 42.

29. Draper, *Life of Daniel Boone*, 213. The European bison (*Bison bonasus*)—whose habitat resembles the eastern North American woodlands—prefers acorns and chestnuts during fall and winter months and they are considered to be the animal's favorite foods. See Zdzislaw Pucek, ed., *European Bison: Status Survey and Conservation Action Plan* (Cambridge: International Union for the Conservation of Nature and Natural Resources, 2004), 22.

30. In individual calendar years, tobacco and rice exports surpassed skins and furs in value in several American colonies during the first half of the eighteenth century. For the entire period, however, furs and skins were the most valued export commodity. See Verner W. Crane, *The Southern Frontier, 1670–1732* (Durham, N.C.: Duke University Press, 1929), 328–31; Paul C. Phillips, *The Fur Trade*, vol. 1 (Norman: University of Oklahoma Press, 1961), 338; James A. Hanson, *When Skins Were Money: A History of the Fur Trade* (Chadron, Neb.: Museum of the Fur Trade, 2005), 49.

31. The aggregate figure of a half million hides comes from several sources. For regional estimates of deerskin exports, see Kathryn E. Holland Braund, *Deerskins and Duffels: Creek Indian Trade with Anglo-America, 1685–1815* (Lincoln: University of Nebraska Press, 1993), 70–72, 97, 226; Eric Hinderaker, *Elusive Empires: Constructing Colonialism in the Ohio Valley, 1673–1800* (Cambridge: Cambridge University Press, 1993), 23–24; Heather A. Lapham, *Hunting for Hides: Deerskins, Status, and Cultural Change in the Protohistoric Appalachians* (Tuscaloosa: University of Alabama Press, 2005), 6–7; Edward B. Barbier, *Scarcity and Frontiers: How Economies Have Developed through Natural Resource Exploitation* (Cambridge: Cambridge University Press, 2011), 317.

32. Eric J. Dolin, *Fur, Fortune, and Empire: The Epic History of the Fur Trade in America* (New York: W. W. Norton, 2010), xv–xvi. Although the fur trade (involving mostly beaver pelts) had an indirect connection to oak-chestnut forests, in New England and the Mid-Atlantic beavers lived in close proximity to chestnut groves and consumed chestnut bark.

33. Silver, *New Face on the Countryside*, 92–93; Holland Braund, *Deerskins and Duffels*, 69–71; Lapham, *Hunting for Hides*, 6–8, 47–51; H. Thomas Foster II and Arthur D. Cohen, "Palynological Evidence of the Effects of the Deerskin

Trade on Forest Fires during the Eighteenth Century in Southeastern North America," *American Antiquity* 72, no. 1 (January 2007): 35–51; Joseph M. Hall, *Zamumo's Gifts: Indian-European Exchange in the Colonial Southeast* (Philadelphia: University of Pennsylvania Press), 2009, 66, 123.

34. E. Lucy Braun, *Deciduous Forests of Eastern North America* (Philadelphia: Blakiston, 1950), 192–258.

35. Albert S. Gatschet, *A Migration Legend of the Creek Indians, with a Linguistic, Historic and Ethnographic Introduction*, vol. 1 (Philadelphia: D. G. Brinton, 1884), 107. Gatschet interviewed dozens of southeastern Indians during the late nineteenth century, so it is assumed that the creation story was told to him by actual Choctaw informants.

36. Henry R. Schoolcraft, *Information Respecting the History, Condition, and Prospects of the Indian Tribes of the United States: Collected and Prepared under the Direction of the Bureau of Indian Affairs*, part 5 (Philadelphia: J. B. Lippincott, 1855), 276. See also Joel W. Martin, "The Green Corn Ceremony of the Muskogees," in *Religions of the United States in Practice*, vol. 1, ed. Colleen McDannell (Princeton, N.J.: Princeton University Press, 2001), 49; Bill Grantham, *Creation Myths and Legends of the Creek Indians* (Gainesville: University Press of Florida, 2002), 65.

37. Horatio B. Cushman, *History of the Choctaw, Chickasaw, and Natchez Indians* (Greenville, Tex.: Headlight Printing House, 1899), 537. See also Charles Gayarré, *History of Louisiana: The French Domination* (New York: William J. Widdleton, 1867), 308; Charles C. Jones Jr., *Antiquities of the Southern Indians, Particularly of the Georgia Tribes* (New York: D. Appleton, 1873), 99–100.

38. Cushman, *History of the Choctaw, Chickasaw, and Natchez*, 537. See also John R. Swanton, *Indian Tribes of the Lower Mississippi Valley and Adjacent Coast of the Gulf of Mexico* (Washington, D.C.: GPO, 1911), 109–10.

39. Antoine-Simon Le Page du Pratz, *History of Louisiana, or of the Western Parts of Virginia and Carolina* (London: T. Becket, 1774), 342; Gordon M. Sayre, "A Newly Discovered Manuscript Map by Antoine-Simon Le Page du Pratz," *French Colonial History* 11 (2010): 41.

40. While most anthropologists accept oral tradition as a valid source of historical data, some argue that "reliable oral historical tradition doesn't necessarily survive beyond the first generation." Clifton Amsbury, "On the Reliability of Oral and Traditional History," *American Ethnologist* 22, no. 2 (May 1995), 412. Anthropologist Ernest S. Burch Jr. has countered this view, stating that oral traditions may, in fact, be passed down for generations, if not centuries. Burch, "More on the Reliability of Oral and Traditional History," *American Ethnologist* 23, no. 1 (February 1996): 130–32.

41. Chestnuts are mentioned five times in James Mooney's *Myths of the Cherokee* and play a major role in two Cherokee stories. Mooney, *Myths of the*

Cherokee: Extract from the Nineteenth Annual Report of the Bureau of American Ethnology (Washington, D.C.: GPO, 1902), 53, 317, 327–29, 439, 516. Mooney's primary informant was A'yunni (Swimmer), a Cherokee elder who spoke little English. Most anthropologists accept the validity of Swimmer's accounts, believing them to be embedded in Cherokee tradition. An informed discussion regarding Mooney and Swimmer is found in Raymond D. Fogelson, "An Analysis of Cherokee Sorcery and Witchcraft," in *Four Centuries of Southern Indians*, ed. Charles M. Hudson (Athens: University of Georgia Press, 2007), 113–31.

42. Mooney, *Myths of the Cherokee*, 327.

43. Ibid., 328–29.

44. Erminnie A. Smith, *Myths of the Iroquois: Extract from the Second Annual Report of the Bureau of Ethnology* (Washington, D.C.: GPO, 1883), 97–99.

45. Henry R. Schoolcraft, *Notes on the Iroquois* (New York: Bartlett and Welford, 1846), 158; William M. Beauchamp, *The Iroquois Trail, or Footprints of the Six Nations, in Customs, Traditions, and History* (Fayetteville, N.Y.: H. C. Beauchamp, 1892), 14; Anthony W. Wonderly, *Oneida Iroquois Folklore, Myth, and History: New York Oral Narrative from the Notes of H. E. Allen and Others* (Syracuse, N.Y.: Syracuse University Press, 2004), 91–95.

46. Theresa Bane, *Encyclopedia of Beasts and Monsters in Myth, Legend and Folklore* (Jefferson, N.C.: McFarland, 2016), 128.

47. Ibid. See also Smith, *Myths of the Iroquois*, 62.

48. Henry R. Schoolcraft, *Archives of Aboriginal Knowledge*, vol. 1 (Philadelphia: J. B. Lippincott, 1860), 429, 430, plate 72. In other versions, the woman is roasting acorns. See David Cusick, *David Cusick's Sketches of Ancient History of the Six Nations* (Lewiston, N.Y.: Tuscarora Village, 1828), 19; William W. Canfield, *Legends of the Iroquois Told by the 'Corn Planter': From Authoritative Notes and Studies* (New York: A. Wessels, 1902), 125–26.

49. Office of Indian Affairs, "Report Concerning Indians in North Carolina," in *Annual Report of the Commissioner of Indian Affairs, for the Year 1902*, part 1 (Washington, D.C: GPO, 1902), 260.

50. Cyrus Byington, *A Dictionary of the Choctaw Language*, ed. John R. Swanton and Henry S. Halbert, Smithsonian Institution Bureau of American Ethnology Bulletin no. 46 (Washington, D.C.: GPO, 1915), 307, 360.

51. Manataka American Indian Council, "Cherokee Dictionary," accessed March 3, 2018, http://www.manataka.org/page122.html. Several sources cite *u-na-gi-na* as the Cherokee word for chestnut, including the online dictionary maintained by the Cherokee Nation, accessed March 3, 2019, https:// language .cherokee.org/word-list. See also Ruth B. Holmes and Betty S. Smith, *Beginning Cherokee* (Norman: University of Oklahoma Press, 1977), 297.

52. James Adair, *The History of the American Indians* (London: Edward and Charles Dilly, 1775), 70.

53. David Zeisberger, *Zeisberger's Indian Dictionary* (Cambridge, Ma.: John Wilson and Son, 1887), 36; William M. Beauchamp, "Onondaga Plant Names," *Journal of American Folklore* 15, no. 57 (April–June 1902): 92.

54. Ives Goddard, "Pidgin Delaware," in *Contact Languages: A Wider Perspective*, ed. Sarah G. Thomason (Amsterdam, Netherlands: John Benjamins, 1997), 54.

55. William W. Tooker, "The Algonquian Terms *Patawomeke* and *Massawomeke*," *American Anthropologist* 7, no. 2 (April 1894): 185.

56. Michael McCafferty, *Native American Place-Names of Indiana* (Urbana: University of Illinois Press, 2008), 92–93.

57. Ibid., 93.

58. Keith A. Baca, *Native American Place Names in Mississippi* (Jackson: University of Mississippi Press, 2007), 36, 121; William Bright, *Native American Placenames of the United States* (Norman: University of Oklahoma Press, 2004), 149, 359; William W. Tooker, *The Indian Place-Names on Long Island and Islands Adjacent*, ed. Alexander F. Chamberlain (New York: G. P. Putnam's Sons, 1911), 270; William M. Beauchamp, *Aboriginal Place Names of New York*, New York State Museum Bulletin no. 108 (Albany: New York State Education Department, 1907), 241, 268.

59. On wild game availability during the frontier period, see David Ramsay, *Ramsay's History of South Carolina, from Its First Settlement in 1670 to the Year 1808*, vol. 2 (Newberry, S.C.: W. J. Duffie, 1858), 305; François André Michaux, "Travels of François André Michaux," in *Early Western Travels, 1748–1846*, vol. 3, ed. Reuben G. Thwaites (Cleveland: Arthur H. Clark, 1904), 199–201; William T. Hornaday, *Our Vanishing Wild Life: Its Extermination and Preservation* (New York: New York Zoological Society, 1913), 42–45; Arnow, *Seedtime on the Cumberland*, 220–21.

60. See, for example, J. Wayne Fears, "Let's Discuss Chestnuts," Whitetails Unlimited, accessed February 8, 2018, https://www.whitetailsunlimited.com/conservation/foodplotdoc/lets-discuss-chestnuts.phtml.

61. Foster and Cohen, "Palynological Evidence," 40, 41.

62. Ibid., 36, 47. See also H. Thomas Foster II, Bryan Black, and Marc D. Abrams, "A Witness Tree Analysis of the Effects of Native American Indians on the Pre-European Settlement Forests in East-Central Alabama," *Human Ecology* 32, no. 1 (February 2004): 37. In New England, fire had the opposite effect on chestnuts, as David Foster and colleagues demonstrated in a 2002 study. In cooler and more northerly latitudes, natural or human-set fires dried out excessively wet soils and opened up dense forest canopies. See David R. Foster et al., "Oak, Chestnut and Fire: Climate and Cultural Controls of Long-Term Forest Dynamics in New England, USA," *Journal of Biogeography* 29, nos. 10–11 (October 2002): 1373–74.

63. Edward K. Faison and David R. Foster, "Did American Chestnut Really Dominate the Eastern Forest?," *Arnoldia* 72, no. 2 (2014): 23.

64. Emily W. B. (Russell) Southgate, "Forest History of the Highlands," in *The Highlands: Critical Resources, Treasured Landscapes*, ed. Richard G. Lathrop Jr. (New Brunswick, N.J.: Rutgers University Press, 2011), 116.

65. Ibid., 117.

66. Melissa A. Thomas-Van Gundy and Michael P. Strager, *European Settlement-Era Vegetation of the Monongahela National Forest, West Virginia*, USDA Forest Service General Technical Report NRS-101 (Newtown Square, Pa.: U.S. Forest Service, Northern Research Station, 2012), 13.

67. Todd F. Hutchinson et al., "History of Forests and Land-Use," in *Characteristics of Mixed-Oak Ecosystems in Southern Ohio Prior to the Reintroduction of Fire*, U.S. Forest Service General Technical Report NE-299, ed. Elaine Kennedy Sutherland and Todd F. Hutchinson (Newtown Square, Pa.: U.S. Forest Service, 2003), 19–20.

68. Isaac Weld Jr., *Travels through the States of North America, and the Provinces of Upper and Lower Canada, during the Years 1795, 1796, and 1797*, 4th ed. (London: John Stockdale, 1800), 151, 204, 232, 295, 422.

69. John R. Stilgoe, *Common Landscape of America, 1580–1845* (New Haven, Conn.: Yale University Press, 1982), 189, 191–92; Terry G. Jordan and Matti Kaups, *The American Backwoods Frontier: An Ethnic and Ecological Interpretation* (Baltimore: Johns Hopkins University Press, 1989), 106–7; Eugene Cotton Mather and John Fraser Hart, "Fences and Farms," *Geographical Review* 44, no. 2 (April 1954): 201–23.

70. John R. Wennersten, *The Chesapeake: An Environmental Biography* (Baltimore: Maryland Historical Society, 2001), 15.

71. Hugh Jones, *The Present State of Virginia* (New York: Joseph Sabin, 1865), 39. In his explorations of New Jersey, New York, and Pennsylvania, Swedish naturalist Pehr Kalm observed that after "cedar-wood," chestnut was most common for fence construction. Kalm, *Travels into North America: Containing Its Natural History, and a Circumstantial Account of Its Plantations and Agriculture in General*, vol. 2 (London: T. Lowndes, 1771), 54.

72. Vanessa Patrick, "Partitioning the Landscape: Fences in Colonial Virginia," *Magazine Antiques* 154, no. 7 (July 1998): 100.

73. Thomas Anburey, *Travels through the Interior Parts of America in a Series of Letters*, vol. 2 (London: William Lane, 1789), 323–24. According to Anburey, New Englanders were fond of saying that when a man is overly intoxicated, "he is making Virginia fences" (324).

74. *Pennsylvania Gazette*, April 25, 1751, 2.

75. Edwin M. Betts, ed., *Thomas Jefferson's Garden Book* (Philadelphia: American Philosophical Society, 1944), 17; James A. Bear and Lucia Stanton, eds., *Jefferson's Memorandum Books: Accounts, with Legal Records and Miscellany, 1767–1826*, vol. 1 (Princeton: Princeton University Press, 1997), 459; Sara Bon-Harper, Fraser

Neiman, and Derek Wheeler, *Monticello's Park Cemetery*, Monticello Department of Archaeology Technical Report Series no. 5 (Charlottesville, Va.: Thomas Jefferson Foundation, 2003), 2.

76. On the initial settlement of the Southwest Mountains, see Edgar Woods, *Albemarle County in Virginia* (Charlottesville, Va.: Michie, 1901), 2–3, 19.

77. Philip Alexander Bruce, *Economic History of Virginia in the Seventeenth Century*, vol. 1 (New York: Macmillan, 1896), 317.

78. James Barbour, "On the Improvement of Agriculture, and the Importance of Legislative Aid on that Object: Description of the South West Mountain Lands," *Farmers' Register* 2, no. 11 (April 1835): 705.

79. *Genesee Farmer and Gardener's Journal* 3, no. 21 (May 25, 1833): 164. For a sampling of eighteenth-century fence laws, see Virgil Maxcy, ed., *The Laws of Maryland*, vol. 2 (Baltimore: Philip H. Nicklin, 1811), 103, 443; John Purdon, *A Digest of the Laws of Pennsylvania (1700–1836)* (Philadelphia: M'Carty & Davis, 1837), 426–27; Charles Z. Lincoln, William H. Johnson, and A. Judd Northrup, eds., *The Colonial Laws of New York from the Year 1664 to the Revolution*, vol. 2 (Albany, N.Y.: James B. Lyon, 1894), 65–67, 326–29.

80. Nora P. Small, *Beauty and Convenience: Architecture and Order in the New Republic* (Knoxville: University of Tennessee Press, 2003), 87. New England barn builders "preferred chestnut . . . at least into the 1840s" (ibid.). Chestnut wood was also used for covered bridges, as their mud sills needed to be weather resistant and long-lasting. See Eric Sloane, *American Barns and Covered Bridges* (New York: Funk and Wagnalls, 1954), 87; Joseph Monninger, *A Barn in New England: Making a Home on Three Acres* (San Francisco: Chronicle Books, 2001), 155.

81. Cornelius Weygandt, *A Passing America: Considerations of Things of Yesterday Fast Fading from Our World* (New York: Henry Holt, 1932), 186.

82. Randy Leffingwell, *The American Barn* (St. Paul, Minn.: Motorbooks International, 2003), 77–78; Bruce Irving, *New England Icons: Shaker Villages, Saltboxes, Stone Walls, and Steeples* (Woodstock, Vt.: Countryman, 2011), 31; Michael J. Till, *Along Massachusetts's Historic Route 20* (Charleston, S.C.: Arcadia, 2012), 125.

83. Heritage Restorations, "Old Scotch Barns in the New World," accessed April 24, 2017, https://www.heritagebarns.com/ blog/old-scotch-barns-in-the-new-world-a-wee-bit-different-of-a-barn. Professional barn restorers use chestnut as a way of dating construction. In New York, hemlock was the predominant wood for barn building until the beginning of the nineteenth century (ibid.).

84. Thomas Fessenden, "Forest Trees," *New England Farmer* 7, no. 18 (November 21, 1828): 138.

85. Benjamin Whitman et al., *History of Erie County, Pennsylvania*, vol. 1 (Chicago: Warner, Beers, 1884), 189. See also Solon J. Buck and Elizabeth H.

Buck, *The Planting of Civilization in Western Pennsylvania* (Pittsburgh, Pa.: University of Pittsburgh Press, 1939), 69.

86. David A. Clary, *George Washington's First War: His Early Military Adventures* (New York: Simon and Schuster, 2011), 47.

87. Jedidiah Morse, *The American Gazetteer* (Boston: Thomas and Andrews, 1797), n.p. According to a 1795 military report, the land surrounding Fort Presque Isle was cultivated or cleared "for a mile and a half for some directions." A century later, the area had "grown up thick with young chestnut and linden." See Whitman, *History of Erie County*, 206.

88. Jefferson memorandum to Samuel Clarkson, September 23, 1792, Manuscript Division, Alderman Library, University of Virginia. See also William M. Kelso, "The Archaeology of Slave Life at Thomas Jefferson's Monticello: 'A Wolf by the Ears,'" *Journal of New World Archaeology* 6, no. 4 (June 1986): 5–20.

89. Lucia C. Stanton, *"Those Who Labor for My Happiness:" Slavery at Jefferson's Monticello* (Charlottesville: University of Virginia Press, 2012), 56.

90. Bear and Stanton, *Jefferson's Memorandum Books*, 1:43.

91. Jack McLaughlin, *Jefferson and Monticello: The Biography of a Builder* (New York: Henry Holt, 1988), 170–71.

92. Marc Leepson, *Saving Monticello: The Levy's Family's Epic Quest to Rescue the House that Jefferson Built* (New York: Free Press, 2001), 121. See also Mesick/Cohen/Waite Architects, "Monticello Roof: Historic Structure Report" (Albany, N.Y.: Mesick/Cohen/Waite Architects, 1990), 5; Hunt McKinnon et al., *The Many Roofs of Monticello: A Case Study* (Raleigh: North Carolina State University College of Design, 2006), 1–37.

93. Quoted in Leepson, *Saving Monticello*, 122. See also James Barron, "Charlottesville Journal: Patching Up the Flaws in Jefferson's Legacy," *New York Times*, December 25, 1990, accessed February 6, 2018, https://www.nytimes.com/1990/12/25/us/charlottesville-journal-patching-up-the-flaws-in-jefferson-s-legacy.html.

94. Charles F. Hobson and Robert Rutland, eds., *Papers of James Madison, Congressional Series*, vol. 13 (Charlottesville: University Press of Virginia, 1981), 304; "James Madison to Mordecai Collins, Lewis Collins, and Sawney (Slave). Instructions." Library of Congress, James Madison Papers, Online collection, accessed May 1, 2018, https://www.loc.gov/item/mjm023773/.

95. James Madison, "Address to the Agriculture Society of Albemarle," *Niles' Weekly Register* 14, no. 359 (July 18, 1818): 354; James Madison, *Letters and Other Writings of James Madison*, vol. 3 (Philadelphia: J. B. Lippincott, 1865), 93.

96. Milton E. Ailes, "Our Farmer Presidents," *Frank Leslie's Popular Monthly* 56, no. 2 (June 1903): 146. William Dupont, "Topography of the Mountain Grounds and Vicinity at Montpelier, Orange County, Virginia, 1908," map in Montpelier Foundation Archives, Montpelier Station, Virginia; personal communication,

C. Thomas Chapman, executive projects manager, James Madison's Montpelier, to Catherine Mayes, Virginia Chapter president, American Chestnut Foundation, August 17, 2012.

97. Daniel L. Druckenbrod and Herman H. Shugart, "Forest History of James Madison's Montpelier Plantation," *Journal of the Torrey Botanical Society* 131, no. 3 (July–September 2004): 209.

98. Thomas Dierauf, *History of the Montpelier Landmark Forest: Human Disturbance and Forest Recovery* (Montpelier Station, Va.: Montpelier Foundation, 2010), 9–10.

99. Ibid.

100. Druckenbrod and Shugart, "Forest History," 215.

101. Gerald K. Kelso, "Palynology in Historical Rural-Landscape Studies: Great Meadows, Pennsylvania," *American Antiquity* 59, no. 2 (April 1994): 369.

102. Richard Moldenke, *Charcoal Iron* (Lime Rock, Conn.: Salisbury Iron Corporation, 1920), 8–32; Paul T. Gundrum, "The Charcoal Iron Industry in Eighteenth-Century America: An Expression of Regional Economic Variation" (PhD diss., University of Wisconsin, 1974), 55–59; John Bezis-Selfa, *Forging America: Ironworkers, Adventurers and the Industrious Revolution* (Ithaca, N.Y.: Cornell University Press, 2004), 27–42.

103. Alfred P. Muntz, "Forests and Iron: the Charcoal Iron Industry of the New Jersey Highlands," *Geografiska Annaler* 42, no. 4 (1960): 320–21; Michael P. Conzen, *The Making of the American Landscape* (New York: Routledge, 1994), 162–63; Robert B. Gordon, *A Landscape Transformed: The Ironmaking District of Salisbury, Connecticut* (New York: Oxford University Press, 2001), 4–7, 54–58.

104. Byrd, *Westover Manuscripts*, 127.

105. Johan David Schoepf, *Travels in the Confederation, 1783–1784* (Philadelphia: William J. Campbell, 1911), 37. See also John Rutherford, "Notes on the State of New Jersey, August 1786," in *Proceedings of the New Jersey Historical Society, Second Series*, vol. 1 (Newark: New Jersey Historical Society, 1867–69), 86.

106. Joshua D. Warfield, *The Founders of Arundel and Howard Counties, Maryland* (Baltimore: Kohn and Pollack, 1905), 502; Ruthella M. Bibbins, "The City of Baltimore, 1797–1850," in *Baltimore: Its History and People*, vol. 1, ed. Clayton C. Hall (New York: Lewis Historical Publishing, 1912), 73–74; Robin S. Doak, *Voices from Colonial America: Maryland, 1634–1776* (Washington, D.C.: National Geographic Society, 2007), 83.

107. H. P. Harris, "Charcoal Manufacture in the Salisbury Iron Region," *Journal of the United States Association of Charcoal Iron Workers* 6, no. 1 (February 1885): 50–51; Bernhard E. Fernow, "Shall Our Charcoal in Future Be Produced from Coppice Growth or Timber Forest?," *Journal of the United States Association of Charcoal Iron Workers* 6, no. 5 (October 1885): 277.

108. See, for example, Austin F. Hawes and Ralph Hawley, *Forest Survey of Litchfield and New Haven Counties, Connecticut*, Connecticut Agricultural Experiment Station Bulletin no. 162, Forestry Publication no. 5 (New Haven, Conn.: Connecticut Agricultural Experiment Station, 1909), 16; William W. Ashe, *Chestnut in Tennessee*, Tennessee Geological Survey Series Bulletin no. 10–B (Nashville, Tenn.: Baird-Ward, 1912), 23, 35.

109. Gordon G. Whitney, *From Coastal Wilderness to Fruited Plain: A History of Environmental Change in Temperate North America from 1500 to the Present* (Cambridge: Cambridge University Press, 1996), 225–26.

110. Brian M. Fagan, *The Little Ice Age: How Climate Made History, 1300–1850* (New York: Basic Books, 2000), 50, 96, 157; Dee C. Pederson et al., "Medieval Warming, Little Ice Age, and European Impact on the Environment during the Last Millennium in the Lower Hudson Valley, New York, USA," *Quaternary Research* 63, no. 3 (May 2005): 238–49; Lesley-Ann Dupigny-Giroux, "Backward Seasons, Droughts and Other Bioclimatic Indicators of Variability," in *Historical Climate Variability and Impacts in North America*, ed. Lesley-Ann Dupigny-Giroux and Cary J. Mock (Heidelberg: Springer, 2009), 231–50.

111. Landon Carter, *The Diary of Colonel Landon Carter of Sabine Hall, 1752–1778*, vol. 1, ed. Jack P. Greene (Charlottesville: University of Virginia Press, 1965), 382.

112. Jefferson is quoted in Barbour, "On the Improvement of Agriculture," 705. Barbour noted that one-fourth of the southwest mountains—110 miles in length and five miles wide—was covered with chestnut (ibid.).

113. Betts, *Thomas Jefferson's Garden Book*, 377; Peter J. Hatch, *The Fruit and Fruit Trees of Monticello* (Charlottesville: University of Virginia Press, 1998), 27.

114. Jefferson to Elisha Watkins, September 27, 1808, quoted in Betts, *Thomas Jefferson's Garden Book*, 327.

115. Betts, *Thomas Jefferson's Garden Book*, 327.

116. Evidence for chestnut decline as a result of colder temperatures is also found in the fossil pollen record. For fluctuating chestnut pollen counts in colonial New England, see James S. Clark, "Late-Holocene Vegetation and Coastal Processes at a Long Island Tidal Marsh," *Journal of Ecology* 74, no. 2 (June 1986): fig. 3; Robert E. Loeb, "Lake Pollen Records of the Past Century in Northern New Jersey and Southeastern New York, U.S.A.," *Palynology* 13 (1989): 3–19; David R. Foster and Tad M. Zebryk, "Long-Term Vegetation Dynamics and Disturbance History of a Tsuga-Dominated Forest in New England," *Ecology* 74, no. 4 (June 1993): 990, 992.

117. In the Appalachian highlands, peak distribution of chestnut occurred around 1200 AD. On one North Carolina mountaintop—Pisgah Ridge—the species occupied nearly 25 percent of the woodlands. By 1800, it represented only 5 percent of the forest. At Horse Cove Bog in Macon County, North Carolina,

peak distribution was longer, as the trees comprised 32 percent of the area a millennium ago and 25 percent in 1830. See David S. Shafer, "Flat Laurel Gap Bog, Pisgah Ridge, North Carolina: Late Holocene Development of a High-Elevation Heath Bald," *Castanea* 51, no. 1 (March 1986): 4; Elizabeth Martin, "Holocene Environmental History of Panthertown Valley in the Blue Ridge Mountains of North Carolina" (master's thesis, Western Carolina University, 2014), 21, 33.

118. On chestnut cold tolerance, see Paul G. Schaberg et al., "Is Nut Cold Tolerance a Limitation to the Restoration of American Chestnut in the Northeastern United States?," *Ecological Restoration* 27, no. 3 (September 2009): 266–68; Kendra M. Gurney et al., "Inadequate Cold Tolerance as a Possible Limitation to American Chestnut Restoration in the Northeastern United States," *Restoration Ecology* 19, no. 1 (January 2011): 55–63; Thomas Saielli et al., "Nut Cold Hardiness as a Factor Influencing the Restoration of American Chestnut in Northern Latitudes and High Elevations," *Canadian Journal of Forest Research* 42, no. 5 (2012): 849–57.

119. Roland M. Harper, *Forests of Alabama*, Geological Survey of Alabama, Monograph 10 (Wetumpka, Ala.: Wetumpka Printing Company, 1943), 25n1. In another study, Harper noted residents of northwest Alabama linked chestnut decline to "a late freeze in May 1854, which killed chestnut trees over considerable areas." Harper, *Economic Botany of Alabama: Catalogue of the Trees, Shrubs and Vines of Alabama, with Their Economic Properties and Local Distribution*, Geological Survey of Alabama, State Commission on Forestry (Birmingham, Ala.: Birmingham Printing Company, 1928), 109. Evidence for chestnuts in southern Alabama during the nineteenth century is found in a monograph by Charles T. Mohr, who wrote in 1901 that "the existence of the chestnut in the Upper Division of the Coast Pine Belt is at present only indicated by the large stumps, which have during the long periods of time resisted decay." Mohr, *Plant Life of Alabama* (Montgomery: Brown Printing Company, 1901), 14n1.

120. Mohr, *Plant Life of Alabama*, 14n1.

121. Navin Ramankutty and Jonathon A. Foley, "Estimating Historical Changes in Global Land Cover: Croplands from 1700 to 1992," *Global Biogeochemical Cycles* 13, no. 4 (December 1999): 997–1027.

122. Earle, *Evolution of a Tidewater Settlement System*, 34. In her study of accumulated ragweed pollen in streambeds, geologist Grace Brush estimated 40 percent of the Chesapeake Tidewater was cleared of standing timber by 1780. Grace S. Brush, "Forests before and after the Colonial Encounter," in *Discovering the Chesapeake: The History of an Ecosystem*, ed. Philip D. Curtin, Grace S. Brush, and George W. Fisher (Baltimore: Johns Hopkins University Press, 2001), 47. See also Wennersten, *Chesapeake*, 27–32.

123. *Pennsylvania Gazette* (Philadelphia), July 7, 1763, 1. A decade earlier, another Chestnut Hill property held sixty acres of woodlands, including "a number of chestnut trees, and other good timber" (*Pennsylvania Gazette* (Philadelphia), June 20, 1755, 1).

124. *Virginia Gazette*, November 7, 1777, A1.

125. William M. Kozlowski, "A Visit to Mount Vernon a Century Ago," *Century* 63, no. 4 (February 1902): 518. Scottish book publisher John Duncan observed two large Spanish chestnuts at Mount Vernon in 1818. The trees, he was told, "sprung from nuts planted by the General's own hand." John M. Duncan, *Travels through Part of the United States and Canada in 1818 and 1819*, vol. 1 (Glasgow, Scotland: Hurst, Robinson, 1823), 290.

126. John Clarke, *Land, Power, and Economics on the Frontier of Upper Canada* (Montreal: McGill-Queens University Press, 2001), 24. In Essex County, American chestnut was the tenth most common tree species, ranking above American sycamore, red maple, tuliptree, and black walnut in frequency (ibid.). See also Jeffrey R. Tindall et al., "Ecological Status of American Chestnut (*Castanea dentata*) in its Native Range in Canada," *Canadian Journal of Forest Research* 34, no. 12 (2004): 2554–63; Michael Henry and Peter Quinby, *Ontario's Old-Growth Forests* (Markham, Ontario: Fitzhenry and Whiteside, 2010), 47–48.

127. Clarke, *Land, Power, and Economics*, 22–26.

128. Gajewski, Swain, and Peterson, "Late Holocene Pollen Stratigraphy," 380–81.

129. Edwin D. Sanborn, *History of New Hampshire, from its First Discovery to the Year 1830* (Manchester, N.H.: John B. Clarke, 1875), 404; Philip L. Buttrick, "Commercial Uses of Chestnut," *American Forestry* 21, no. 262 (October 1915): 962; Julia Lichtblau, "Sterling Heights: A House and its Family," *Old-House Journal* 12, no. 8 (October 1984): 167–71. The home of my former professor Lewis Sumberg—built in 1755 in Wayland, Massachusetts—contained considerable chestnut timber, including framing beams and joists (personal communication, April 7, 2013).

130. Barry A. Greenlaw, *New England Furniture at Williamsburg* (Williamsburg, Va.: Colonial Williamsburg Foundation, 1974), 67, 119, 131–32, 159; John Kassay, *The Book of American Windsor Furniture: Styles and Technologies* (Amherst: University of Massachusetts Press, 1998), 9, 32, 56, 64, 69, 77, 91, 127; Frances G. Safford, *American Furniture in the Metropolitan Museum of Art*, vol. 1 (New Haven, Conn.: Yale University Press, 2007), 142, 199, 212, 233, 260, 330.

131. Nancy G. Evans, *Windsor Chair-Making in America: From Craft Shop to Consumer* (Hanover, N.H.: University Press of New England, 2006), 89.

132. Ebenezer Tracy Sr., Estate Records, New London County, Lisbon, Connecticut, 1803.

133. *Pennsylvania Evening Post*, May 7, 1776, 230.

134. It is impossible to cite all news articles and advertisements related to chestnut sales during the nineteenth century. A broad sample—representing numerous cities and different geographic regions—is found in the endnotes to chapter 5.

CHAPTER 5. Cash Will Be Paid If Delivered Soon

1. United States Census Bureau, "1810 Fast Facts: 10 Largest Urban Places," accessed April 22, 2017, http://www.census.gov/history/www/through_the_decades/fast_facts/1810_fast_facts.html.

2. Advertised imported foods included salmon from Canada, figs from Turkey, raisins from Spain, olives from Italy, sugar from the Caribbean, spices from the Near East, and wine, brandy, and spirits from several European countries. See, for example, *Wilmington (N.C.) Gazette*, April 1, 1806; *New York Gazette*, January 25, 1819, 1; *Essex Register* (Salem, Mass.), January 30, 1819, 1; *Connecticut Courant* (Hartford), November 2, 1819, 4; *Providence (R.I.) Patriot*, December 19, 1821, 3. On food imports in the late eighteenth century, see David Klingaman, "Food Surpluses and Deficits in the American Colonies, 1768–1772," *Journal of Economic History* 31, no. 3 (September 1971): 553–69.

3. For discussions on forest clearance in early nineteenth-century America, see Henry M. Miller, "Transforming a 'Splendid and Delightsome Land': Colonists and Ecological Change in the Chesapeake 1607–1820," *Journal of the Washington Academy of Sciences* 76, no. 3 (September 1986): 173–87; Ted Steinberg, *Down to Earth: Nature's Role in American History* (New York: Oxford University Press, 2002), 45–61; Michael Williams, *Deforesting the Earth: From Prehistory to Global Crisis* (Chicago: University of Chicago Press, 2003), 222–30, 301–15; Brian Donahue, "Another Look from Sanderson's Farm: A Perspective on New England Environmental History and Conservation," *Environmental History* 12, no. 1 (January 2007): 18.

4. *Connecticut Courant* (Hartford), January 11, 1796, 1.

5. *Centinel of Freedom* (Newark, N.J.), October 16, 1798, 3.

6. *Providence (R.I.) Phoenix*, October 25, 1806, 4.

7. François André Michaux, *Histoire des Arbres Forestiers de l'Amérique Septentrionale*, vol. 2 (Paris: L. Haussmann, 1812), 159. The 1819 English edition of the volume puts the price of chestnuts at three dollars per bushel. F. André Michaux, *The North American Sylva*, vol. 3 (Philadelphia: Thomas Dobson, 1819), 11.

8. *Nantucket Gazette*, May 20, 1816, 1.

9. *New York Gazette & General Advertiser*, December 22, 1818, 3.

10. *Connecticut Courant* (Hartford), October 27, 1818, 1.

11. *American Advocate* (Hallowell, Maine), November 27, 1819, 3.

12. *Daily National Intelligencer* (Washington, D.C.), April 24, 1820, 1; *New York Gazette & General Advertiser*, January 2, 1821, 1; *Providence (R.I.) Patriot*, December 19, 1821, 3; *Newport (R.I.) Mercury*, March 9, 1822, 4; *Hallowell (*Maine) *Gazette*, November 26, 1823, 3.

13. See, for example, *New York Daily Advertiser*, August 17, 1819, 1. The original ad appeared May 18 and lists Spanish chestnuts among the items sold at the Broadway marketplace. According to the nineteenth-century British horticulturalist John Claudius Loudon, chestnuts imported from Spain or Italy were kiln-dried "to prevent germination on their passage." Loudon, *Arboretum et Fruticetum Brittannicum, or The Trees and Shrubs of Britain*, vol. 3 (London: Longman, Orme, Brown, Green, and Longmans, 1838), 1990.

14. The farm and house were located on the old Bloomfield Road, "2 1/2 miles from Newark and 1 1/2 miles from Belleville and the Passaic River." *Evening Post* (New York, N.Y.), March 11, 1815, 4.

15. Benjamin Hawkins, *Letters of Benjamin Hawkins, 1796–1806*, ed. Stephen B. Weeks (Savannah: Georgia Historical Society, 1916), 18. The letter is dated November 28, 1796. The 340-mile roundtrip journey took seventeen days and the rucksack filled with chestnuts likely weighed sixty-five pounds. The Pine Log settlement, added Hawkins, was situated "on the richest lands I have seen, . . . the growth [of] poplar and chestnut very large without any undergrowth" (ibid.).

16. Lewis Eldon Atherton, *The Southern Country Store, 1800–1860* (Westport, Conn.: Greenwood, 1968), 90. Atherton's observations were taken from barter and day books recorded by Hogg between 1803 and 1808 (92).

17. On the importance of roads and other transportation networks on the early American economy, see Paul A. Groves, "The Northeast and Regional Integration, 1800–1860," in *North America: The Historical Geography of a Changing Continent*, 2nd ed., ed. Thomas F. McIlwraith and Edward K. Muller (Lanham, Md.: Rowman & Littlefield, 2001), 189–206.

18. See, for example, George Washington to Dr. William Thornton, October 7, 1797, in *The Writings of George Washington from the Original Manuscript Sources, 1745–1799*, vol. 36, ed. John C. Fitzpatrick (Washington, D.C.: GPO, 1941), 39n21.

19. Loudon, *Arboretum et Fruticetum Brittannicum*, 1984. Regarding the marrons of France, Loudon added that they were "much preferred, being larger, more farinaceous, and sweeter" (ibid.). See also Michaux, *Histoire des Arbres Forestiers*, 163–64.

20. Madame de Tessé to Thomas Jefferson, December 8, 1811, in *The Papers of Thomas Jefferson, Retirement Series*, vol. 4, ed. Susan H. Perdue and Robert F. Haggard (Princeton, N.J.: Princeton University Press, 2008), 322–23. Thomas Jefferson to Madame de Tessé, December 8, 1813, in *The Papers of Thomas Jefferson, Retirement Series*, vol. 7, ed. Perdue and Haggard (Princeton, N.J.: Princeton University Press, 2010), 34.

21. Thomas Jefferson to John Patton Emmett, April 27, 1826, in *The Writings of Thomas Jefferson, Monticello Edition*, vol. 16, ed. Andrew A. Lipscomb (Washington, D.C.: Thomas Jefferson Memorial Association, 1904), 167.

22. Correspondence from M. F. Wheeler to editor John S. Skinner, printed as "French Chestnuts," *American Farmer* 9, no. 34 (November 9, 1827): 272. For an account of the importance of French and Italian marrons in European horticulture, including their impact on village life, see James Villas, *Villas at Table: A Passion for Food and Drink* (New York: Harper & Row, 1988), 245–50; Vieri Bufalari, Andrea Semplici, and Massimo Ricciolini, "The Marron from Mugello," in *Journey through Tuscany to Discover Typical Products*, ed. Giovanni Breschi (Florence: Giunti Gruppo Editoriale, 2001), 136–45.

23. Wheeler to Skinner, "French Chestnuts," 272.

24. Thomas Jefferson, *Notes on the State of Virginia* (London: John Stockdale, 1787), 107.

25. On the impact of Buffon's theory of natural degeneracy on Jefferson and other early American thinkers, see Lee A. Dugatkin, *Mr. Jefferson and the Giant Moose: Natural History in Early America* (Chicago: University of Chicago Press, 2009), ix–xii, 1–30, 42–69, 108–14, 153–60; James D. Drake, *The Nation's Nature: How Continental Presumptions Gave Rise to the United States of America* (Charlottesville: University of Virginia Press, 2011), 18–20, 58–60, 231–42, 286, 346.

26. Philip J. Pauly, *Fruits and Plains: The Horticultural Transformation of America* (Cambridge, Mass.: Harvard University Press, 2008), 11.

27. Although Dupont supposedly grew chestnuts the year he moved to the Powder Mill estate, correspondence between him and a neighbor reveals this was not the case. According to Dupont, all chestnuts sent from France were spoiled the first year. See George H. Powell, *The European and Japanese Chestnuts in the Eastern United States*, Delaware College Agricultural Experiment Station Bulletin no. 42 (Newark, Del.: Delaware College Agricultural Experiment Station, 1898), 3.

28. Jefferson, *Notes on the State of Virginia*, 48–50.

29. Frederick Ungeheuer, Lewis Hurlbut, and Ethel Hurlbut, *Roxbury Remembered* (Oxford, Conn.: Connecticut Heritage Press, 1989), 45; Eldred N. Woodcock, *Fifty Years a Hunter and Trapper: Experiences and Observations of E. N. Woodcock*, ed. A. R. Harding (St. Louis, Mo.: A. R. Harding, 1941), 107.

30. Campbell Gibson, "Population of the 100 Largest Cities and Other Urban Places in the United States: 1790–1990," U.S. Census Bureau, Population Division Working Paper no. 27 (June 1998), table 6: Population of the 90 Urban Places, accessed May 2, 2018, https://www2.census.gov/library/working-papers/1998/demo/pop-twps0027/tab06.txt.

31. Lewis Cecil Gray, *History of Agriculture in the Southern United States to 1860*, vol. 2 (Gloucester, Mass.: Peter Smith, 1958), 884; Tyrel G. Moore, "Economic Development in Appalachian Kentucky, 1800–1860," in *Appalachian Frontiers:*

Settlement, Society and Development in the Preindustrial Era, ed. Robert D. Mitchell (Lexington: University Press of Kentucky, 1991), 223; Mary B. Pudup, "Town and Country in the Transformation of Appalachian Kentucky," in *Appalachia in the Making: The Mountain South in the Nineteenth Century*, ed. Mary B. Pudup, Dwight B. Billings, and Altina L. Waller (Chapel Hill: University of North Carolina Press, 1995), 280.

32. *New-Hampshire Patriot and State Gazette* (Concord), October 17, 1831, 3.

33. *Independent Inquirer* (Brattleboro, Vt.), October 19, 1833, 3.

34. *Vermont Phoenix* (Brattleboro), October 15, 1839, 3.

35. In 1830, the price for chestnuts in the Boston Vegetable Market was $1.75 per bushel. *Hampshire (Mass.) Gazette*, November 24, 1830, 3. However, that same year the treasurer of the American Baptist Foreign Mission Society recorded a five-dollar gift from a Mr. Jonathon Whitney of Conway, Massachusetts, "received from the proceeds of four bushels of chestnuts." Heman Lincoln, "Treasurer's Report," in *The American Baptist Magazine*, vol. 11, ed. Board of Managers of the Baptist General Convention (Boston: Lincoln & Edmands, 1831), 189.

36. Mollie Ellen Bowles, "Interactions between *Phytophthora* Species and *Castanea* species and the Creation of a Genetic Linkage Map for the F1 Parent in a First-Generation Backcross Family of *Castanea* species" (master's thesis, North Carolina State University, 2006), 3; Yilmaz Balci and John C. Bienapfl, "*Phytophthora* in U.S. Forests," in *Phytophthora: A Global Perspective*, ed. Kurt Lamour (Oxfordshire, UK: CAB International, 2013), 135–45.

37. Adrienne R. Hardham, "*Phytophthora cinnamomi*," *Molecular Plant Pathology* 6, no. 6 (November 2005): 589–604.

38. George A. Zentmyer, "Origin of *Phytophthora cinnamomi*: Evidence That It Is Not an Indigenous Fungus in the Americas," *Phytopathology* 67, no. 11 (November 1977): 1373–77. See also Francine Govers, "Foreword," in *Oomycete Genetics and Genomics: Diversity, Interactions and Research Tools*, ed. Kurt Lamour and Sophien Kamoun (Hoboken, N.J.: John Wiley & Sons, 2009), ix–xi.

39. John T. Kliejunas et al., *Review of Literature on Climate Change and Forest Diseases of Western North America*, USDA General Technical Report PSW-GTR-225 (Albany, Calif.: USDA, U.S. Forest Service, Pacific Southwest Research Station, 2009), 14–15. See also Thomas Jung et al., "The Impact of Invasive *Phytophthora* Species on European Forests," in Lamour, *Phytophthora: A Global Perspective*, 146–58.

40. Dorothy Shaw, E. Gwenda Cartledge, and D. Jean Stamps, "First Records of *Phytophthora Cinnamomi* in Papua, New Guinea," *Papua New Guinea Agricultural Journal* 23, nos. 1–2 (1972): 46; Marie-Laure Desprez-Loustau, "Alien Fungi of Europe," in Delivering Alien Invasive Species Inventories for Europe (DAISIE), *Handbook of Alien Species in Europe*, vol. 3, *Invading Nature* (Sao Paulo, Brazil: Springer, 2009), 20.

41. Isabel Azcárate Luxán, *Plagas agrícolas y forestales en España en los siglos XVIII y XIX* (Madrid: Ministerio de Agricultura, Pesca y Alimentación, Secretaría General Técnica, 1996), 338–39.

42. Antoinette Fauve-Chamoux, "Chestnuts," in *The Cambridge World History of Food*, vol. 1, ed. Kenneth F. Kiple and Kriemhild C. Ornelas (Cambridge: Cambridge University Press, 2000), 359–63; Andrea Vannini and Anna M Vettraino, "Ink Disease in Chestnuts: Impact on the European Chestnut," *Forest Snow and Landscape Research* 76, no. 3 (January 2001): 345–46.

43. Clive M. Brasier, "The Role of *Phytophthora* Pathogens in Forests and Semi-natural Communities in Europe and Africa," in *Phytophthora Diseases of Forest Trees*, ed. Everett M. Hansen and Wendy Sutton (Corvallis, Ore.: Forest Research Laboratory, Oregon State University, 2000), 10. See also A. Ruth Finlay and A. R. McCracken, "Microbial Suppression of *Phytophthora cinnamomi*," in *Phytophthora: Symposium of the British Mycological Society*, ed. John A. Lucas et al. (Cambridge: Cambridge University Press, 1991), 384–98; K. L. McCarren et al., "The Role of Chlamydosphores of *Phytophthora cinnamomi*: A Review," *Australasian Plant Pathology* 34, no. 3 (September 2005): 333–38; Courtney Reuter, "*Phytophthora cinnamomi* Rands," pathogen profile, North Carolina State University, Department of Plant Pathology, accessed May 21, 2018, http://www.cals.ncsu.edu/course/pp728/cinnamomi/p_cinnamomi.htm.

44. Donald C. Erwin and Olaf K. Ribeiro, *Phytophthora Diseases Worldwide* (St. Paul, Minn.: American Phytopathological Society, 1996), 59, 272, 406; Kliejunas et al., *Review of Literature on Climate Change*, 15; Balci and Bienapfl, "*Phytophthora* in U.S. Forests," 135. A study by D. Michael Benson found the reproductive spores of the disease are inactivated if subsurface soil temperatures fall below freezing for several consecutive days. See Benson, "Cold Inactivation of *Phytophthora cinnamomi*," *Phytopathology* 72, no. 5 (May 1982): 561.

45. Barbara Wilson, Janine Kinloch, and Marnie Swinburn, "Distribution and Impacts of *Phytophthora cinnamomi*," in *Biodiversity Values and Threatening Processes of the Gnangara Groundwater System*, ed. Barbara A. Wilson and Leonie E. Valentine (Perth, W.A.: Gnangara Sustainability Strategy Taskforce, 2009), 14.

46. Bowen S. Crandall, G. F. Gravatt, and Margaret M. Ryan, "Root Disease of *Castanea* Species and Some Coniferous and Broadleaf Nursery Stocks, Caused by *Phytophthora Cinnamomi*," *Phytopathology* 35 (1945): 163. See also George A. Zentmyer, "Origin of *Phytophthora Cinnamomi*," *California Avocado Society Yearbook* 60 (1976), 154–56; Everett M. Hansen, "*Phytophthora* in the Americas," in *Phytophthora Diseases of Forest Trees: Proceedings from the First International Meeting on Phytophthora in Forest and Wild Ecosystems*, ed. Everett M. Hansen and Wendy Sutton (Corvallis: Oregon State University, 2000), 24.

47. William L. Jones, "On the Death and Disappearance of Some Trees and Shrubs," *American Journal of Science and Arts* 1, no. 5 (May 1846): 450.

48. Ibid. Jones also observed moisture-tolerant pondspice (*Litsea aestivalis*) and loblolly bay (*Gordonia lasianthus*) dying in the same vicinity, suggesting the die-off was not caused by flooding (ibid.). See also Leonard Holmes Healey, *Forty-Fifth Annual Report of the Secretary of the Connecticut Board of Agriculture, 1912* (Hartford: Connecticut Board of Agriculture, 1913), 408–9.

49. Jones, "On the Death and Disappearance," 450.

50. See, for example, Philip L. Buttrick, "Chestnut in North Carolina," in *Chestnut and the Chestnut Blight in North Carolina*, North Carolina Geologic and Economic Survey Economic Paper no. 56 (Raleigh: North Carolina Geological and Economic Survey, 1925), 7; G. F. Gravatt, "The Chestnut Blight in North Carolina," in *Chestnut and the Chestnut Blight*, 15.

51. Franklin B. Hough, *Report upon Forestry*, vol. 1, USDA, U.S. Forest Service (Washington, D.C.: GPO, 1878), 470.

52. George F. Gravatt, *The Chestnut Blight in Virginia*, (Richmond: Virginia Department of Agriculture and Immigration, 1914), 3.

53. James W. Holland, "The Beginning of Public Agricultural Experimentation in America: The Trustees' Garden in Georgia," *Agricultural History* 12, no. 3 (July 1938): 278; Mart A. Stewart, *"What Nature Suffers to Groe": Life, Labor, and Landscape on the Georgia Coast, 1680–1920* (Athens: University of Georgia Press, 2002), 46–47, 60–61; David L. Cowan, "Trustee Garden," *New Georgia Encyclopedia*, Georgia Humanities Council, accessed May 22, 2018, https://www.georgiaencyclopedia.org/articles/history-archaeology/trustee-garden.

54. In a September 6, 1765, diary entry, botanist John Bartram claimed chestnut and other "northward trees" were not found below New Ebenezer, a town seventy miles north of Savannah. John Bartram and Frances Harper, "Diary of a Journey through the Carolinas, Georgia, and Florida from July 1, 1765 to April 10, 1766," *Transactions of the American Philosophical Society* 33, no. 1 (December 1942): 24.

55. Frederick Watts, *Report of the Commissioner of Agriculture for the Year 1875* (Washington, D.C.: GPO, 1875), 271. Georgia counties identified as having large stands of chestnut with no signs of disease or die-off include Bartow, Dade, Dawson, Forsyth, and Rabun.

56. Ibid., 272.

57. Ibid.

58. Ibid., 273.

59. Ibid., 275.

60. Witness-tree surveys of the area were compiled from early land lottery maps housed in the Georgia State archives. The maps show concentrations of chestnut trees in the lower Piedmont, particularly along the borders of Talbot, Upson, and Taylor Counties. University of Georgia botanist Gayther L. Plummer found few chestnuts in the Georgia coastal plain, although his data set only included Wayne, Appling, Early, and Irwin Counties. Plummer, "18th Century

Forests in Georgia," *Bulletin of the Georgia Academy of Sciences* 33, no. 1 (1975): 10, 12. See also C. Mark Cowell, "Presettlement Piedmont Forests: Patterns of Composition and Disturbance in Central Georgia," *Annals of the Association of American Geographers* 85, no. 1 (March 1995): 71; Mark Davis, "Key to Reviving Georgia's Chestnut Trees May Lie in the Past," *Atlanta Journal-Constitution*, Wednesday, February 11, 2009, A1.

61. Roland M. Harper, *Economic Botany of Alabama*, part 1, *Geographical Report* (Montgomery: Brown Printing Company, 1913), 109.

62. Hilgard is quoted in Andrew D. Hopkins, "Relation of Insects to the Death of Chestnut Trees," *American Forestry* 18, no. 4 (April 1912): 221.

63. Caroline Rumbold, "A New Record of a Chestnut-Tree Disease in Mississippi," *Science* 34, no. 887 (December 29, 1911): 917.

64. When American chestnuts were killed by late freezes, some trees continued sprouting from their stumps. However, this was not the case with mortality due to ink disease. As plant pathologist Sandra Anagnostakis noted, "trees killed by the [*Phytophthora*] organism do not sprout and never recover." See Sandra L. Anagnostakis, "Chestnut Breeding in the United States for Disease and Insect Resistance," *Plant Disease* 96, no. 10 (October 2012): 1395.

65. Watts, *Report of the Commissioner of Agriculture*, 272.

66. Ibid. The conversion of forests into cotton plantations required the cutting of tens of thousands of chestnut trees. However, in the drier soils of the foothills, southern pine species often replaced chestnut after the forest was cut down. In Walton County, commissioner Watts observed that after the chestnuts were felled, the land took on "a heavy growth of old-field pine" (272).

67. By 1760, Charleston was awash with gardeners, nurserymen, and seed dealers, all of whom sold nonnative trees, plants, and shrubs to prospective buyers. A comprehensive summary of those in the nursery trade is found in Barbara W. Sarudy, "Garden History: Seed Dealers & Nursery Owners of South Carolina," accessed June 7, 2018, http://americangardenhistory.blogspot.com/2013/04/garden-history-seed-dealers-nursery.html.

68. Harry Hammond, *South Carolina: Resources and Population, Institutions and Industry* (Charleston: State Board of Agriculture of South Carolina, 1883), 146.

69. Benjamin L. C. Wailes, *Report on the Agriculture and Geology of Mississippi: Embracing a Sketch of the Social and Natural History of the State* (Jackson, Miss.: E. Barksdale, State Printer, 1854), 354. See also John L. Campbell and William H. Ruffner, *A Physical Survey Extending from Atlanta, Ga., across Alabama and Mississippi to the Mississippi River* (New York: E. F. Weeks, 1883), 92, 93.

70. The most likely culprit was late-May freezes, and not *Phytophthora*, given the suddenness and pervasiveness of the die-off. In 1856, the editorial pages of

Andrew Jackson Downing's *The Horticulturist* noted that "all chestnut-trees throughout Rockingham County, N.C., and the surrounding counties, have died this season," *Horticulturist* 6, no. 2 (February 1856), 97.

71. Crandall, Gravatt, and Ryan, "Root Disease of *Castanea* Species," 162–80. For climate conditions along the coastal plain of North Carolina and Virginia during the nineteenth century, see Jason T. Ortegren, "Tree-Ring Based Reconstruction of Multi-Year Summer Droughts in Piedmont and Coastal Plain Climate Divisions of the Southern U.S., 1690–2006" (PhD diss., University of North Carolina at Greensboro, 2008), 59–63; Kathryn Perkins Wolff, "A 1,461-Year Growing Season Precipitation Reconstruction for the Carolina Coastal Plain" (master's thesis, University of Arkansas, 2012), 13–22.

72. "American Chesnut [*sic*]," *American Magazine of Useful and Entertaining Knowledge* 1, no. 11 (November 1834): 102. The writer also appears to have plagiarized Daniel Jay Browne's description of the American chestnut in *The Sylva Americana*, published in 1832. See Browne, *The Sylva Americana, or A Description of the Forest Trees Indigenous to the United States* (Boston: William Hyde, 1832), 131–32.

73. See "On Planting Chestnuts: When and How Done?," *American Farmer* 18, no. 7 (July 29, 1825): 149; Ira Hopkins, "Experiments in Planting Chestnuts for Fencing Timber," *American Farmer* 8, no. 21 (August 11, 1826): 162; D. T., "Chestnut Timber," *Genesee Farmer and Gardener's Journal* 3 (May 25, 1833): 164; Henry Colman, "Tree Planting," *New England Farmer and Horticultural Register* 18, no. 10 (September 11, 1839): 88.

74. Andrew Jackson Downing, "Large Chestnuts," *Horticulturist and Journal of Rural Art and Rural Taste* 2 (November 1847): 240 (italics in original).

75. Theodore K. Long, *Forty Letters to Carson Long, 1904–1910* (New Bloomfield, Pa.: Carson Long Institute, 1931), 72.

76. William Walker, "Chestnuts Successful," *Prairie Farmer* 8, no. 8 (August 1848): 249.

77. Ibid.

78. Ibid (parenthesis in original).

79. Noah Webster, *An American Dictionary of the English Language*, vol. 1 (New York: S. Converse, 1828), s.v. "chestnut-tree."

80. Michaux, *North American Sylva* (1819), 3:11.

81. Arthur H. Graves, "The Future of the Chestnut Tree in North America," *Popular Science Monthly* 84, no. 6 (June 1914): 553; Donald C. Peattie, *A Natural History of Trees of Eastern and Central North America* (Boston: Houghton Mifflin, 1991), 189–90; Robert A. Mohlenbrock and Paul M. Thomson, *Flowering Plants: Smartweeds to Hazelnuts*, Illustrated Flora of Illinois (Carbondale: Southern Illinois University Press, 2009), 118–19.

82. See, for example, William Lord, "A Fenced-In Forest," in *Mighty Giants: An American Chestnut Anthology*, ed. Chris Bolgiano and Glenn Novak (Bennington, Vt.: American Chestnut Foundation, 2007), 131; George C. Kingston, *James Madison Hood: Lincoln's Consul to the Court of Siam* (Jefferson, N.C.: McFarland, 2013), 105.

83. J. McCan Davis, "Origin of the Lincoln Rail: As Related by Governor Oglesby," *Century Illustrated Monthly Magazine* 60 (n.s. 38) (June 1900): 271–75; Jane Martin Johns, *Personal Recollections of Early Decatur, Abraham Lincoln, Richard J. Oglesby, and the Civil War* (Decatur, Ill.: Decatur Chapter, Daughters of the American Revolution, 1912), 80–81; Mark A. Plummer, *Lincoln's Rail-Splitter: Governor Richard J. Oglesby* (Urbana: University of Illinois Press, 2001), 41.

84. According to his former law partner William Herndon, the young Lincoln "made a frontiersman's living by hard work, poling a flat-boat, getting out cedar and chestnut rails, even sawing wood. The scene of his early struggles was Indiana." George A. Townsend, *The Real Life of Abraham Lincoln: A Talk with Mr. Herndon, His Late Law Partner* (New York: Bible House, 1867), 6.

85. T. K. Burrill, *Ninth Report of the Board of Trustees of the Illinois Industrial University* (Urbana: Illinois Industrial University, 1878), 268.

86. In 1900, a chestnut in the same grove measured "four-and-a-half or five feet in diameter." William Trelease, "The Chestnut in Illinois," in *Transactions of the Illinois Academy of Science*, vol. 10, *1917*, ed. J. L. Pricer (Springfield: State of Illinois, 1917), 143.

87. Burrill, *Ninth Report*, 268. See also Trelease, "Chestnut in Illinois," 144; Mohlenbrock and Thomson, *Flowering Plants*, 11; Peattie, *Natural History of Trees*, 189; John E. Schwegman, *The Natural Heritage of Illinois: Essays on Its Lands, Waters, Flora, and Fauna* (Carbondale: Southern Illinois University Press, 2016), 217.

88. Dennis quoted in "Maryland Farming," *Dairy Farmer* 1, no. 11 (November 1860): 216.

89. Ted Olsen, *Blue Ridge Folklife* (Jackson: University of Mississippi Press, 1998), 145–46; Anne M. Whisnant, *Super-Scenic Motorway: A Blue Ridge Parkway History* (Chapel Hill: University of North Carolina Press, 2006), 2, 7, 221.

90. Martin Van Buren, *Message from the President of the United States to the Two Houses of Congress at the Commencement of the Third Session of the Twenty-Fifth Congress* (Washington, D.C.: Thomas Allen, 1838), 258.

91. Ibid. For the important role picket fences played in the architectural development of the United States, see Richard L. Bushman, *The Refinement of America: Persons, Houses, Cities* (New York: Vintage, 1993), 133–35, 256–60, 471; James R. Cothran, *Gardens and Historic Plants of the Antebellum South* (Columbia: University of South Carolina Press, 2003), 47, 65–69, 72, 90.

92. Elias Nason, *A Gazetteer of the State of Massachusetts* (Boston: B. B. Russell, 1874), 17, 24, 55, 292, 523; Henry David Thoreau, *The Journal: 1837–1861*, ed. Damion

Searls (New York: New York Review Books, 2009), 331, 495; David R. Foster, *Thoreau's Country: Journey through a Transformed Landscape* (Cambridge, Mass.: Harvard University Press, 2001), 177.

93. Henry David Thoreau, *The Writings of Henry David Thoreau*, vol. 20, *Journal XIV: August 1, 1860–November 3, 1861*, ed. Bradford Torrey (Boston: Houghton Mifflin, 1906), 137.

94. Thoreau quoted in Leo Stoller, *After Walden: Thoreau's Changing Views on Economic Man* (Stanford, Calif.: Stanford University Press, 1957), 86.

95. See, for example, Foster, *Thoreau's Country*, 176.

96. Henry David Thoreau, *Faith in a Seed: The Dispersion of Seeds and Other Late Natural History Writings*, ed. Bradley P. Dean (Washington, D.C.: Island Press, 1993), 114, 126–31, 143, 147, 260; Henry David Thoreau, *Wild Fruits: Thoreau's Rediscovered Last Manuscript*, ed. Bradley P. Dean (New York: W. W. Norton, 2000), 209–16, 256.

CHAPTER 6. Placed There by a Quadruped or Bird

1. Henry David Thoreau, *Walden, or Life in the Woods* (Boston: Ticknor and Fields, 1854), 257.

2. Ibid.

3. Ibid., 256. A detailed description of the Walden Pond environs, with references to Thoreau's own writings and letters, is found in William B. Maynard, *Walden Pond: A History* (New York: Oxford University Press, 2004), 1–28.

4. Jeffrey S. Cramer, ed., *Walden: A Fully Annotated Edition* (New Haven, Conn.: Yale University Press, 2004), 228. Before the first heavy frost, chestnut burrs required prying open with the soles of shoes, gloved hands, or wooden sticks. It is thus not surprising that Thoreau mentions, in several journal entries, the impact of cold evenings on his chestnut-gathering activities. See, for example, Thoreau, *Writings of Henry David Thoreau*, vol. 16, *Journal X: August 8, 1857–June 29, 1858*, ed. Bradford Torrey (Boston: Houghton Mifflin, 1906), 121–22. All subsequently cited volumes of *The Writings of Henry David Thoreau* were also edited by Torrey and published by Houghton Mifflin in 1906.

5. Thoreau, *Walden, or Life in the Woods*, 257. Although Thoreau lived and wrote at the cabin for a two-year period beginning in 1845, portions of the *Walden* manuscript were written during the early 1850s, long after he had left his life of solitude for Concord proper.

6. Thoreau, *Writings of Henry David Thoreau*, vol. 20, *Journal XIV: August 1, 1860–November 3, 1861*, 137. In the same entry, Thoreau wrote, "the trees which *with us* grow in masses, i.e., not merely scattering, are:— 1, 2., White and pitch pine; 3. Oaks; 4. White birch; 5. Red maple; 6. Chestnut; 7. Hickory," followed by other species also listed in item 7 (134). The list of species appears to be in descending

order of frequency, as Thoreau adds, "of these only white and pitch pine, oaks, white birch, and red maple are *now* both important and abundant. (Chestnut and hickory have become rare.)" Italics and parenthesis in original.

7. Thoreau, *Walden, or Life in the Woods*, 256. Ralph Waldo Emerson, Thoreau's friend and neighbor, was also aware the Lincoln chestnut trees were being felled for railroad construction. See Emerson, *The Journals and Miscellaneous Notebooks of Ralph Waldo Emerson, vol. 7, 1838–1842*, ed. A. W. Plumstead and Harrison Hayford (Cambridge, Mass.: The Belknap Press of Harvard University Press, 1969), 482.

8. Thoreau's last mention of chestnuts occurs in the journal entry dated December 2, 1860, a day before he contracted the bronchitis that, combined with his tuberculosis, confined him at home until his death in May 1862. On that day, Thoreau observed a large chestnut stump with several large coppice sprouts. The sprouts had been cut three years earlier, allowing him to count forty-two growth rings, an indication the parent tree was harvested in 1815. Thoreau, *Writings of Henry David Thoreau*, vol. 20, *Journal XIV: August 1, 1860–November 3, 1861*, 290.

9. Thoreau, *Writings of Henry David Thoreau*, vol. 10, *Journal IV: March 1, 1852–February 27, 1853*, 434–35; Thoreau, *Writings of Henry David Thoreau*, vol. 16, *Journal X: August 8, 1857–June 29, 1858*, 64–65; Thoreau, *Writings of Henry David Thoreau*, vol. 18, *Journal XII: March 2, 1859–November 30, 1859*, 125; Thoreau, *Walden, or Life in the Woods*, 257.

10. *The Dispersion of Seeds* was one of three unfinished manuscripts in Thoreau's possession at the time of his death. It was published for the first time—with minor revisions—in 1993. A discussion of Thoreau's contribution to our understanding of forest succession, as well as his impact on American natural history writing, is found in the introduction. See Robert D. Richardson Jr., "Introduction: Thoreau's Broken Task," in Henry David Thoreau, *Faith in a Seed: The Dispersion of Seeds and Other Late Natural History Writings*, ed. Bradley P. Dean (Washington, D.C.: Island Press, 1993), 3–17.

11. Thoreau, *Faith in a Seed*, 128.

12. Ibid., 127. In Thoreau's opinion, the animals traveled from the woodlands near Flint Pond, where the trees commonly grew. Thoreau, *Writings of Henry David Thoreau*, vol. 20, *Journal XIV: August 1, 1860–November 3, 1861*, 137.

13. Thoreau, *Writings of Henry David Thoreau*, vol. 20, *Journal XIV: August 1, 1860–November 3, 1861*, 136.

14. Ibid. See also Thoreau, *Faith in A Seed*, 127.

15. Thoreau, *Faith in a Seed*, 126. In his journal, Thoreau noted that "squirrels, etc., went further for chestnuts than for acorns in proportion as they were a greater rarity. I suspect that a squirrel may convey them sometimes a quarter or a half a mile even." Thoreau, *Writings of Henry David Thoreau*, vol. 20, *Journal XIV: August 1, 1860–November 3, 1861*, 137.

16. Thoreau, *Writings of Henry David Thoreau*, vol. 20, *Journal XIV: August 1, 1860–November 3, 1861*, 138. Thoreau thought the killing of squirrels important enough to chestnut health that he recommended educating farmers about the practice. Thoreau's remarks do not make him an anti-hunting advocate, although he took that position in later writings and did not hunt or fish as an adult. See Forrest E. Wood Jr., "Thoreau: Vegetarian Hunter and Fisherman," in *Addresses of the Mississippi Philosophical Association*, ed. Bennie R. Crockett Jr. (Atlanta, Ga.: Rodopi, 2000), 251–60; Thoreau, *Walden: A Fully Annotated Edition*, 202–4, 203n8, 204n12, 204n14.

17. Bernd Heinrich, "Revitalizing Our Forests," *New York Times*, December 20, 2013, http://www.nytimes.com/2013/12/21/opinion/revitalizing-our-forests.html.

18. Thoreau, *Writings of Henry David Thoreau*, vol. 16, *Journal X: August 8, 1857–June 29, 1858*, 179–80. See also Thoreau, *Faith in a Seed*, 147–48.

19. Thoreau, *Writings of Henry David Thoreau*, vol. 10, *Journal IV: March 1, 1852–February 27, 1853*, 462.

20. On October 6, 1857, Thoreau recalled seeing "one or two chestnut burs open on the trees. The squirrels, red and gray, are on all sides throwing them down. You cannot stand long in the woods without hearing one fall." Thoreau, *Writings of Henry David Thoreau*, vol. 16, *Journal X: August 8, 1857–June 29, 1858*, 70.

21. Ibid., 69–70.

22. Thoreau, *Writings of Henry David Thoreau*, vol. 15, *Journal IX: August 16, 1856–August 7, 1857*, 120. Thoreau adds in the same entry that "the chestnuts are not so ready to fall as I expected. Perhaps the burs require to be dried now after the rain. In a day or two they will nearly all come down" (119).

23. Thoreau, *Writings of Henry David Thoreau*, vol. 10, *Journal IV: March 1, 1852–February 27, 1853*, 388.

24. The first appearance of the term in a major U.S. periodical occurred as early as 1836. See Henry William Herbert, "Scenes and Stories of the Hudson," *American Monthly Magazine* 2 (July 1836): 14–21.

25. Thoreau, *Writings of Henry David Thoreau*, vol. 15, *Journal IX: August 16, 1856–August 7, 1857*, 130.

26. Thoreau, *Walden, or Life in the Woods*, 211.

27. Harriet Beecher Stowe, *Poganuc People: Their Loves and Lives* (New York: Fords, Howard & Hulbert, 1878).

28. Harriet Beecher Stowe, "Introductory Remarks," in *The Writings of Harriet Beecher Stowe*, vol. 11 (Boston: Houghton Mifflin, 1898), ix.

29. Stowe, *Poganuc People*, 223.

30. Ibid., 224.

31. Ibid., 227.

32. Henry Ward Beecher, *Pittsfield (Mass.) Sun*, September 22, 1870, 1.

33. On the loss of the New England woodland commons, see Richard W. Judd, *Common Lands, Common People: The Origins of Conservation in Northern New England* (Cambridge, Mass.: Harvard University Press, 1997), 40–122; Brian Donahue, *Reclaiming the Commons: Community Farms and Forests in a New England Town* (New Haven, Conn.: Yale University Press, 1999), 115–62; Andrew M. Barton, with Alan S. White and Charles V. Cogbill, *The Changing Nature of the Maine Woods* (Lebanon: University of New Hampshire Press, 2012), 100–17.

34. Winslow Homer, *Chestnutting*, in *Every Saturday: A Journal of Choice Reading*, n.s., 1 (October 29, 1870): 700.

35. Ibid., 691.

36. Ibid. Although Homer drew the image, the wood engraving was done by John Andrew & Son of Boston in 1869. See Marc Simpson, *Winslow Homer: The Clark Collection* (New Haven, Conn.: Yale University Press, 2013), 193.

37. Most injuries went unreported in village newspapers. To be newsworthy, victims had to be severely injured. Such accidents occurred with regularity from the 1860s forward, which is not surprising, since tens of thousands of individuals were annually involved in gathering chestnuts. For examples of such accidents see *Massachusetts Spy* (Worcester), October 21, 1861, 1; *Daily Albany (N.Y.) Argus*, October 18, 1872, 2; *Indianapolis News*, September 27, 1872, 2; *Connecticut Courant* (Hartford), October 10, 1875, 2; *Middlebury (Vt.) Register*, November 9, 1875, 2; *Hicksville (N.Y.) News*, October 23, 1884, 3; *Vermont Phoenix* (Brattleboro), October 7, 1887, 2; *Scranton (Pa.) Republican*, September 26, 1892, 5; *Middleton (N.Y.) Daily Argus*, October 5, 1895, 5.

38. *New Haven (Conn.) Palladium*, October 27, 1863, 1.

39. *Norwich (Conn.) Aurora*, October 29, 1873, 3.

40. Cornelius Weygandt, *A Passing America: Considerations of Things of Yesterday Fast Fading from Our World* (New York: Henry Holt, 1932), 180. Born in 1871, Weygandt's observations are from his boyhood days in the Germantown section of Philadelphia.

41. Ibid., 181. Adds Weygandt, "none of the other ways of chestnutting compare to this clubbing. . . . Once, in boyhood, I climbed a great-headed tree in the open, and by shaking and whipping strewed down a whole bushel of nuts; but to clubbing it was what casting for trout in a stocked pond is to casting for them in a golden sanded pool of a tumbling run high in the mountains" (ibid.).

42. Henry David Thoreau, *Wild Fruits: Thoreau's Rediscovered Last Manuscript*, ed. Bradley P. Dean (New York: W. W. Norton, 2000), 212. See also Thoreau, *Writings of Henry David Thoreau*, vol. 9, *Journal III: September 16, 1851–April 30, 1852*, 122.

43. Thoreau, *Writings of Henry David Thoreau*, vol. 10, *Journal IV: March 1, 1852–February 27, 1853*, 125.

44. Thoreau penned the remarks October 23, 1855. Henry David Thoreau, *The Writings of Henry David Thoreau*, vol. 13, *Journal VII: September 1, 1854–October 30,*

1855, 514. In *Wild Fruits* he wrote, "is it not a barbarous way, to jar the tree? I trust I *do repent of it*. Gently shake it only, or better, let the wind shake it for you" (215; Thoreau's emphasis).

45. Thoreau, *Writings of Henry David Thoreau*, vol. 10, *Journal IV: March 1, 1852–February 27, 1853*, 427. In *Wild Fruits*, drawing on his journal entry of October 24, 1857, Thoreau wrote about "picking chestnuts all the afternoon. . . . It is as good as a journey; I seem to have been somewhere and done something. . . . As I go stooping and brushing the leaves aside by the hour, I am not thinking of chestnuts merely, but I find myself humming a thought of more significance. This occupation affords a certain broad pause and opportunity to start again afterward,—*turn over a new leaf*" (Thoreau's emphasis)." Thoreau, *Wild Fruits*, 215.

46. For commentary regarding Thoreau's last months at Concord, with references to his views on the Civil War, see Robert D. Richardson Jr., *Henry Thoreau: A Life of the Mind* (Berkeley: University of California Press, 1986), 386–88; Walter R. Harding, *The Days of Henry Thoreau* (New York: Knopf, 1965), 447–56.

47. Chestnuts are not listed among official rations as they were likely collected or purchased in the theater of combat. Writing from the Greenbriar Bridge Confederate camp in West Virginia, Private John H. Kimzey told his father "we scarcely get enough to eat, particularly in the Bread Line. Consequently we have to apply almost daily to our purses for assistance . . . the citizens keeps up a market for us. Apples $3, Irish potatoes $1.50 per bushel, Cheese $5 per cake, Butter 50cts per lb., Chestnuts 30cts a quart." Christopher M. Watford, ed., *The Civil War in North Carolina: Soldiers' and Civilians' Letters and Diaries, 1861–1865*, vol. 2, *The Mountains* (Jefferson, N.C.: McFarland, 2003), 27.

48. Tens of thousands of acres of timber were felled by Confederate and Union soldiers when constructing bridges, fortifications, breastworks, campgrounds, canals, corrals, and hitching posts. Damaging forest fires resulted from exploding munitions during hot and dry weather and considerable amounts of wood were also used for cooking and warming fires. See Jack Temple Kirby, "The American Civil War: An Environmental View," National Humanities Center, accessed January 22, 2018, http://nationalhumanitiescenter.org/tserve/nattrans/ntuseland/essays/amcwar.htm; Mark Fiege, "Gettysburg and the Organic Nature of the American Civil War," in *Natural Enemy, Natural Ally: Toward an Environmental History of War*, eds. Richard P. Tucker and Edmund Russell (Corvallis: Oregon State University Press, 2004), 93–109; Lisa M. Brady, *War upon the Land: Military Strategy and the Transformation of Southern Landscapes during the American Civil War* (Athens: University of Georgia Press, 2012); Erin Stewart Mauldin, *Unredeemed Land: An Environmental History of Civil War and Emancipation in the Cotton South* (Oxford: Oxford University Press, 2018), 11–41.

49. Kelby Ouchley, *Flora and Fauna of the Civil War: An Environmental Reference Guide* (Baton Rouge: Louisiana State University Press, 2010), 33.

50. Ibid., 34.

51. Ibid.

52. Ibid.

53. *North Carolina Standard* (Raleigh), November 17, 1861, 1.

54. Berry Benson, *Berry Benson's Civil War Book: Memoirs of a Confederate Scout and Sharpshooter* (Athens: University of Georgia Press, 2007), 157.

55. Ibid., 158.

56. Ibid., 159. Benson mentions gathering or consuming chestnuts six of the twelve days it took him—on foot and by boat—to reach the Virginia border.

57. George B. McClellan, *Report of Major General George McClellan upon the Organization of the Army of the Potomac and Its Campaigns in Virginia and Maryland from July 26, 1861, to November 7, 1862* (Chicago: Times Steam Book and Job Printing Establishment, 1864), 7–8. See also Jeffry D. Wert, *The Sword of Lincoln: The Army of the Potomac* (New York: Simon and Shuster, 2005), 33.

58. Thomas H. Mann, *Fighting with the Eighteenth Massachusetts: The Civil War Memoir of Thomas H. Mann*, ed. John J. Hennessy (Baton Rouge: Louisiana State University Press, 2000), 24.

59. "The Tory Women of East Tennessee," *Charlotte Democrat*, November 24, 1863, 2. Although the names of the Confederate officers were abbreviated, the pair were likely members of the forces defending Knoxville, Tennessee, from General Burnside's Army of the Ohio in November 1863.

60. Ibid.

61. Ibid. The editor was probably Alexander Hamilton Jones, a Union sympathizer who wrote for the *Hendersonville Times* during the Civil War. See William S. Powell, ed., *Dictionary of North Carolina Biography*, vol. 3, *H–K* (Chapel Hill: University of North Carolina Press, 1988), 312–13.

62. "Tory Women of East Tennessee," 2.

63. United States War Department, *The War of the Rebellion: A Compilation of the Official Records of the Union and Confederate Armies*, vol. 39, part 3 (Washington, D.C.: GPO, 1892), 358. For the geographic location and strategic mission of Sherman's final campaign, with references to its impact on the environment, see Edward Caudill and Paul Ashdown, *Sherman's March in Myth and Memory* (Lanham: Md.: Rowman and Littlefield, 2009), 65–88, 179–85; Megan Kate Nelson, *Ruin Nation: Destruction and the American Civil War* (Athens: University of Georgia Press, 2012), 46–70, 101, 115–16.

64. See, for example, Richard H. Steckel, "A Peculiar Population: The Nutrition, Health and Mortality of American Slaves from Childhood to Maturity," *Journal of Economic History* 46, no. 3 (September 1986): 721–41; Wilma A. Dunaway, *The African American Family in Slavery and Emancipation* (Cambridge: Cambridge University Press, 2003), 100–13; Herbert C. Covey and Dwight Eisnach, *What*

the Slaves Ate: Recollections of African American Foods and Foodways from the Slave Narratives (Santa Barbara, Calif.: ABC-CLIO, 2009), 2–4, 10–38.

65. For a summary of the interpretive shortcomings of WPA Slave narratives, see Paul D. Escott, "The Art and Science of Reading WPA Slave Narratives," in *The Slave's Narrative*, ed. Charles T. Davis and Henry Louis Gates Jr. (Oxford: Oxford University Press, 1985), 40–48. See also Covey and Eisnach, *What the Slaves Ate*, 7–9.

66. Norman R. Yetman, "An Introduction to the WPA Slave Narratives: A Collective Portrait," Born in Slavery: Slave Narratives from the Federal Writer's Project, 1936–1938 (collection), Manuscript Division, Library of Congress, accessed March 11, 2018, https://www.loc.gov/collections/slave-narratives-from-the-federal-writers-project-1936-to-1938/articles-and-essays/introduction-to-the-wpa-slave-narratives.

67. Works Project Administration, *Slave Narratives: A Folk History of Slavery in the United States from Interviews with Former Slaves*, vol. 2, *Arkansas Narratives*, part 7 (Washington, D.C.: Library of Congress, Manuscript Division, 1941), 33.

68. Works Project Administration, *Slave Narratives: A Folk History of Slavery in the United States from Interviews with Former Slaves*, vol. 9, *Mississippi Narratives*, part 2 (Washington, D.C.: Library of Congress, 1941), 343.

69. Covey and Eisnach, *What the Slaves Ate*, 179. See also Appendix K, "Nuts," 280.

70. Works Project Administration, *Slave Narratives: A Folk History of Slavery in the United States from Interviews with Former Slaves*, vol. 4, *Georgia Narratives*, part 4 (Washington, D.C.: Library of Congress, 1941), 158.

71. Works Project Administration, *Slave Narratives: A Folk History of Slavery in the United States from Interviews with Former Slaves*, vol. 1, *Alabama Narratives* (Washington, D.C.: Library of Congress, 1941), 243.

72. William J. Clarke, *North Carolina Standard* (Raleigh), November 20, 1861, 1.

73. Emil Rosenblatt and Ruth Rosenblatt, eds., *Hard Marching Every Day: The Civil War Letters of Private Wilbur Fisk, 1861–1865* (Lawrence: University Press of Kansas, 1994), 8.

74. Charles Larimer, ed., *Love and Valor: Intimate Civil War Letters between Captain Jacob and Emeline Ritner* (Chicago: Sigourney Press, 2000), 280. Writing near Rome, Georgia, on October 13, 1864, Ritner noted that "'the boys' are out in the woods gathering chestnuts and persimmons—they are very plenty here" (374).

75. Josie Underwood, *Josie Underwood's Civil War Diary*, ed. Nancy Disher Baird (Lexington: University Press of Kentucky, 2009), 105.

76. Hampton Smith, ed. *Brother of Mine: The Civil War Letters of Thomas and William Christie* (St. Paul: Minnesota Historical Society, 2011), 74. A similar observation was made by Cyrus F. Boyd as his regiment marched toward Corinth,

Mississippi. Recording their entry into the town of Ripley, he wrote, "When we halted this evening the rain was pouring down and we are wet through and mud from head to foot. Are burning some chestnut rails to-night to dry our clothes." Mildred Throne, ed., *The Civil War Diary of Cyrus F. Boyd, Fifteenth Iowa Infantry, 1861–1863* (Baton Rouge: Louisiana University Press, 1998), 78.

77. Smith, *Brother of Mine*, 74.

78. Frederick Tilp, *This Was Potomac River* (Bladenburg, Md.: Frederick Tilp, 1978), 156.

79. English botanist John Hill described European chestnuts as having "a rough prickly Shell, and under that, each particular Chesnut, has its firm brown Coat and a thin Skin." According to Hill, the skin "is a very fine Astringent; it stops Purgings and Overflowing of the Menses," Hill, *The Useful Family Herbal* (London: W. Johnston, 1754), 76. The American chestnut is also mentioned in the 1812 edition of William Cullen's *Materia Medica*. See Cullen, *Professor's Cullen's Treatise of the Materia Medica*, vol. 1 (Philadelphia: Edward Parker, 1812), 182, 282. For the medicinal use of chestnut leaves by Cherokees, see Paul B. Hamel and Mary Chiltoskey, *Cherokee Plants: Their Uses: A 400 Year History* (Sylva, N.C.: Herald, 1978), 29.

80. Francis P. Porcher, *Resources of the Southern Fields and Forests* (Charleston: Steam-Power Press of Evans and Cogswell, 1863), 238. A former slave from Georgia mentions the use of dried chestnut leaves in making a tea for asthma in Works Project Administration, *Slave Narratives: A Folk History of Slavery in the United States from Interviews with Former Slaves*, vol. 4, *Georgia Narratives*, part 1 (Washington, D.C.: Library of Congress, 1941), 145.

81. Porcher, *Resources of the Southern Fields and Forests*, 238.

82. George C. Close, "On Chestnut Leaves in Whooping-Cough," *American Journal of Pharmacy* (January 1863): 35; "Cure for the Whooping Cough," *Atlanta Constitution*, June 16, 1870, 2; Lewis J. Steltzer, "On Chestnut Leaves," *American Journal of Pharmacy* 10, no. 6 (June 1880): 292; F. L. Sim, "Chestnut Leaves in Whooping Cough," *Mississippi Valley Medical Monthly* 4, no. 5 (May 1884): 238.

83. See Steven Foster, *Forest Pharmacy: Medicinal Plants in American Forests* (Durham, N.C.: Forest History Society, 1995), 42; Guy R. Hasegawa, "Pharmacy in the American Civil War," *American Journal of Health-System Pharmacy* 57, no. 5. (March 2000): 475–89; Steven R. Moore and John Parascandola, "The Other Pharmacists in the American Civil War," *American Journal of Health-System Pharmacy* 57, no. 13 (July 2000): 1276.

84. "Ziegler's Grove: Battle of Gettysburg," American Civil War Homepage, accessed February 5, 2018, http://thomaslegion.net/zieglersgrovebattleofgettysburg.html. See also John T. Trowbridge, "The Field of Gettysburg," *Atlantic Monthly* 16, no. 98 (November 1865): 619.

85. John T. Trowbridge, *The South: A Tour of Its Battlefields and Ruined Cities* (Hartford, Conn.: L. Stebbins, 1866), 24. See also Clifford Dowdey, *Death of a Nation: The Story of Lee and His Men at Gettysburg* (New York: Alfred A. Knopf, 1958), 293; Ronald S. Coddington, *Faces of the Civil War: An Album of Union Soldiers and Their Stories* (Baltimore: Johns Hopkins University Press, 2004), 109.

86. John David Hoptak, *The Battle of South Mountain* (Charleston, S.C.: History Press, 2011), 83.

87. Benson J. Lossing, *Pictorial History of the Civil War in the United States of America*, vol. 2 (Hartford, Conn.: Thomas Belknap, 1874), 470, n1. See also Trowbridge, *The South*, 24.

88. Mann, *Fighting with the Eighteenth Massachusetts*, 90–91.

89. Brady, *War upon the Land*, 23. See also Joan E. Cashin, *War Stuff: The Struggle for Human and Environmental Resources in the American Civil War* (Cambridge: Cambridge University Press, 2018), 89–97, 107.

90. For a summary of the collateral environmental damage caused by Civil War campaigns, including from road construction and forest fires, see Cashin, *War Stuff*, 98–107.

91. Several areas suffered economically well into the 1870s, such as northwest Georgia and the Cumberland Plateau of Tennessee and Kentucky, slowing the pace of forest clearance. Elsewhere, wooded landscapes continued to be converted into fields for agriculture. See Steven Hahn, *The Roots of Southern Populism: Yeoman Farmers and the Transformation of the Georgia Upcountry, 1850–1890* (New York: Oxford University Press, 1983), 137–69; Stephen V. Ash, *Middle Tennessee Society Transformed, 1860–1870: War and Peace in the Upper South* (Knoxville: University of Tennessee Press, 2006), 175–91.

92. Nelson, *Ruin Nation*, 114–16, 151–52; Brady, War upon the Land, 22–23, 57; Cashin, *War Stuff*, 100, 191, Mauldin, *Unredeemed Land*, 87–88. At least one thousand miles of rails and crossties were destroyed by Union armies alone. Repairing the lines required an exorbitant amount of wood, as three thousand crossties were needed to build a single mile of track. In southern states, however, oak, hemlock, and cypress were most often used for that purpose.

93. "The Aorangi Collie Kennels," *Dogdom Monthly* 8, no. 10 (December 1907): 1012.

94. On the decay of chestnut crossties, see Franklin Benjamin Hough, *Report upon Forestry*, vol. 1, USDA, U.S. Forest Service (Washington, D.C.: GPO, 1878), 116; Nathaniel H. Egleston, *Report upon Forestry*, vol. 4, USDA, U.S. Forest Service (Washington, D.C.: GPO, 1884), 140; John C. Trautwine, *The Civil Engineer's Pocket-Book*, 15th ed. (New York: John Wiley & Sons, 1891), 759.

95. "Railroad Sleepers—How Shall We Get Them?," *Merchants' Magazine and Commercial Review* 38 (January–June 1858): 384.

96. Henry Hall, *Report of the Ship-Building Industry in the United States*, U.S. Department of the Interior, Census Office (Washington, D.C.: GPO, 1884), 28, 48, 64, 119, 159, 227; Glenn A. Knoblock, *The American Clipper Ship, 1845–1920: A Comprehensive History, with a Listing of Builders and Their Ships* (Jefferson, N.C.: McFarland, 2014), 40.

97. *New York Times*, June 20, 1865, 2.

98. Thoreau, *Writings of Henry David Thoreau*, vol. 20, *Journal XIV: August 1, 1860–November 3, 1861*, 133.

99. Ibid., 135. Although pines and birches often replaced oaks and chestnuts, they were "of feebler growth" as a result of soil depletion caused by livestock grazing, injudicious timbering, and indiscriminate fires (ibid.).

100. Brian Donahue, *The Great Meadow: Farmers and the Land in Colonial Concord* (New Haven, Conn.: Yale University Press, 2004), 212. See also Brian Donahue, "Another Look from Sanderson's Farm: A Perspective on New England Environmental History and Conservation," *Environmental History* 12, no. 1 (January 2007): 19–20; David Foster et al., "New England's Forest Landscape," in *Agrarian Landscapes in Transition: Comparisons of Long-term Ecological and Cultural Change*, ed. Charles L. Redman and David R. Foster (New York: Oxford University Press, 2008), 59–62. Before reforestation, the Concord landscape remained a patchwork of weedy pastures, heath shrubs, brushy copses, and grassy meadows. Although Donahue believes this was "the common condition of much of Concord's pastureland during the nineteenth century and would lead in time to the return of the forest," the woodlands that returned were largely void of chestnut (Donahue, *Great Meadow*, 212).

101. Edward K. Faison and David R. Foster, "Did American Chestnut Really Dominate the Eastern Forest?," *Arnoldia* 72, no. 2 (2014): 18–32.

102. George B. Emerson, *A Report of the Trees and Shrubs Growing Naturally in the Forests of Massachusetts*, 3rd ed. (Boston: Little, Brown, 1878), 188; George F. Schwarz, "The Sprout Forests of the Housatonic Valley of Connecticut," *Forestry Quarterly* 5, no. 2 (June 1907): 121–53; Raphael Zon, *Chestnut in Southern Maryland*, USDA Bureau of Forestry Bulletin no. 53 (Washington, D.C.: GPO, 1904), 8–9; Emily W. B. Russell, *People and the Land through Time: Linking Ecology and History* (New Haven, Conn.: Yale University Press, 1997), 121.

103. Frederick Watts, *Report of the Commissioner of Agriculture for the Year 1876* (Washington, D.C.: GPO, 1876), 302.

104. *Atlanta Constitution*, October 11, 1871, 1.

105. Samuel B. Buckley, "Mountains of North Carolina and Tennessee," *American Journal of Science and Arts*, 2nd ser., 27 (March 1850): 292; Matthew F. Maury and William M. Fontaine, *Resources of West Virginia* (Wheeling, W.V.: Register Company, 1876), 116; J. T. Rothrock, *Annual Report of the Pennsylvania Department of*

Agriculture, part 2 (Harrisburg, Pa.: State Printer of Pennsylvania, 1896), 261; Gifford Pinchot and W. W. Ashe, *Timber Trees and Forests of North Carolina*, North Carolina Geological Survey Bulletin no. 6 (Raleigh, N.C.: M. I. & J. C. Stewart, 1897), 109; Jason Duke, *Tennessee Coal Mining, Railroading, and Logging in Cumberland, Fentress, Overton, and Putnam Counties* (Paducah, Ky.: Turner, 2003), 13, 60.

106. Thoreau, *Writings of Henry David Thoreau*, vol. 20, *Journal XIV: August 1, 1860–November 3, 1861*, 184–85.

107. William W. Ashe, *Chestnut in Tennessee*, Tennessee Geological Survey Series Bulletin no. 10–B (Nashville, Tenn.: Baird-Ward, 1912), 23; John S. Holmes, *Forest Conditions in Western North Carolina*, North Carolina Geological and Economic Survey Bulletin no. 23 (Raleigh, Va.: Edwards and Broughton, 1911), 21–23, 53, 66; John Preston Arthur, *A History of Watauga County, North Carolina* (Richmond, Va.: Everett Waddey, 1915), 24, 26–28, 126, 238; Society of American Foresters, *Forest Cover Types of the Eastern United States* (Washington, D.C.: Society of American Foresters, 1932), 22.

CHAPTER 7. Along All Prominent Thoroughfares

1. U.S. Bureau of the Census, "Population of the 100 Largest Urban Places, Listed Alphabetically by State: 1790–1990," accessed April 30, 2017, https://www.census.gov/library/working-papers/1998/demo/POP-twps0027.html.

2. "The Nut Trade: How Chestnuts and Pea-Nuts Are Sold—The Profits and Losses of the Business," *New York Times*, December 22, 1872, 5.

3. "The Nut Trade," 5. If chestnut roasters were on the corners of "every sixth block," as claimed by the *New York Times*, the number of Manhattan vendors (from Harlem to Lower Manhattan) would have exceeded two hundred in 1872. In 1866, the number of chestnut vendors in Paris, France, was four hundred. See Frank Leslie, *Frank Leslie's Illustrated Newspaper*, January 13, 1866, 3.

4. Junius Henri Browne, *The Great Metropolis: A Mirror of New York* (Hartford, Conn.: American Publishing Company, 1869), 135.

5. "The Nut Trade," 5.

6. Ibid. Two years later, the price for a bushel of chestnuts in New York City dropped to four dollars. *Wheeling (Va.) Register*, October 15, 1874, 1. The decline in chestnut prices was linked to the arrival of railroads in remote rural areas where the trees were more prevalent, thus increasing supply.

7. *Massachusetts Spy* (Worcester), February 23, 1866, 8; *National Aegis* (Worcester, Mass.), June 18, 1870, 5; *Evening Gazette* (Port Jervis, N.Y.), April 9, 1870, 4; *Brooklyn (N.Y.) Daily Eagle*, October 16, 1873, 1; H. C. Bunner, "The Chestnut Trade," *Puck* 14, no. 342 (September 26, 1883): 70.

8. "The Nut Trade," p. 5.

9. *Atlanta Constitution*, October 20, 1871, 4. An exception was the Cincinnati, Ohio, market, where chestnuts were advertised for $6.50 per bushel in 1872. *Cincinnati Daily Enquirer*, October 16, 1872, 6.

10. Charles H. Haswell, *Reminiscences of an Octogenarian of the City of New York, 1816 to 1860* (New York: Harpers and Brothers, 1896), 229.

11. *New York Tribune*, February 6, 1856, 8. In the 1850s and 1860s, newspaper advertisements for Italian chestnuts were published in late winter or early spring, presumably as American chestnuts became scarce or unavailable.

12. *Commercial Advertiser* (New York City), April 23, 1866, 4. "Eighty barrels and fourteen cases of chestnuts" were unloaded from the Genoan "Bark Gulia" in a single shipment.

13. Federal Writers' Project, Works Progress Administration, *The Italians of New York* (New York: Random House, 1938), 184–85.

14. *Hinds County Gazette* (Raymond, Mich.), March 6, 1878, 1.

15. A. W. Roberts, "Chestnuts," *Forest and Stream: A Journal of Outdoor Life, Travel, Nature Study, Shooting, Fishing, Yachting* 17, no. 16 (November 17, 1881): 306.

16. *New York Sun*, October 10, 1883, 3.

17. Ibid. Italian Americans also gathered and sold native chestnuts. As reported in the magazine *Puck*, a pedestrian walking along Park Place asked an Italian vendor where he got his chestnuts. "My wifa and family go out in Jersey and chestnut picka in the daytime," he replied, "and I roasta them here, and sella them for five cents." "The Chestnut Trade," *Puck* 14, no. 343 (October 3, 1883): 70.

18. *Atlanta Constitution*, November 10, 1870, 2. The figure assumes the purchase price for chestnuts was two dollars per bushel and each bushel weighed fifty-four pounds. Although Virginia, Tennessee, and South Carolina had passed laws standardizing the weight of chestnut bushels (57, 50, and 50 pounds, respectively), Georgia had not done so by that date. See Treasury Department, *Quarterly Reports of the Chief of the Bureau of Statistics, Showing the Imports and Exports of the United States* (Washington, D.C.: GPO, 1878), 378.

19. *Atlanta Constitution*, October 31, 1879, 1.

20. "Beating the Road: Smuggling Chestnuts from Gainesville to Atlanta," *Atlanta Constitution*, October 30, 1885, 7.

21. *Wheeling (W.Va.) Register*, October 21, 1875, 3; *Richmond (Va.) Whig*, November 22, 1870, 3; *Alexandria (Va.) Gazette*, November 16, 1888, 2; *New York Tribune*, October 12, 1887, 4.

22. For a comprehensive study of the economic impact of the chestnut trade on southwest Virginia, see Ralph H. Lutts, "Like Manna from God: The American Chestnut Trade in Southwestern Virginia," *Environmental History* 9, no. 3 (July 2004): 497–525.

23. Virginia Railroad Commissioner, *Twelfth Annual Report of the Railroad*

Commissioner of the State of Virginia (Richmond: J. H. O'Bannon, Superintendent of Public Printing, 1888), xxix.

24. Ibid., xxx. Rangeley's initial monetary request is omitted from the report, although it was obviously higher than the payoff recommendation submitted by the claims agent.

25. Ibid. Freight rates varied according to shipping distances, as perishability had a direct impact on the profit margins of buyers and sellers. In Georgia, chestnuts were perishable third-class freight and were shipped at a rate of 13 cents per hundred pounds for distances of ten miles or less, a rate that became incrementally more expensive as destinations became longer. In 1881, the cost of shipping 3,080 pounds of chestnuts 250 miles was $16.94, nearly three dollars less than the Virginia rate. See Railroad Commission of Georgia, *Third Semi-annual Report of the Railroad Commission of the State of Georgia* (Atlanta: Constitution Publishing Company, 1881), 112, 119.

26. Virginia Railroad Commissioner, *Twelfth Annual Report*, xxx. Satisfied with the payment received for the delayed and canceled sale, Rangeley wrote to the railroad commissioner, thanking him for the "prompt attention to the claims I have sent you" (xliv). For chestnut wholesale prices in 1887, see *Wheeling (W.V.) Register*, October 19, 1887, 4; *Cleveland Plain Dealer*, November 16, 1887, 7; *New York Tribune*, October 12, 1887, 4; *Daily Inter Ocean* (Chicago, Ill.), October 26, 1887, 9.

27. Patrick County Historical Society, *History of Patrick County, Virginia* (Stuart, Va.: Patrick County Historical Society, 1999), 359. A report in the *Asheville Citizen* noted that 152 bushels of chestnuts were sent from there to Detroit, Michigan, on November 25, 1887. Shipped by Hall, Smith & Company, the nuts weighed nearly seven thousand pounds. The freight costs were $78.95. *Asheville Citizen*, November 26, 1887, 2.

28. Patrick County Historical Society, *History of Patrick County, Virginia*, 359. Unscheduled trains, or "extras," were used to handle large volumes of freight, including chestnuts and apples, during peak harvest season. See Shirley Grose, ed., *Patrick County, Virginia, Heritage Book*, vol. 1, *1791–1999* (Summersville, W.Va.: Walsworth, 2000), 5.

29. *Atlanta Constitution*, October 25, 1886, 4. In 1884, the subscription rate for the *North Georgia Times* was one dollar. Department of the Interior, Census Office, *History and Present Condition of the Newspaper and Periodic Press of the United States* (Washington, D.C.: GPO, 1884), 217.

30. "A Big Chestnut Crop: It Will Be Sufficient to Pay This Year's Taxes in Lumpkin County," *Atlanta Constitution*, September 7, 1891, 2.

31. *New York Times*, October 16, 1892, 1. Hamburg is presently located in the township of Lyme, Connecticut. Population estimates for Hamburg are published in State of Connecticut, *Register and Manual for the State of Connecticut* (Hartford, Conn.: Case, Lockwood, and Brainard, 1893), 304.

32. *New York Times*, October 16, 1892, 1. The estimate for skilled-labor wages for New England in 1892 was taken from Melita Podesta, "1890s Family," educational brochure (Boston: Federal Reserve Bank of Boston, 2004).

33. *New York Times*, October 16, 1892, 1. Although Hamburg merchants also purchased chestnuts by the quart, the average price paid for a single bushel was $2.80.

34. Two species of weevils commonly attacked chestnuts: *Curculio proboscideus* and the smaller *Curculio armiger*. By 1904, damage done by chestnut weevils fell short of 25 percent, although in some years it might be as high as 40 to 50 percent. Frank H. Chittenden, "The Nut Weevils," *Yearbook for the Department of Agriculture, 1904* (Washington, D.C.: GPO, 1905), 300.

35. Fred E. Brooks, "Nut Weevils in West Virginia," *First Annual Report of the West Virginia Agricultural Experiment Station, 1887–1888* (Charleston: Moses W. Donnally, 1890), 25–26.

36. Fred E. Brooks, *The Chestnut Curculios*, USDA Technical Bulletin no. 130 (Washington, D.C.: GPO, 1929), 3.

37. U.S. Food and Drug Administration, *Notices of Judgment under the Food and Drugs Act, Nos. 5000–6000* (Washington, D.C.: GPO, 1918), 390.

38. W. G. Campbell, acting chief of the Bureau of Chemistry, quoted in Brooks, *Chestnut Curculios*, 3.

39. Notice no. 5344, "Adulteration of Chestnuts," in USDA Bureau of Chemistry, Service and Regulatory Announcements, Supplement 37 (February 23, 1918), 390.

40. Ibid.

41. In 1894, a survey conducted by the North Carolina Experiment Station found those living in areas where chestnuts were abundant often scalded them in boiling water. After boiling—which killed the weevil larvae and their eggs—the nuts were laid in the sun to dry. See Gerald McCarthy, "The Chestnut and Its Weevil," in *Annual Report of the North Carolina Agricultural Experiment Station* (Raleigh: North Carolina College of Agriculture and Mechanic Arts, 1895), 268.

42. William W. Ashe, *Forest Fires: Their Destructive Work, Causes and Prevention*, North Carolina Geological Survey Bulletin no. 7 (Raleigh, N.C.: Josephus Daniels, State Printer, 1895), 49; James Mooney, *Myths of the Cherokee: Extract from the Nineteenth Annual Report of the Bureau of American Ethnology* (Washington, D.C.: GPO, 1902), 317.

43. See George H. Powell, *The European and Japanese Chestnuts in the Eastern United States*, Delaware College Agricultural Experiment Station Bulletin no. 42 (Newark, Del.: Delaware College Agricultural Experiment Station, 1898), 19; Chittenden, "Nut Weevils," 300.

44. Mrs. D. A. [Mary Johnson] Lincoln, *Mrs. Lincoln's Boston Cook Book: What to Do and What Not to Do in Cooking* (Boston: Roberts Brothers, 1884), 192, 256,

258, 356. Lincoln was the first principal of the Boston Cooking School and the culinary editor of *American Kitchen Magazine.*

45. Ibid., 258.

46. Ladies' Benevolent Society of Lowry Hill Congregational Church, *Cook Book of Tried Recipes* (Minneapolis: Byron and Willard Printers, 1897), 131. For additional chestnut croquette recipes, see Maria Parloa, *Miss Parloa's Kitchen Companion: A Guide for All Who Would Be Good Housekeepers* (New York: Clover Publishing Company, 1887), 382; Maud C. Cooke, *Breakfast, Dinner and Supper, or What to Eat and How to Prepare It* (Philadelphia: J. H. Moore, 1897), 266; Fannie Merritt Farmer, *The Boston Cooking-School Cook Book* (Boston: Little, Brown, 1900), 308.

47. *Brooklyn Eagle*, November 15, 1902, 10.

48. Almeda Lambert, *Guide for Nut Cookery: Together with a Brief History of Nuts and Their Food Value* (Battle Creek, Mich.: Joseph Lambert, 1899), 298.

49. Ibid., 330, 373–74, 408. See also Jules A. Harder, *The Physiology of Taste: Harder's Book of Practical American Cookery*, vol. 2 (San Francisco: n.p., 1885). 488; Archie C. Hoff, *International Dessert and Pastry Specialties of the World Famous Chefs: United States, Canada, Europe* (Los Angeles: International Publishing Company, 1913), 29, 33, 59.

50. Mary Ronald, *The Century Cook Book* (New York: Century Company, 1895), 410.

51. On the origin and cultivation of Paragon chestnuts, see J. Russell Smith, *Tree Crops, A Permanent Agriculture* (Washington, D.C.: Island Press, 1987), 139; William Lord, "The Paragon Chestnut: Pedigree and History," *Chestnut: The New Journal of the American Chestnut Foundation* 29, no. 2 (Summer 2015): 19–27.

52. Addison S. Flowers, *Nut Menu: A Treatise on the Preparation of Nuts for the Palate* (Mount Joy, Pa.: [Glen Orchard Hotel], 1903), 21.

53. Ibid., 21–22. In Lambert's *Guide for Nut Cookery* another recipe for chestnut cake calls for chestnut flour ground in the "family grist-mill or coffee mill" (215).

54. Flowers, *Nut Menu*, 26–27.

55. Alphonse Bertillon, "Anthropometrical Descriptions: A New Method of Determining Individual Identity," *Atlantic Medical Weekly* 1, no. 10 (October 1893): 489; Paolo Mantegazza, "Mantegazza on Expressions in the Face," *Werner's Magazine* 17, no. 1 (January 1895): 7.

56. Gardner Dexter Hiscox, ed., *Henley's Twentieth Century Book of Recipes, Formulas and Processes* (New York: N. W. Henley, 1907), 267, 391, 556.

57. Charles Miesse, *Points on Coal and the Coal Business* (Myerstown, Pa.: Feese and Uhrich, 1887), 131.

58. William Bainbridge, *A Treatise on the Law of Mines and Minerals: First American Edition from the Third London Edition of George M. Dallas, Esq.*

(Philadelphia: John Campbell, 1871), 474. See also Frederick E. Saward, ed., *The Coal Trade: A Compendium of Valuable Information Relative to Coal Production, Prices, Transportation, etc.* (New York: N.p., 1880), 12.

59. The expression was part of the popular lexicon after the early 1880s. In 1886, numerous newspapers published explanations for its origin, but there is little consensus about who first coined the phrase. See Robert A. Palmatier, *Food: A Dictionary of Literal and Nonliteral Terms* (Westport, Conn.: Greenwood, 2000), 257.

60. William Dimond, *The Broken Sword: A Grand Melodrama, Interspersed with Songs, Choruses, etc.* (London: J. Barker, 1816), 13.

61. Untitled and unsigned article beginning "Mr. Martin Hanley, manager of Harrigan's Theatre, relates the origin of 'Chestnut,'" in *Theatre* 2, no. 4 (October 11, 1886): 95. See also *Boston Globe*, April 10, 1885, 6.

62. Earl Shinn and Francis Hopkins Smith, *A Book of the Tile Club* (New York: Houghton Mifflin, 1886), 31–35.

63. Ibid., 32.

64. Edward V. Lucas, *Edwin Austin Abbey, Royal Academician: The Record of His Life and Work*, vol. 1, *1852–1893* (New York: Charles Scribner's Sons, 1921), 52.

65. Ibid., 52–53. See also Elisa Tamarkin, "The Chestnuts of Edwin Austin Abbey: History Painting and the Transference of Culture in Turn-of-the-Century America," *Prospects* 24 (October 1999): 417–47.

66. Lucas, *Edwin Austin Abbey*, 53.

67. Edward Payson Roe, *Opening a Chestnut Burr* (New York: Dodd, Mead, 1874). A gardener and nurseryman, Roe was among the most popular American writers of the Reconstruction era. A "country ballad" about chestnut gathering also appears in the book, although it is unclear if Roe wrote the song lyrics himself or adapted them for his own use (76).

68. E. Clark Reed, "Opening the Chestnut Burr" (New York: Central Publishing Company, 1884), Library of Congress, Historic Sheet Music Collection, 1800 to 1922, Music Division, box 64, file 4358, item #100008612, 5, https://www.loc.gov/item/ihas.100008612.

69. Mulbro [no first name], "Chestnuts on the Brain" (N.p.: William F. Shaw, 1885), Library of Congress, Historic Sheet Music Collection, electronic resource, accessed June 30, 2018, https://www.loc.gov/item/sm1885.25306.

70. "Scenes in Fairmount Park," *Art Journal*, n.s., vol. 4 (1878): 2.

CHAPTER 8. The Wonder and Admiration of All

1. William P. Corsa, comp., *Nut Culture in the United States, Embracing Native and Introduced Species* (Washington, D.C.: GPO, 1896); Andrew S. Fuller, *The Nut Culturist: A Treatise on the Propagation, Planting, and Cultivation of Nut-Bearing*

Trees and Shrubs (New York: Orange Judd, 1896); William A. Buckhout, *Chestnut Culture for Fruit*, Pennsylvania State College Agricultural Experiment Station Bulletin no. 36 (State College: Pennsylvania State College, 1896).

2. See, for example, M. T. Georgeson, "Autumn in Japan," *Orchard and Garden* 11, no. 1 (January 1899): 16–17.

3. Corsa, *Nut Culture in the United States*, 85–88. Readers of the report were told orchards of "100 to 300 trees" had been established in a dozen states (79).

4. George H. Powell, *The European and Japanese Chestnuts in the Eastern United States*, Delaware College Agricultural Experiment Station Bulletin no. 42 (Newark, Del.: Delaware College Agricultural Experiment Station, 1898), 26. See also Nelson F. Davis, *Chestnut Culture in Pennsylvania*, Pennsylvania Department of Agriculture Bulletin no. 123 (Harrisburg, Pa.: William Stanley Ray, 1904), 14–15.

5. Corsa, *Nut Culture in the United States*, 87. See also Joseph L. Lovett, *The Paragon Chestnut* (Emilie, Pa.: Joseph L. Lovett, 1896), 1–7; G. Harold Powell, "The Chestnut Grove and the Chestnut Orchard," *American Gardening: A Weekly Illustrated* 20 (April 15, 1899): 280.

6. Corsa, *Nut Culture in the United States*, 87.

7. Ibid. John R. Parry, William's son, edited a volume of reports and essays about commercial chestnut growing that mentions their New Jersey orchard and Japanese chestnut cultivation. See John R. Parry, ed., *Nuts for Profit: A Treatise on the Propagation and Cultivation of Nut-Bearing Trees* (Camden, N.J.: Sinnickson Chew, 1897), 66, 78, 80, 88–91.

8. Ernest A. Sterling, *Chestnut Culture in the Northeastern United States: Reprinted from Seventh Report of the Forest, Fish and Game Commission, State of New York* (Albany, N.Y.: J. B. Lyon, 1903), 100.

9. G. Harold Powell, "The Historical Status of Commercial Chestnut Culture in the United States," *American Gardening* 20, no. 220 (March 11, 1899): 178.

10. Richard Edwards, *Industries of New Jersey*, part 1, *Trenton, Princeton, Hightstown, Pennington, and Hopewell* (New York: Historical Publishing, 1882), 78; Powell, *European and Japanese Chestnuts*, 15; New York Forest, Fish, and Game Commission, *New York Forest, Fish, and Game Commission Eighth Annual Report* (Albany, N.Y.: Argus, 1903), 104–5.

11. Charles S. Sargent, "Notes," *Garden and Forest* 8 (November 13, 1895): 460. See also Powell, *European and Japanese Chestnuts*, 32; Parry, *Nuts for Profit*, 75.

12. Henry S. Williams, *Luther Burbank: His Life and Work* (New York: Hearst's International Library, 1915), 207. Burbank's account book for 1885 records the sale of twenty-five "Japan Chestnut trees" and "2lbs Japan Mammoth chestnuts" to Perry Hulbert. Luther Burbank Papers, 1830–1989, Manuscript Division, Library of Congress, box 21.

13. Luther Burbank, *How Plants Are Trained to Work for Man*, vol. 8, *Trees, Biography, Index* (New York: P. F. Collier & Son, 1921), 55. On the importance

of Burbank's experiments in the development of American horticulture, see Noel Kingsbury, *Hybrid: The History and Science of Plant Breeding* (Chicago: University of Chicago Press, 2009), 187–96; Jane S. Smith, *The Garden of Invention: Luther Burbank and the Business of Breeding Plants* (New York: Penguin, 2009), 1–34.

14. Powell, *European and Japanese Chestnuts*, 12. In a stock notebook dated 1887–88, Burbank claims to have 1,140 "China Chestnut, Japan Chestnut" trees on hand. Luther Burbank Papers, 1830–1989, Manuscript Division, Library of Congress, box 25, folder 2: "Stock Notebook, 1887–1888," n.p.

15. Powell, *European and Japanese Chestnuts*, 29, 31, 32; Sterling, *Chestnut Culture*, 99–100; Burbank, *How Plants Are Trained*, 63.

16. George H. Powell, "The Chestnut Industry," in *Transactions of the Peninsula Horticultural Society, Eleventh Annual Session* (Dover, Del.: Press of the Delawarean, 1898), 67; Powell, *European and Japanese Chestnuts*, 6; Parry, *Nuts for Profit*, 79–80.

17. Parry, *Nuts for Profit*, 79.

18. Powell, "Chestnut Industry," 67; Powell, *European and Japanese Chestnuts*, 5. The Mammoth Chestnut Company was incorporated in the state of New Jersey on August 24, 1891. New Jersey Department of State, *Corporations of New Jersey: List of Certificates Filed in the New Jersey Department of State, 1846–1891* (Trenton, N.J.: Naar, Day & Naar, 1892), 143.

19. U.S. House of Representatives, *Report of the Industrial Commission on Agriculture and Agricultural Labor*, 57th Congress, 1st Session, doc. no. 179, vol. 10 (Washington, D.C.: GPO, 1901), 394. When Congressman A. L. Harris asked Hale how many chestnuts he produced per acre, he replied, "60 to 75 bushels" (ibid.).

20. Ibid.

21. According to a plan book dated 1884–88, Burbank was growing "9 Japan chestnuts, 8 China chestnuts, and 1 Italian chestnut" at his Whittier property near Sebastopol. See Luther Burbank Papers, 1830–1989, Manuscript Division, Library of Congress, box 24, folder 1: "Plan Book, 1884–1888," n.p. In 1893, he was crossing "Hybrid American" with "Japan Chestnuts" and "Hybrid [American]" with "Chinquapin" and "Japan chestnuts," Luther Burbank Papers, 1830–1989, Manuscript Division, Library of Congress, box 25, folder 3: "Stock Records, 1887–94," n.p.

22. The Prolific, a Japanese chestnut imported by John T. Lovett, was introduced by the Parry brothers in 1897. It was one of twenty-three different Japanese varieties available in 1898 and one of four imported by Lovett. Powell, *European and Japanese Chestnuts*, 30–32.

23. John T. Lovett, *Lovett's Guide to Fruit Culture* (Little Silver, N.J.: Monmouth Nursery, 1901), 17. In 1901, two- to three-foot saplings sold for $3.50 per dozen and trees five to six feet sold for $5.00 a dozen. Lovett also

sold American chestnuts, advertising them as "the well-known chestnut of the forest. Its sweetness and delicacy of flavor or as a shade tree is unsurpassed" (ibid.).

24. Corsa, *Nut Culture in the United States*, 84.

25. Ibid. A total of seventeen named American varieties are described in the text, from locations ranging from Mountainville, New York, to Otto, Tennessee. The USDA ranked the Otto chestnut "the highest flavored chestnut" (84).

26. Fuller, *Nut Culturist*, 9. The United States annually imported, on average, 2,147,636 pounds of chestnuts valued at $78,658 ($2.7 million in today's currency).

27. Fifty-Ninth Congress, 2nd Session, *Commercial Relations of the United States with Foreign Countries during the Year 1906*, House of Representatives Document no. 354 (Washington, D.C.: GPO, 1907), 593, 600, 604, 607, 668.

28. Sterling, *Chestnut Culture*, 93.

29. Davis, *Chestnut Culture in Pennsylvania*, 19. This "remark" is a paraphrase of comments made by Coleman K. Sober, who grew Paragon chestnuts near Lewisburg, Pennsylvania, in the 1890s and early 1900s. See Sober, "Chestnut Culture in Pennsylvania," in *Report of the Forty-Sixth Annual Meeting of the State Horticultural Association of Pennsylvania, Harrisburg, Pennsylvania, January 17–18, 1905*, State Horticultural Association of Pennsylvania (Harrisburg, Pa.: William Stanley Ray, State Printer, 1905), 54–58.

30. Davis, *Chestnut Culture in Pennsylvania*, 21–22.

31. Recent studies challenge earlier beliefs regarding the supposed inability of grafted plants to share genetic material, as well as pass DNA to offspring. See Sandra Stegemann and Ralph Bock, "Exchange of Genetic Material between Cells in Plant Tissue Grafts," *Science* 324, no. 5927 (May 2009): 649–51; Victor M. Haroldsen et al., "Mobility of Transgenic Nucleic Acids and Proteins within Grafted Rootstocks for Agricultural Improvement," *Frontiers in Plant Science* 3, no. 39 (March 2, 2012): 39–47; Rui Wu et al., "Inter-Species Grafting Caused Extensive and Heritable Alterations of DNA Methylation in *Solanaceae* Plants," *PLoS ONE* 8, no. 4 (April 16, 2013): 1371–77.

32. The numbers are approximations using data from the following sources: Powell, *European and Japanese Chestnuts*, 23–34; U.S. House of Representatives, *Report of the Industrial Commission*, 394–95; Sterling, *Chestnut Culture*, 98–104; Sandra L. Anagnostakis, "A Historical Reference for Chestnut Introductions into North America," Connecticut Agricultural Experiment Station, accessed July 28, 2018, http://www.ct.gov/caes/cwp/view.asp?a=2815&q=376740.

33. William T. Hornaday, *Popular Official Guide to the New York Zoological Park, as Far as Completed* (New York: New York Zoological Society, 1900), 4, 9–10, 13.

34. Hermann W. Merkel, "A Deadly Fungus on the American Chestnut," in *Tenth Annual Report of the New York Zoological Society* (New York: New York Zoological Society, 1906): 97–103.

35. Ibid., 101, 103; William A. Murrill, "A Serious Chestnut Disease," *Journal of the New York Botanical Garden* 7, no. 78 (June 1906): 143; Paul J. Anderson and William H. Rankin, *Endothia Canker of Chestnut*, Cornell University Agricultural Experiment Station Bulletin 347 (Ithaca, N.Y.: Cornell University College of Agriculture, 1914), 538–39.

36. Merkel, "Deadly Fungus on the American Chestnut," 101; Murrill, "Serious Chestnut Disease," 153; G. G. Copp, "A Disease Which Threatens the American Chestnut Tree," Scientific American 95, no. 24 (December 1906): 451.

37. William A. Murrill, "A New Chestnut Disease," *Torreya* 6, no. 9 (September 1906): 186–89; William A. Murrill, "Further Remarks on a Serious Chestnut Disease," *Journal of the New York Botanical Garden* 7, no. 81 (September 1906): 203–11; John Mickleborough, *A Report on the Chestnut Tree Blight: The Fungus, Diaporthe Parasitica, Murrill* (Harrisburg, Pa.: Commonwealth of Pennsylvania Department of Forestry, 1909), 9.

38. Elmer R. Hodson, *Extent and Importance of the Chestnut Bark Disease* (Washington, D.C.: GPO, 1908), 1–7; Haven Metcalf and J. Franklin Collins, *The Present Status of the Chestnut Bark Disease*, USDA Bureau of Plant Industry Bulletin no. 141, part 5 (Washington, D.C.: GPO, 1909), 45; Marieka Gryzenhout, Brenda D. Wingfield, and Michael J. Wingfield, *Taxonomy, Phylogeny, and Ecology of Bark-Inhabiting and Tree-Pathogenic Fungi in the Cryphonectriaceae* (St. Paul, Minn.: American Phytopathological Society, 2009), 3–10.

39. Murrill, "Serious Chestnut Disease," 146. Murrill initially referred to the disease as "chestnut canker." See Murrill, "The Chestnut Canker," *Torreya* 8, no. 5 (May 1908): 111–12.

40. Metcalf and Collins, *Present Status of the Chestnut Bark Disease*, 7–10; Frank W. Rane, *The Chestnut Bark Disease: A Grave Danger Which Threatens Our Forest Trees, with Its Remedy* (Boston: Massachusetts State Forester's Office, 1911), 3–7; Haven Metcalf, "The Chestnut Tree Blight," *Scientific American* 106, no. 11 (March 16, 1912): 241–42.

41. Anderson and Rankin, *Endothia Canker of Chestnut*, 549; Frederick Deforest Heald, *The Symptoms of Chestnut Tree Blight and a Brief Description of the Blight Fungus* (Philadelphia: Pennsylvania Chestnut Tree Blight Commission, 1913), 3–11, plates 1–16; Merkel, "Deadly Fungus on the American Chestnut," 98.

42. Metcalf and Collins, *Present Status of the Chestnut Bark Disease*, 48.

43. Murrill, "Serious Chestnut Disease," 148, 150.

44. Merkel, "Deadly Fungus on the American Chestnut," 97, 101–2. Besides the Bordeaux Mixture, Merkel used a liquid solution of copper sulfate, but found it difficult to spray the trees "more than once during the past season" (ibid., 102).

45. Murrill, "Further Remarks on a Serious Chestnut Disease," 207. See also Anderson and Rankin, *Endothia Canker of Chestnut*, 539; Mickleborough, *Report on the Chestnut Tree Blight*, 9.

46. Nathaniel L. Britton, "Report of the Secretary and Director-in-Chief for the Year 1907," *Bulletin of the New York Botanical Garden* 6, no. 19 (February 24, 1908): 11. According to Britton, chestnut blight was "impossible to combat" (ibid.).

47. "Chestnut Trees Face Destruction," *New York Times*, May 21, 1908, 4. Murrill stated that "there should be a law to prevent the shipping of our chestnut trees now to other States," adding that "they have been shipped recently to California" (ibid.).

48. William A. Orton, "Plant Diseases in 1907," in *Yearbook of the United States Department of Agriculture: 1907* (Washington, D.C.: GPO, 1907): 587; Anderson and Rankin, *Endothia Canker of Chestnut*, 540; George P. Clinton, "Report of the Botanist for 1911–1912," in *Thirty-Sixth Annual Report of the Connecticut Agricultural Experiment Station: Being the Annual Report for the Year Ending October 31, 1912* (Hartford, Conn.: Tuttle, Morehouse & Taylor, 1913): 370.

49. This argument is based on the belief the disease moved southward and northward at the same rate. A point of origin farther south seems more logical, however. One of the earliest published maps of chestnut blight puts its epicenter near Flushing, Queens. See Metcalf and Collins, *Present Status of the Chestnut Bark Disease*, 46, fig. 2.

50. William A. Murrill, "The Spread of the Chestnut Disease," *Journal of the New York Botanical Garden* 9, no. 98 (February 1908): 27.

51. Clinton, "Report of the Botanist for 1911 and 1912," 361, 374.

52. Murrill, "Spread of Chestnut Disease," 27. Japanese chestnuts demonstrated a range of susceptibility to the blight and could, in rare instances, die from the attack.

53. Nathaniel L. Britton, "Appendix 3: List of Plants in the Grounds of the New York Botanical Temporary Garden and in the Greenhouse in 1897," in *Bulletin of the New York Botanical Garden* 1, no. 3 (February 15, 1898): 141.

54. No records in the Botanical Garden archives document a purchase of Japanese chestnut trees prior to 1904.

55. Anderson and Rankin, *Endothia Canker of Chestnut*, 565–73. See also Frank H. Tainter and Fred A. Baker, *Principles of Forest Pathology* (New York: John Wiley & Sons, 1996), 577–79.

56. Powell, *European and Japanese Chestnuts*, 21, fig. 7. Others noted chestnut blight initially appeared on the southern or western sides of the infected trees. In Connecticut, George P. Clinton witnessed numerous cankers, "and by far the larger part of these were on the south or southwest sides of the trees." Clinton, "Report of the Botanist for 1908," in *Thirty-First and Thirty-Second Annual Reports of the Connecticut Agricultural Experiment Station: Being the Biennial Report for the Two Years Ended October 31, 1908* (Hartford: Connecticut Agricultural Experiment Station, 1913): 882.

57. Powell, *European and Japanese Chestnuts*, 21. Powell's fieldwork took him to both New Jersey and Delaware, where presumably he observed the pathogen on European and American chestnut seedlings.

58. Frederick D. Chester, "Report of the Mycologist," in *Fourteenth Annual Report of the Delaware College Agricultural Experiment Station*, Arthur T. Neale (Wilmington, Del.: Mercantile Printing Company, 1903), 44.

59. Ibid. Early drawings of the blight fungus are found in John Mickleborough, *Report on the Chestnut Tree Blight*, 8–9; Ralph A. Waldron, "Physiological Studies on the Chestnut Blight Disease," in *Annual Report of the Pennsylvania State College for the Year 1912–1913*, Part 2 (Harrisburg, Pa.: William Stanley Ray, State Printer, 1914): 152–53, figs. 1, 2.

60. Metcalf and Collins, *Present Status of the Chestnut Bark Disease*, 46–47; Clinton, "Report of the Botanist for 1911–1912," 360.

61. Haven Metcalf, "Diseases of the Chestnut and Other Trees," in *Transactions of the Massachusetts Horticultural Society for the Year 1912*, part 1 (Boston: Massachusetts Horticultural Society, 1912), 78. See also Anderson and Rankin, *Endothia Canker of Chestnut*, 555; Haven Metcalf and J. Franklin Collins, *The Control of the Chestnut Bark Disease*, USDA Farmer's Bulletin no. 467 (Washington, D.C.: GPO, 1911), 20.

62. Haven Metcalf, *The Immunity of the Japanese Chestnut to the Bark Disease*, USDA Bureau of Plant Industry Bulletin no. 121, part 6 (Washington, D.C: GPO, 1908), 3–4.

63. Ibid., 4.

64. Cornelius L. Shear and Neil E. Stevens, "The Discovery of the Chestnut-Blight Parasite (*Endothia parasitica*) and Other Chestnut Fungi in Japan," *Science* 43, no. 1101 (February 4, 1916): 173–74.

65. Ibid., 175. Meyer previously observed *Cryphonectria parasitica* in Northern China, perhaps the first American to do so. On June 13, 1913, the U.S. State Department received a cablegram from him stating that he had "discovered the chestnut bark fungus." David Fairchild, "The Discovery of the Chestnut-Bark Disease in China," *Science* 38, no. 974 (August 29, 1913): 297; Cornelius L. Shear and Neil Stevens, "The Chestnut-Blight Parasite (*Endothia parasitica*) from China," *Science* 38, no. 974 (August 29, 1913): 295–97.

66. Shear and Stevens, "Discovery of the Chestnut-Blight Parasite," 175.

67. Ibid., 175–76. In 1914, Y. Kozai of the Imperial Agricultural Station near Tokyo wrote, "the chestnut blight is found to some extent in the Provinces of Tamba, Ise, Suruga and Shimotsuke (Nikko is in the latter). This disease is limited to the seedlings in the nursery and the young trees (three or four years old) in the field and may be prevented by spraying with Bordeaux mixture" (174).

68. Clinton, "Report of the Botanist for 1911–1912" 407–13.

69. Ibid., 390–407.

70. George P. Clinton, "Report of the Botanist for 1908," 362–63; George P. Clinton, "Some Facts and Theories Concerning Chestnut Blight," in *The Conference Called by the Governor of Pennsylvania to Consider Ways and Means for Preventing the Spread of the Chestnut Tree Bark Disease, Harrisburg, Pennsylvania, February 20 and 21, 1912*, Pennsylvania Chestnut Tree Blight Commission (Harrisburg, Pa.: C. E. Aughinbaugh, 1912), 78–80.

71. William G. Farlow, "Paper by Professor W. G. Farlow" (untitled, read by George P. Clinton in conference session), in Pennsylvania Chestnut Tree Blight Commission, *Conference Called by the Governor of Pennsylvania*, 71; William G. Farlow, "The Fungus of the Chestnut-Tree Blight," *Science* 35, no. 906 (May 10, 1912): 717–21; George P. Clinton, "The Relationships of the Chestnut Blight Fungus," *Science* 36, no. 939 (December 27, 1912): 907–14.

72. Murrill, "Serious Chestnut Disease," 153.

73. Raphael Zon, *Chestnut in Southern Maryland*, USDA Bureau of Forestry Bulletin no. 53 (Washington, D.C.: GPO, 1904), 14. Zon defined southern Maryland as Anne Arundel, Calvert, Charles, and Prince George Counties.

74. Ibid., 29. According to Zon, not only was it true that the "productive capacity of stumps decreases with their age" (31), but "there is an age when trees cease entirely to produce sprouts" (7).

75. Ibid., 29–30.

76. Ibid. (my emphasis). *Cryphonectria parasitica* was observed in Glenn Dale, Maryland, in 1905, at the breeding orchards of Walter Van Fleet. Van Fleet was growing both foreign and domestic chestnuts there as early as 1894 and had a working relationship with nurseryman John T. Lovett of Little Silver, New Jersey. See Walter Van Fleet, "Chestnut Work at Bell Experiment Plot," in *Report of the Proceedings at the Eleventh Annual Meeting of the Northern Nut Growers Association, Washington, D.C., October 7–8, 1920* (N.p.: Northern Nut Growers Association, 1921), 16–21.

77. David C. Marshall, "Periodical Cicadas, *Magicicada* spp. (Hemiptera: Cicadidae)," in *Encyclopedia of Entomology*, vol. 2, ed. John L. Capinera (New York: Springer, 2008), 2785–94.

78. Charles V. Riley, *The Periodical Cicada: An Account of Cicada Septendecim and its Tredecim Race, with a Chronology of All Broods Known*, USDA Division of Entomology Bulletin no. 8 (Washington, D.C: GPO, 1885), 10. In 1913, forest pathologist Samuel B. Detwiler observed blight infections "in wounds made by cicadas in 1911." See Samuel B. Detwiler, "Observations on Sanitation Cutting in Controlling the Chestnut Blight in Pennsylvania," in Chestnut Tree Blight Commission, *Final Report of the Pennsylvania Chestnut Tree Blight Commission, January 1 to December 15, 1913* (Harrisburg, Pa.: William Stanley Ray, 1914), 67.

79. "Periodical Cicadas," Magicicada Mapping Project, accessed August 23, 2018, http://www.magicicada.org/about/brood_pages/broodX.php.

80. In 1970, folksinger Bob Dylan witnessed the emergence of this brood when receiving his honorary degree from Princeton University. The experience inspired the song "Day of the Locusts" on the album *New Morning* (1970). In 1902, photographic evidence formed the basis for an estimate that "from 30,000 to 40,000 Cicada pupae" emerged from beneath a single large tree. Charles L. Marlatt, *The Periodical Cicada*, USDA Bureau of Entomology Bulletin no. 71 (Washington, D.C: GPO, 1907), 99.

81. Nelson F. Davis, "Chestnut Culture," in *The Publications of the Pennsylvania Chestnut Tree Blight Commission, 1911–1913* (Harrisburg, Pa.: William Stanley Ray, State Printer, 1915), 95. See also Altus Lacy Quaintance, *The Periodical Cicada, and Its Occurrence in Maryland in 1902*, Maryland Agricultural Experiment Station Bulletin no. 87 (College Park, Md.: Maryland Agricultural Experiment Station, 1902). The Quaintance report contains dozens of letters from farmers who witnessed cicada damage on or near their homesteads. George W. Main wrote, for example, "Oak and Chestnut trees are . . . greatly damaged on both Frederick and Washington County sides of the Mountain; it looks as though a fire had passed over killing the limbs" (111).

82. Robert A. Haack and Robert E. Acciavatti, "Twolined Chestnut Borer," Forest Insect and Disease Leaflet 168, accessed August 27, 2018, http://www.na.fs.fed.us/spfo/pubs/fidls/chestnutborer/chestnutborer.html.

83. Frank H. Chittenden, *The Two-Lined Chestnut Borer (Agrilus bilineatus Weber.)*, USDA Division of Entomology Circular no. 24, second series (Washington, D.C.: GPO, 1897), 2.

84. *Macon (Ga.) Telegraph*, August 54, 1912, 4.

85. "Chestnut Timberworm: *Melittomma sericeum* (Harris, 1841)," Center for Invasive Species and Ecosystem Health, accessed August 27, 2018, http://www.invasive.org/browse/subthumb.cfm?sub=2377. The impact of the insects on commercial lumber is found in Anthony D. Hopkins, "Insect Injuries to Hardwood Forest Trees," in *Yearbook of the United States Department of Agriculture, 1903* (Washington, D.C: GPO, 1904), 326; Philip L. Buttrick, "Commercial Uses of Chestnut," *American Forestry* 21, no. 262 (October 1915): 965–66.

86. Buttrick, "Commercial Uses of Chestnut," 965; Eric Sloan, *Our Vanishing Landscape* (New York: Funk and Wagnalls, 1955), 12–13; Constance E. Richards and Kenneth L. Richards, *North Carolina's Mountains: Including Asheville, Biltmore Estate, Cherokee, and the Blue Ridge Parkway* (Guilford: Morris Book Publishing, 2008), 66, 88.

87. Anthony D. Hopkins, *Defects in Wood Caused by Insects*, West Virginia Agricultural Experiment Station Bulletin no. 35 (Charleston: Moses W. Donnally, W.V., 1894), 292–93; Ephraim Porter Felt, *New York State Museum Memoir 8: Insects Affecting Park and Woodland Trees* (Albany, N.Y.: New York State Education Department, 1905), 7, collected in New York State Assembly,

Documents of the Assembly of the State of New York, One Hundred and Thirtieth Session, 1907, vol. 22, no. 66, part 3 (Albany, N.Y.: J. B. Lyon, 1907), 7; Agricultural Service Company, "Insect Injuries to the Wood of Living Trees," in *Abridged Agricultural Records*, vol. 6, *Insects Affecting Vegetation* (Washington, D.C.: Agriculture Service Company, 1912), 311.

88. U.S Senate, *Report of the National Conservation Commission*, vol. 2, 60th Congress, 2d Session (Washington, D.C.: GPO, 1909), 471. The National Conservation Commission claimed timber worms destroyed "not far from 30 percent" of the "average lumber product" (ibid.).

89. See, for example, Chittenden, *Two-Lined Chestnut Borer*, 2–3. This idea was also promoted by George P. Clinton of the Connecticut Agricultural Experiment Station. Clinton, "Some Facts and Theories Concerning Chestnut Blight," 76, 81.

90. Mickleborough, *Report on the Chestnut Tree Blight*, 10; Thomas E. Snyder, "Pole Damage by Insects," *Telephony* 60, no. 3 (January 21, 1911): 59–63; Mark Carleton, "The Fight to Save the Chestnut Trees: Final Report of the General Manager," in Chestnut Tree Blight Commission, *Final Report of the Pennsylvania Chestnut Tree Blight Commission*, 38–39; Frank W. Rane, quoted in meeting minutes, in Pennsylvania Chestnut Tree Blight Commission, *Conference Called by the Governor of Pennsylvania*, 152; Anthony D. Hopkins, "Relation of Insects to the Death of Chestnut Trees," *American Forestry* 18, no. 4 (April 1912): 221.

91. John Gifford, "A Preliminary Report on the Forest Conditions of South Jersey," in *Annual Report of the State Geologist for the Year 1894*, Geological Survey of New Jersey (Trenton, N.J.: John L. Murphy, 1895), 270. A single fire in Ocean and Burlington Counties consumed 125,000 acres, 271.

92. F. R. Meier, "Forest Fires in New Jersey during 1904," in *Annual Report of the State Geologist for the Year 1904*, Geological Survey of New Jersey (Trenton, N.J.: MacCrellish & Quigley , 1905), 288.

93. Ibid., 285.

94. Ibid., 286.

95. Ibid., 285.

96. Ibid., 277, 289. In 1902, sixty-five fires burned 98,850 acres of timber in New Jersey alone. F. R. Meier, "Forest Fires in New Jersey during 1902," in *Annual Report of the State Geologist for the Year 1902*, Geological Survey of New Jersey (Trenton, N.J.: John L. Murphy, 1903), 106.

97. In 1880, New Jersey railroad operators burned 36,948 acres of timberland out of the 71,074 acres destroyed by fire. Nathaniel H. Egleston, "Appendix A: Connection of Railroads with Forest Fires," in *Report on the Relation of Railroads to Forest Supplies and Forestry: Together with Appendices*, USDA Forestry Division Bulletin no. 1, ed. Bernhard E. Fernow (Washington, D.C.: GPO, 1887), 129.

98. Mickleborough, *Report on the Chestnut Tree Blight*, 10; Clinton, "Some Facts and Theories Concerning Chestnut Blight," 81; Paul J. Anderson and

D. C. Babcock, *Field Studies on the Dissemination and Growth of the Chestnut Blight Fungus*, Pennsylvania Chestnut Tree Blight Commission Bulletin no. 3 (Harrisburg, Pa.: C. E. Aughinbaugh, 1913), 16.

99. Detwiler, "Observations on Sanitation Cutting," 66. See also Philip L. Buttrick, "The Effect of Forest Fires on Trees and Reproduction in Southern New England," *Forestry Quarterly* 10, no. 2 (June 1912): 198–99.

100. Clinton, "Report of the Botanist for 1911–1912," 401.

101. Frank W. Rane, "Report of State Foresters or Other Officials on the Present Extent of the Bark Disease: An Estimate of the Present and Possible Future Losses," in *The Chestnut Blight Disease: Means of Identification, Remedies Suggested and Need of Co-operation to Control and Eradicate the Blight*, Pennsylvania Chestnut Tree Blight Commission Bulletin no. 1 (Harrisburg, Pa.: C. E. Aughinbaugh, 1912), 152.

102. See Commonwealth of Pennsylvania, *The Publications of the Pennsylvania Chestnut Tree Blight Commission, 1911–1913* (Harrisburg, Pa.: William Stanley Ray, State Printer, 1915).

103. The Pennsylvania Chestnut Blight Conference was held in the main chamber of the state house of representatives on February 20 and 21, 1912. Governor John Tener opened the proceedings, which included a written message from President William Taft, who told participants they had his "earnest sympathy" in fighting the disease, calling it "a destructive enemy of one of our most beautiful trees." Commonwealth of Pennsylvania, *Conference Called by the Governor of Pennsylvania*, 176.

104. Detwiler, "Observations on Sanitation Cutting," 77.

105. Irvin C. Williams, "A History of the Early Pennsylvania Effort to Combat the Chestnut Disease," in Chestnut Tree Blight Commission, *Final Report of the Pennsylvania Chestnut Blight Commission*, 23.

106. Detwiler, "Observations on Sanitation Cutting," 65.

107. The Plant Quarantine Act of 1912 placed tighter controls on the import of foreign nursery stock, but did not restrict the transport of trees within the United States, although it gave the USDA legal authority to do so. In Pennsylvania, legislation passed in 1911 placed a quarantine on the shipment of chestnut nursery stock, but it was not rigorously enforced. In practice, there was no prohibition on the shipment of trees in or out of state, only a mandate that they be inspected and sprayed with fungicide. Samuel B. Detwiler, "The Pennsylvania Programme," in Pennsylvania Chestnut Tree Blight Commission, *Conference Called by the Governor of Pennsylvania*, 132–33.

108. Federal Horticultural Board, "Notice of Proposed Quarantine on Account of the Chestnut-Bark Disease (*Endothia parasitica*)," in USDA *Service and Regulatory Announcements, April 1915* (Washington, D.C.: Federal Horticultural Board, 1915), 25. In attendance at a hearing of the board was

Coleman K. Sober, who convinced the USDA he could control the blight's spread even where it had been observed nearby. Because of Sober's testimony, the agency ruled "no action will be taken, therefore, pending further investigation of the possibility of the control of the disease under orchard conditions." Federal Horticultural Board, "Report on the Chestnut-Bark Disease Hearing," in USDA *Service and Regulatory Announcements, May 1915* (Washington, D.C.: Federal Horticultural Board, 1915), 39.

109. An excellent summary of the orchardist and nurserymen lobby in U.S. agricultural policy, including their role in importing foreign plants, is found in Philip J. Pauly, *Fruits and Plains: The Horticultural Transformation of America* (Cambridge, Mass.: Harvard University Press, 2007), 131–63.

110. See John B. Smith, *The Pernicious or San José Scale*, New Jersey Agricultural Experiment Stations Bulletin no. 116 (Trenton: New Jersey Agricultural Experiment Stations, 1896); William B. Alwood, *The Distribution of San José Scale in Virginia*, Virginia Agricultural Experiment Station Bulletin no. 66 (Blacksburg: Virginia Agricultural Experiment Station, 1896); Theodore D. A. Cockerell, *San José Scale and Its Nearest Allies*, USDA Division of Entomology Bulletin no. 6 (Washington, D.C.: GPO, 1897).

111. Leland O. Howard and Charles L. Marlatt, *The San José Scale: Its Occurrences in the United States*, USDA Division of Entomology Bulletin no. 3 (Washington, D.C.: GPO, 1896), 9–14, 36–38; Willis G. Johnson, *Report on the San José Scale in Maryland and Remedies for Its Suppression and Control*, Maryland Agricultural Experiment Station Bulletin no. 57 (College Park, Md.: Maryland Agricultural Experiment Station, 1898), 11–17.

112. Howard and Marlatt, *San José Scale*, 15.

113. Charles V. Riley, *The San José Scale*, Maryland Agricultural Experiment Station Bulletin no. 32 (College Park, Md.: Maryland Agricultural Experiment Station, 1895), 93–94.

114. Ibid., 94.

115. Ibid.

116. *Lovett's Catalogue*, Spring 1900 edition (Little Silver, N.J.: Monmouth Nursery, 1900), 2. Above the certification letter, Lovett wrote, "In order to have virgin soil upon which to grow Fruit Trees and Plants, and in a place removed from all infestation of the dreaded San José Scale, I am now growing my Fruit Trees and Small Fruit Plants . . . at Deal, Monmouth County, N.J." (ibid.)

117. Riley, *San José Scale*, 92–93, 110; Howard and Marlatt, *San José Scale*, 28; Smith, *Pernicious or San José Scale*, 4–5; Johnson, *Report on the San José Scale*, 14.

118. Francis M. Webster, "The San José Scale," in "Twenty-Seventh Annual Report of the Entomological Society of Ontario" (separately paginated), in *Annual Report of the Department of Agriculture of the Province of Ontario*, vol. 2 (Toronto: Warwick Bros. and Rutter, 1897), 93–94.

119. Ibid. The extent to which nursery owners escaped liability from government legislation and took no responsibility for their actions is addressed in an uncharacteristically frank paper delivered by entomologist Francis Webster at the twentieth annual meeting of the American Association of Nurserymen in 1895. In referencing the San José scale epidemic, he applauded a *Rural New Yorker* report exposing the carelessness of the Parsons and Sons nursery, stating "it was no excuse at all for them to plead ignorance of the dangerous character of the pest and neglect the repeated warnings that have been given" (94) Webster saw the epidemic as "a test case, as it were, as well as a reminder that the coming twentieth century would bring to us problems which we had not previously been called upon to solve" (86).

120. Corsa, *Nut Culture in the United States*, 88; Powell, *European and Japanese Chestnuts*, 4–5; George H. Powell, "Some Climatic and Fungus Troubles of the Chestnuts," *American Gardening* 20 (August 12, 1899): 559.

121. Tokyo Nursery Company, *General Catalogue of Plants, Bulbs, Seeds, etc.* (Tokyo: Aoyama Industrial Press, 1898), 18; Yokohama Nursery Company, *Catalogue of the Yokohama Nursery Company* (Yokohama, 1898), 40; Isabel S. Cunningham, *Frank N. Myer: Plant Hunter in Asia* (Ames: Iowa State University Press, 1984), 203; Bureau of Plant Industries, Foreign Seed and Plant Introductions Photo Album: Chili, Kiansu, and Chikiang Provinces, China and Nikko, Japan, April 23—September 14, 1915, National Agricultural Library, Special Collections, Frank. N. Meyer Collection, Plant Exploration, image no. 12347.

122. Frederick D. Heald, M. W. Gardner, and R. A. Studhalter, "Wind Dissemination of Ascospores of the Chestnut Blight Fungus," *Phytopathology* 4 (1914): 51.

123. In 1913, forest pathologist Frederick D. Heald wrote that "rain and wind are undoubtedly the most important natural agents in the dissemination of spores," Heald, *Symptoms of Chestnut Tree Blight*, 11.

124. E. B. Garriott, "Forecasts and Warnings," *Monthly Weather Review* 27 (October 1899): 449; James E. Hudgins, *Tropical Cyclones Affecting North Carolina since 1586: An Historical Perspective*, NOAA Technical Memorandum NWS ER92 (Bohemia, N.Y.: NOAA Scientific Services Division, 2000), 22.

125. Clinton, "Report of the Botanist for 1908," 888; Clinton, "Facts and Theories Concerning Chestnut Blight," 81.

126. The view that there were multiple sources of the blight in the Bronx is also advanced in Nicholas P. Money, The Triumph of the Fungi*: A Rotten History* (Oxford: Oxford University Press, 2007), 4. Several studies find birds to be reliable carriers of the fungus spores. See Frederick D. Heald and R. A. Studhalter, "Preliminary Note on Birds as Carriers of the Chestnut Blight Fungus," *Science* 38, no. 973 (August 22, 1913): 278–80; Craig S. Scharf and Nancy K. DePalma, "Birds and Mammals as Vectors of the Chestnut Blight Fungus

(*Endothia parasitica*)," *Canadian Journal of Zoology* 59, no. 9 (February 2011): 1647–50.

127. E. B. Garriott, "Forecasts and Warnings," *Monthly Weather Review* 32, no. 9 (September 1904): 401–2; Willis Moore, "Report of the Chief of the Weather Bureau," in *Annual Reports of the Department of Agriculture for the Fiscal Year Ended June 30, 1905*, U.S. Department of Agriculture (Washington, D.C.: GPO, 1905), 12–13; Emery R. Boose, Kristen E. Chamberlin, and David R. Foster, "Landscape and Regional Impacts of Hurricanes in New England," *Ecological Monographs* 71, no. 1 (February 2001): 27–48.

128. John Bancroft Bevins, "Editorial Notes," *New York Observer* 84, no. 34 (August 23, 1906), 234.

129. "The Costly Blight of the Chestnut Canker," *New York Times Magazine*, May 31, 1908, 9.

130. "The Chestnut Bark Disease," *New York Times*, August 25, 1909, 8.

131. Ibid. The recommendation resulted from the work of entomologist Charles Marlatt, who fought for the passage of stricter federal quarantine legislation after his experience with San José scale. See Philip J. Pauly, "The Beauty and Menace of the Japanese Cherry Trees: Conflicting Visions of American Ecological Independence," in *Science and the American Century: Readings from "Isis"*, ed. Sally Gregory Kohlstedt and David Kaiser (Chicago: University of Chicago Press, 2013), 49–50.

132. "The Chestnut Bark Disease," *New York Times*, August 25, 1909, 8.

CHAPTER 9. To Maintain the Balance of Nature

1. Roy G. Pierce, "The Need of State and National Quarantines to Prevent Rapid Spread of Chestnut Blight to Southern States," *Forestry Quarterly* 17, no. 1 (January 1919): 449–51. Pierce was concerned about the effectiveness of quarantines and noted North Carolina, South Carolina, Kentucky, Tennessee, Georgia, and Alabama had not adopted "measures for preventing blighted nursery stock from coming into their states" (450).

2. *New York Times*, October 2, 1910, section 5, 2. These are conservative figures as the U.S. chestnut crop alone was valued at $20 million in 1909. C. S. Knapp, "The Chestnut Blight," in *Transactions of the Warren Academy of Sciences, 1912–1913*, vol. 2, part 2 (Warren, Pa.: Warren Academy of Sciences, 1915), 65–66.

3. Samuel B. Detwiler, "Pennsylvania Programme," in *The Conference Called by the Governor of Pennsylvania to Consider Ways and Means for Preventing the Spread of the Chestnut Tree Bark Disease, Harrisburg, Pennsylvania, February 20 and 21, 1912*, Pennsylvania Chestnut Tree Blight Commission (Harrisburg, Pa.: C. E. Aughinbaugh, 1912), 130; Samuel B. Detwiler, "The Farmer and the Chestnut Blight," in *Proceedings of the Farmers' Annual Normal Institute and Spring*

Meeting, Commonwealth of Pennsylvania Bulletin no. 229 (Harrisburg, Pa.: C. E. Aughinbaugh, 1912), 71–72.

4. *New York Times*, July 17, 1910, 5. More than a thousand trees at Roosevelt's Sagamore Hill estate were infected with the blight before 1910 (*New York Times Magazine*, October 2, 1910, section 5, 2). The blight was also discovered in Petersham, Massachusetts, that year by John G. Jack of the Harvard Forest School. By 1912, 122 trees were severely infected, so much so that one forester predicted they would "die in two years." Joseph Kittredge Jr., "Notes on the Chestnut Bark Disease (*Diaporthe parasitica* Murrill) in Petersham, Massachusetts," *Bulletin of the Harvard Forestry Club* 2 (1913): 22.

5. *New York Times*, July 17, 1910, 5. Other New York estates suffered damage as a result of chestnut blight, including the John D. Rockefeller home in Pocantico Hills and the Helen Gould estate on the Hudson River. See also *New York Times*, October 2, 1910, section 5, 2.

6. *New York Times*, July 17, 1910, 5.

7. Oliver Peck Newman, "Foreign Plant Pests: A Narcissus Embargo Follows Success with Our Quarantine Policy," *American Review of Reviews* 72, no. 10 (October 1925): 412–13. Newman was referring to the Japanese chestnuts imported to America in 1876 by Samuel B. Parsons. He did so for plant collector Thomas Hogg, who needed the trees to complete his orchard of chestnut species. Other commentators suggested those trees were the source of the blight, but it is doubtful that this early shipment carried the disease. If they were infected with *Cryphonectria parasitica*, the pathogen would have spread to nearby trees in two or three years. The Parsons nursery may have been a source of the blight, but only after 1896.

8. Samuel B. Detwiler, "Some Benefits of the Chestnut Blight," *Forest Leaves* 13 (October 1912): 163.

9. Ibid.

10. Ibid., 163–64.

11. David G. White, "The Trend in Chestnut Production and Consumption," *Southern Lumberman* 139, no. 1778 (April 15, 1930): 46, table 1; Henry B. Steer, *Lumber Production in the United States: 1799–1946*, USDA Miscellaneous Publication no. 669 (Washington, D.C.: GPO, 1948), 136–37, table 57.

12. Steer, *Lumber Production*, 136–37, table 57.

13. Ibid. The figure was derived using an estimated four thousand board feet to the acre. Because chestnut did not grow at such densities across the entire state, more land surface was likely impacted as a result of the logging activities. For timber volume levels in Connecticut during the mid-twentieth century, see USDA, *Trends in Connecticut's Forests: A Half-Century of Change*, U.S. Forest Service Document NE-INF-143–01 (Newtown Square, Pa.: U.S. Forest Service, n.d.), 6.

14. See Tom Standage, *The Victorian Internet: the Remarkable Story of the Telegraph and the Nineteenth Century's On-Line Pioneers* (New York: Walker, 2007).

15. James D. Reid, *The Telegraph in America: Its Founders, Promoters, and Noted Men* (New York: Derby Brothers, 1879), 575; Gregory J. Downey, *Telegraph Messenger Boys: Labor, Communication and Technology, 1850–1950* (New York: Routledge, 2002), 15–38, 63–67; Annteresa Lubrano, *The Telegraph: How Technology Innovation Caused Social Change* (New York: Routledge, 2012), 71–72.

16. "Report of the Post-Master General," *Journal of the Telegraph* 6 (December 16, 1872): 27. In 1872, the Western Union Company transmitted 12.4 million messages and accumulating a profit of $2.8 million. Joshua D. Wolff, *Western Union and the Creation of the American Corporate Order, 1845–1893* (New York: Cambridge University Press, 2013), 180.

17. William A. Buckhout, *Chestnut Culture for Fruit*, Pennsylvania State College Agricultural Experiment Station Bulletin no. 36 (State College: Pennsylvania State College, 1896), 5; Bernhard E. Fernow, *Forestry for Farmers*, USDA Farmer's Bulletin no. 67 (Washington, D.C.: GPO, 1898), 36; Carl A. Schenck, *Biltmore Lectures on Sylviculture* (Albany, N.Y.: Brandow Printing Company, 1905), 160.

18. Howard F. Weiss, *Progress in Chestnut Pole Preservation*, USDA Forest Service Circular no. 147 (Washington, D.C.: GPO, 1908), 4–5.

19. Taltavall, "Pole Statistics," 295. See also H. P. Folsom, "The Preservation of Poles," *Telegraph Age* 26 (July 1909): 476; Thomas E. Snyder, *Damage to Chestnut Telephone and Telegraph Poles by Wood-Boring Insects*, USDA Bureau of Entomology Bulletin no. 94 (Washington, D.C.: GPO, 1910), 8–9.

20. John Moody, *The Truth about Trusts: A Description and Analysis of the American Trust Movement* (New York: Moody Publishing Company, 1904), 384. This is a conservative estimate, since chestnut in some years comprised 40 percent of all pole purchases. As more poles were erected west of the Mississippi River, the percentage dropped, falling to 16 percent by 1909. Olin J. Ferguson, *The Elements of Electrical Transmission* (New York: Macmillan, 1911), 47.

21. *New York Times*, November 12, 1887, 8. Written anonymously, the article is cleverly titled "Chestnuts, Most of Them: Gathering Facts about Telephone Poles."

22. Ibid.

23. Gifford Pinchot, *Forest Products of the United States: 1906*, USDA Forest Service Bulletin no. 77 (Washington, D.C.: GPO, 1908), 84.

24. Interstate and Foreign Commerce Committee, *The Transmission of Telegrams: Hearings before the Committee on Interstate and Foreign Commerce, Sixty-second Congress, H.R. 3010* (Washington, D.C.: GPO, 1912), 38, 79.

25. Henry Grinnell, *Seasoning of Telephone and Telegraph Poles*, USDA Forest Service Circular 103 (Washington, D.C.: GPO, 1907), 7. In 1909, of the 875,000 chestnut poles used by the telegraph and telephone industries, 10 percent were replaced annually due to damage or insects. Royal S. Kellogg, *The Timber Supply*

of the United States, USDA Forest Service Circular no. 166 (Washington, D.C.: GPO, 1909), 20–21.

26. Commission on Public Ownership and Operation, *Municipal and Private Operation of Public Utilities*, vol. 2 (New York: National Civic Federation, 1907), 1005–13.

27. Bernhard E. Fernow, "Consumption of Forest Supplies by Railroads and Practicable Economy in Their Use," in *Report on the Substitution of Metal for Wood in Railroad Ties*, ed. Edward E. Russell Tratman, USDA Forestry Division Bulletin no. 4 (Washington, D.C.: GPO, 1890), 13–39; Walter G. Berg, *Buildings and Structures of American Railroads* (New York: John Wiley & Sons, 1893), 118, 352; John Perlin, *A Forest Journey: The Role of Wood in the Development of Civilization* (Cambridge, Mass.: Harvard University Press, 1991), 346–55.

28. Jesse Hardesty, *Railroads: Their Construction, Cost, Operation, and Control* (Topeka, Kans.: Crane, 1898), 7–9; William S. Cappeller, *Annual Report of the Commissioner of Railroads and Telegraphs . . . for the Year 1887* (Columbus, Ohio: Westbote, 1888), 16.

29. Armin E. Shuman, "Statistical Report of the Railroads of the United States, 1880," in U.S. Census Office, *Tenth Census Reports*, vol. 4, *Report on the Agencies of Transportation in the United States* (Washington, D.C.: GPO, 1883), 467–71..

30. Andrew Morrison, "Essay on Track-Work," in *Proceedings of the Seventeenth Meeting of the Association of North American Railroad Superintendents, Held at New York, April 8, 1889*, American Railway Association, Operating Division (Boston: Rand Avery Supply, 1899), separately paginated, 11–13; Michael Williams, *Americans and Their Forests: A Historical Geography* (Cambridge: Cambridge University Press, 1992), 347–49.

31. M. N. Forney, "New Railroads Built in 1886," *Railroad and Engineering Journal* 61 (March 1887): 143. On real and perceived wood shortages in the late nineteenth century, see Edward A. Bowers, "The Present Condition of the Forests on Public Lands," in American Economic Association, *Report of the Proceedings of the American Economic Association at the Fourth Annual Meeting, Washington, D.C., December 26–30, 1890* (Baltimore: Guggenheimer, Weil, 1891), 239–58; Eric Rutkow, *American Canopy: Trees, Forests, and the Making of a Nation* (New York: Scribner, 2012), 99–168.

32. Michael Williams, *Deforesting the Earth: From Prehistory to Global Crisis* (Chicago: University of Chicago Press, 2003), 315.

33. See Bernhard E. Fernow, ed., *Report on the Relation of Railroads to Forest Supplies and Forestry*, Department of Agriculture Forestry Division Bulletin no. 1 (Washington, D.C.: GPO, 1887), 34, 38, 40, 55–58, 60.

34. Ibid., 7.

35. Harry M. Hale, "Cross-Ties Purchased by the Steam Railroads of the

United States in 1905," in Royal S. Kellogg and Harry M. Hale, *Forest Products of the United States: 1905*, USDA Forest Service Bulletin no. 74 (Washington, D.C.: GPO, 1907), 34.

36. Gifford Pinchot, *Forest Products of the United States: 1906*, USDA Forest Service Bulletin no. 77 (Washington, D.C.: GPO, 1908), 50.

37. Ibid.

38. Ernest A. Sterling, "The Development and Status of the Wood Preserving Industry," *Scientific American Supplement* 76, no. 1958 (July 12, 1913): 24. According to Sterling, of the 148 million railroad ties laid in 1910, "18% received preservative treatment," 25. See also Weiss, *Progress in Chestnut Pole Preservation*, 13.

39. George E. Walsh, "Metal and Wood Railroad Ties," *Scientific American* 81, no. 9 (August 26, 1899): 134.

40. This is not a direct quote from a single individual; I derived this formulation from several sources: George E. Walsh, "Forest Culture of To-day," *New England Magazine* 16, no. 4 (June 1897): 408–14; George A. Shepard, "Town of Sandisfield," in *History of Berkshire County, Massachusetts*, vol. 2, ed. Joseph E. Adams Smith (New York: J. B. Beers, 1885), 502; Austin F. Hawes and Ralph C. Hawley, *Forest Survey of Litchfield and New Haven Counties, Connecticut*, Connecticut Agricultural Experiment Station Forestry Publication no. 5 (Hartford, Conn.: Hartford Printing Company, 1909), 44–45.

41. Some late nineteenth-century commentators believed the railroad industry accounted for only 20 percent of the total timber consumption in any given year. See, for example, Bernhard E. Fernow, "Introductory," in *Report on the Substitution of Metal for Wood*, 7. The telegraph and telephone pole industries harvested more trees, but they were smaller in size, making the overall volume lower. The same tree producing a single telegraph pole made only three crossties. See Raphael Zon, *Chestnut in Southern Maryland*, USDA Bureau of Forestry Bulletin no. 53 (Washington, D.C.: GPO, 1904), 27.

42. After 1910, states reporting on the economic impact of wood industries specifically mentioned the value and quantity of chestnut products. In 1912, Connecticut placed the annual harvest of chestnut at 7,244,700 board feet. Albert H. Pierson, *Wood-Using Industries of Connecticut*, Connecticut Agricultural Experiment Station Bulletin no. 174 (New Haven, Conn.: Connecticut Agricultural Experiment Station, 1913), 15–16.

43. White, "Trend in Chestnut Production," 46.

44. The figures are based on those of William W. Ashe, who calculated the yield of "second quality" trees to be 7,400 board feet per acre and "third quality" to be 2,800 board feet per acre (a 5,100 bf per acre mean). Mature stands of timber yielded as much as 15,000 board feet per acre and the very poorest 2,000 board feet per acre. See Ashe, *Chestnut in Tennessee*, Tennessee Geological Survey Series Bulletin no. 10–B (Nashville, Tenn.: Baird-Ward, 1912), 32.

45. In terms of the total volume of U.S. lumber production, chestnut ranked eighteenth in 1899 and fifteenth in 1904. However, the figures do not reflect trees harvested for pole and crosstie construction or tanbark use. Chestnut did not have a major impact on the industry until 1907, when the annual cut increased to 653,239,000 board feet. See White, "Trend in Chestnut Production," 46.

46. Ibid., 47, table 3.

47. John Murdoch Jr., comp., *Chestnut: Its Market in Massachusetts* (Boston: Wright and Potter Printing Company, 1912), 6.

48. Albert H. Pierson, *Wood Using Industries of New Jersey*, Reports of the Forest Park Reservation Commission of New Jersey (Union Hill, N.J.: Dispatch Printing Company, 1914), 21. According to Pierson, "a greater quantity of chestnut is used in New Jersey than of any other hardwood, its rank being sixth among the fifty-six kinds reported" (ibid.).

49. Pierson, *Wood-Using Industries of Connecticut*, 6, 15.

50. Hu Maxwell, "The Uses of Wood: Wood for Musical Instruments," *American Forestry* 26, no. 321 (September 1920): 537.

51. Edwin Meeker, "Some Facts Regarding Pianos," *Hardwood Record* 44, no. 11 (March 25, 1918): 30. A more contemporary appreciation for the role chestnut lumber played in the construction of early twentieth-century pianos is found in Jon Taylor, "One Man's Trash," *Journal of the American Chestnut Foundation* 27, no. 1 (January/February 2013): 26–27.

52. R. V. Reynolds, *Wood-Using Industries of New York*, New York State College of Forestry Technical Publication no. 14 (Syracuse, N.Y.: Syracuse University, 1921), 44.

53. Royal S. Kellogg, *Lumber and Its Uses*, 2nd ed. (New York: U.P.C. Book Company, 1919), 261–62; Maxwell, "Uses of Wood," 535; White, "Trend in Chestnut Production," 47.

54. White, "Trend in Chestnut Production," 47. At that time, the word "coffin" referred to hexagonal or "body"-shaped burial boxes, and the word "casket" referred to rectangular ones. See Martha Pike, "In Memory Of: Artifacts Related to Mourning in Nineteenth Century America," in *Rituals and Ceremonies in Popular Culture*, ed. Ray B. Browne (Bowling Green, Ky.: Bowling Green University Popular Press, 1980), 312.

55. Reynolds, *Wood-Using Industries of New York*, 75.

56. Carroll W. Dunning, *Wood-Using Industries of Ohio*, U.S. Forest Service/Ohio Agricultural Experiment Station Joint Publication (Wooster: Ohio Experiment Station Press, 1912), 83–84.

57. Ibid., 84.

58. Reynolds, *Wood-Using Industries of New York*, 75–77, table 11. In 1911, the New York coffin and casket industry consumed 9.1 million board feet of chestnut. See also John T. Harris, *Wood-Using Industries of New York*, New York

State College of Forestry Bulletin no. 1 (Albany, N.Y.: Syracuse University, 1921), 61, table 17.

59. U.S. Department of Health and Human Services, "The Great Pandemic: The United States in 1918–1919," U.S. Department of Health and Human Services, accessed December 21, 2018, http://www.flu.gov/pandemic/history/1918/the_pandemic/index.html. See also Alfred W. Crosby, *America's Forgotten Pandemic: The Influenza of 1918*, 2nd ed. (Cambridge: Cambridge University Press, 2003), 206–7.

60. Roger E. Simmons, *Wood-Using Industries of North Carolina*, North Carolina Geological and Economic Survey Economic Paper no. 20 (Raleigh, N.C.: E. M. Uzzell, 1910), 19, 42, 62; J. C. Nellis, "Indiana's Wood-Using Industries," *Hardwood Record* 40, no. 12 (October 10, 1915): 17; James K. Crissman, *Death and Dying in Central Appalachia: Changing Attitudes and Practices* (Urbana: University of Illinois Press, 1994), 47–50.

61. Rufus K. Helphenstine Jr., *Wood-Using Industries of North Carolina*, North Carolina Geological and Economic Survey Bulletin no. 30 (Raleigh, N.C.: Mitchell Printing Company, 1923), 70–71.

62. Ibid., 27, 40.

63. Alternative scenarios regarding chestnut blight control are discussed in chapter 10. See also Money, *Triumph of the Fungi*, 7–10; Susan Freinkel, *American Chestnut: The Life, Death, and Rebirth of a Perfect Tree* (Berkeley: University of California Press, 2007), 48–51, 58, 60–64; Rutkow, *American Canopy*, 215–18.

64. Frank W. Rane, "Report of State Foresters or Other Officials on the Present Extent of the Bark Disease: An Estimate of the Present and Possible Future Losses," in *The Chestnut Blight Disease: Means of Identification, Remedies Suggested and Need of Co-operation to Control and Eradicate the Chestnut Blight*, Pennsylvania Chestnut Tree Blight Commission Bulletin no. 1 (Harrisburg, Pa.: C. E. Aughinbaugh, 1912), 152.

65. Mark A. Carleton, "The Fight to Save the Chestnut Trees: Final Report of the General Manager," in Chestnut Tree Blight Commission, *Final Report of the Pennsylvania Chestnut Tree Blight Commission, January 1 to December 31, 1913* (Harrisburg, Pa.: William Stanley Ray, 1914), 29–30; Caroline Rumbold, "Report of the Physiologist," in Commonwealth of Pennsylvania, *Report of the Chestnut Tree Blight Commission, July 1 to December 31, 1912* (Harrisburg, Pa.: C. E. Aughinbaugh, 1913), 45–47; Caroline Rumbold, "The Injection of Chemicals into Chestnut Trees," *American Journal of Botany* 7, no. 1 (January 1920): 1–10.

66. Caroline Rumbold, "Giving Medicine to Trees," *Scientific American Monthly* 2, no. 2 (October 1920): 116.

67. Ibid. Chemical injections have been used to treat Dutch Elm disease as well as ink disease in European chestnuts. See Linda Haugen and Mark Stennes, "Fungicide Injection to Control Dutch Elm Disease: Understanding

the Options," *Plant Diagnostics Quarterly* 20, no. 2 (June 1999): 29–38; S. Gentile, D. Valentino, and G. Tamietti, "Control of Ink Disease by Trunk Injection of Potassium Phosphate," *Journal of Plant Pathology* 91, no. 3 (November 2009): 565–71.

68. Rumbold, "Giving Medicine to Trees," 116.

69. Russell B. Clapper, "Chestnut Breeding, Techniques and Results: I. Breeding Material and Pollination Techniques," *Journal of Heredity* 45, no. 3 (May/June 1954): 107.

70. D. Cunningham, "The American Chestnut of the Future," *Country Life in America* 30, no. 6 (October 1916): 98–100. See also Walter Van Fleet, "Chestnut Breeding Experience," *Journal of Heredity* 5, no. 1 (January 1914): 19–25; Charles R. Burnham, "Historical Overview of Chestnut Breeding in the United States," *Journal of the American Chestnut Foundation* 2, no. 1 (December 1987): 9–11.

71. David Fairchild, "The Discovery of the Chestnut Bark Disease in China," *National Nurseryman* 22, no. 9 (September 1913): 349–50; H. L. Crane, C. A. Reed, and M. N. Wood, "Nut Breeding," in United States Department of Agriculture, *Yearbook of Agriculture* (Washington, D.C.: GPO, 1937), 827–89; John W. McKay and H. L. Crane, "Chinese Chestnut—A Promising New Orchard Crop," *Economic Botany* 7, no. 3 (July–September 1953): 228.

72. National Nut-Growers Association, "Breeding Chestnuts for Disease Control," *Nut-Grower* 15, no. 3 (March 1916): 37.

73. Ibid. See also [Ralph T. Olcott], "Disease-Resistant Chestnuts," *American Nut Journal* 6, no. 4 (April 1917): 59.

74. Clarence A. Reed, "The 1946 Status of Chinese Chestnut Growing in the Eastern United States," *National Horticultural Magazine* 26, no. 4 (April 1947): 88.

75. William A. Murrill, "Hybrid Chestnuts and Other Hybrids," *Journal of the New York Botanical Garden* 28, no. 214 (October 1917): 214.

76. Ibid., 215.

77. Amanda Ulm, "Remember the Chestnut," in *Annual Report of the Board of Regents of the Smithsonian Institution for the Year Ended 1948* (Washington, D.C.: Smithsonian Institution, 1949), 381; Jesse D. Diller, Russell B. Clapper, and Richard A. Jaynes, *Cooperative Test Plots Produce Some Promising Chinese and Hybrid Chestnut Trees*, U.S. Forest Service Research Note NE-25 (Upper Darby: Pa.: U.S. Forest Service, Northeastern Forest Experiment Station, 1964), 5–6.

78. Jesse D. Diller, *Growing Chestnuts for Timber*, Forest Pathology Special Release no. 31 (Beltsville, Md.: USDA Bureau of Plant Industry, 1947), 1.

79. Jesse D. Diller and Russell B. Clapper, "A Progress Report on Attempts to Bring Back the Chestnut Tree in the Eastern United States , 1954–1964," *Journal of Forestry* 63, no. 3 (March 1965): 186–87.

80. Russell B. Clapper and George Gravatt, "Status of Work with Blight Resistant Chestnuts," in *Twenty-Seventh Annual Report of the Northern*

Nut-Growers Association (Geneva, N.Y.: Northern Nut-Growers Association, 1936), 58–60; Russell B. Clapper and George Gravatt, "The American Chestnut: Its Past, Present, and Future," *Southern Lumberman* 167, no. 2105 (1943): 227–29.

81. Russell B. Clapper, "Chestnut Breeding, Techniques and Results, II: Inheritance of Characters, Breeding for Vigor, and Mutations," *Journal of Heredity* 45, no. 4 (July 1954): 201. See also Clapper and Gravatt, "Status of Work with Blight Resistant Chestnuts," 59; Diller and Clapper, "Progress Report," 187.

82. Crane, Reed, and Wood, "Nut Breeding," 834.

83. *New York Times*, July 30, 1911, 6. In 1911, only two large surviving American chestnuts could be found in the New York Botanical Garden. "They are fine old trees," stated the report, "the last of the garden's beautiful chestnut groves, seven feet in diameter and over twenty-two in circumference" (ibid.).

84. Ibid.

85. "Timber Resources of the Southern Appalachians Rapidly Being Developed," *American Lumberman*, part 2, no. 1788 (August 28, 1909): 26; White, "Trend in Chestnut Production," 47; Steer, *Lumber Production*, 136–37.

86. Steer, *Lumber Production*, 137.

87. Ibid., 136–38. In 1911, the U.S. Department of Commerce noted West Virginia was "coming to the front in the production of chestnut lumber, and the output in that state in 1909 exceeded by some 12,000,000 feet the cut in Pennsylvania, which had previously ranked first." U.S. Department of Commerce and Labor, *Forest Products of the United States: 1909* (Washington, D.C.: GPO, 1911), 30.

CHAPTER 10. Grandfather Had Lived in a Log

1. [Wesley T. Christine], "Whiting and Timber Mill Interests," *American Lumberman* 1808 (January 15, 1910): 51–114. Although Streator identifies the location as "Poplar Cove," other captions suggest the cove encompassed the entire "headwaters of Little Santeetlah Creek," 67. Today, only a small portion of the watershed bears the name Poplar Cove, including the highly visited Loop Trail in the Joyce Kilmer Memorial Forest. Owen L. McConnell, personal communication, November 8, 2014. See also Owen L. McConnell, *Unicoi Unity: A Natural History of the Unicoi and Snowbird Mountains and their Plants, Fungi, and Animals* (Bloomington, Ind.: AuthorHouse, 2014), 7–9, 157–59.

2. The two men were possibly employed by *American Lumberman* as backcountry guides, since the photography shoot involved taking hundreds of photographs in rugged terrain. In referencing the photograph collection, the editor claimed they "were the most remarkable series of timber pictures ever taken." Christine, "Whiting and Timber Mill Interests," 32.

3. Ibid., 74. When Craig Lorimer measured the trees in the mid-1970s, several decades after they lost their bark and fell to the forest floor, the trees in

the foreground were 5'7" and 5'10" inches in diameter. Craig G. Lorimer, "Age Structure and Disturbance History of a Southern Appalachian Virgin Forest," *Ecology* 61, no. 5 (October 1980): 1174.

4. McConnell, *Unicoi Unity*, 92. The Whiting brothers owned an additional twenty-three thousand acres in Graham County, but much of that timber had been cut or was in the process of being harvested. [James E. Defebaugh], "Pertaining to the Manufacture of Lumber from Southern Appalachian Hardwood and Softwood Timber," *American Lumberman*, part 1, no. 1716 (April 11, 1908): 1, 63–74; "W. S. Whiting," in *History of North Carolina*, vol. 6, *North Carolina Biography* (Chicago: Lewis, 1919), 347–48.

5. Christine, "Whiting and Timber Mill Interests," 98.

6. Henry B. Steer, *Lumber Production in the United States: 1799–1946*, USDA Miscellaneous Publication no. 669 (Washington, D.C.: GPO, 1948), 137–38, table 57.

7. Christine, "Whiting and Timber Mill Interests," 68. In 1910, Graham County possessed the largest concentration of mature chestnut trees in the United States, possessing as much as twenty-five thousand board feet of timber per acre. See "Land Classification Map of Part of the Southern Appalachian Region, 1904," insert in Horace B. Ayers and William W. Ashe, *The Southern Appalachian Forests*, U.S. Geological Survey Professional Paper no. 37 (Washington, D.C.: GPO, 1905), n.p.

8. Christine, "Whiting and Timber Mill Interests," 66.

9. William W. Ashe, *Chestnut in Tennessee*, Tennessee Geological Survey Series Bulletin no. 10–B (Nashville, Tenn.: Baird-Ward, 1912), 16. Ashe wrote that "chestnut suffers severely from fire because of its thin bark. . . . Since the sprout stands are those left after lumbering, they have been frequently burned, and the trees are either hollow or defective at the base" (10).

10. " 'Twas All Moonshiner: Smoking Tree Mystery in North Carolina a Case for Revenue Only," *New York Times*, July 22, 1894, 8. The incident was reported in dozens of American newspapers, including the *San Francisco Chronicle* (November 5, 1894), 9; *Cincinnati Enquirer* (July 28, 1894), 14; *Los Angeles Herald* (November 7, 1894), 10.

11. "'Twas All Moonshiner." In the account in the Jeffersonville, Virginia, *Clinch Valley News*, the distillery was not inside the chestnut but in a "tunnel-like cavern" thirty feet from the tree. In that version, a small tunnel had been dug into the "eight feet in diameter" chestnut, "thus allowing the smoke to enter the hollow beneath the ground." *Clinch Valley News* (Tazewell, Va.), April 6, 1894, 1.

12. " 'Twas All Moonshiner."

13. Ibid. A prominent member of the Pisgah community, Owens was known for making a "cherry bounce" cordial crafted from mountain honey, wild cherries, and illicit moonshine. For more on Owens, see M. L. White, *A History of the Life of Amos Owens, the Noted Blockader of Cherry Mountain, N.C.* (Shelby, N.C.:

Cleveland Star Printers, 1901); Anita Price Davis, *Legendary Locals of Rutherford County, North Carolina* (Charleston, S.C.: Arcadia, 2013), 22–24. "Blockade" was one of many contemporary terms for moonshine.

14. Carson Brewer, *Just Over the Next Ridge* (Knoxville, Tenn.: Knoxville News-Sentinel, 1992), 100. See also Ted Olson, *Blue Ridge Folklife* (Jackson: University Press of Mississippi, 1998), 59–60; Christopher Camuto, *Another Country: Journeying toward the Cherokee Mountains* (Athens: University of Georgia Press, 2000), 293.

15. Brewer, *Just Over the Next Ridge*, 100.

16. Nelms is quoted in Tim Homan, *Hiking Trails of the Joyce Kilmer–Slickrock and Citico Creek Wildernesses* (Atlanta: Peachtree), 2008, 118.

17. Grossman quoted in Margaret Lynn Brown, *The Wild East: A Biography of the Great Smoky Mountains* (Gainesville: University Press of Florida, 2001), 21. Frank W. Woods, a University of Tennessee forestry professor, believed the chestnut was the same one another Cosby farmer used "as a barn for a pig and a cow." David Wheeler, "Where There Be Mountains, There Be Chestnuts," *Katuah Journal* 21, no. 3 (Fall 1988): 28.

18. Joseph S. Illick, *Tree Habits: How to Know the Hardwoods* (Washington, D.C.: American Tree Association, 1924), 101. Illick, a Pennsylvania State forester, noted the tree's height was "more than 100 feet" (ibid.). See also Samuel B. Detwiler, "The American Chestnut Tree," *American Forestry* 21, no. 262 (October 1915): 957; William Carey Grimm, *Familiar Trees of America* (New York: Harper & Row, 1967), 109.

19. Colby B. Rucker, "Great Eastern Trees, Past and Present," Native Trees Society, February 2004, accessed November 17, 2018, http://www.nativetreesociety.org/bigtree/great_eastern_trees.htm.

20. Randy Cyr, "Francis Cove's Huge American Chestnuts," Native Trees Society, January 18, 2004, accessed November 18, 2018, http://www.nativetreesociety.org/fieldtrips/north_carolina/francis_cove.htm. The Francis Cove tree was a rare and exceptional anomaly. If removed for firewood, the measurement was likely taken at the very base of the stump, which may explain its exceptional diameter. A more skeptical assessment of the tree's size is found in Rachel J. Collins et al., "American Chestnut: Re-examining the Historical Attributes of a Lost Tree," *Journal of Forestry* 116, no. 1 (January 2018): 68–75. As the authors note, the historical record is replete with examples of trees eight, nine, and ten feet in diameter. If the Francis Cove tree was only seventeen feet in circumference (5.4 feet in diameter), as they claim, it would hardly be worthy of mention.

21. Cyr, "Francis Cove's Huge American Chestnuts."

22. See Henry S. Graves, *The Use of Wood for Fuel*, U.S. Forest Service Office of Forest Investigations, comp., USDA Bulletin no. 753 (Washington, D.C.: GPO,

1919), 38; Carl A. Schenck, *Logging and Lumbering, or Forest Utilization: A Textbook for Forest Schools* (Darmstadt, Germany: L. C. Wittich, 1912), 109.

23. Graves, *The Use of Wood for Fuel*, 26. For additional commentary on chestnut firewood, see Philip L. Buttrick, "Commercial Uses of Chestnut," *American Forestry* 21, no. 262 (October 1915): 964; Bethany N. Baxter, "An Oral History of the American Chestnut in Appalachia" (master's thesis, University of Tennessee at Chattanooga, 2009), 36–37; Jay Erskine Leutze, *Stand Up that Mountain: The Battle to Save One Small Community in the Wilderness along the Appalachian Trail* (New York: Scribner, 2013), 79.

24. Emma Bell Miles, *Once I Too Had Wings: The Journals of Emma Belle Miles , 1908–1918*, ed. Steven Cox (Athens: University of Ohio Press, 2014), 28. Miles wrote of the typical East Tennessee distillery, "The still proper is a copper vessel almost round, with a capacity of from 100 to 300 gallons. . . . Chestnut wood, which makes next to no smoke, is preferred for stoking" (28).

25. Thomas T. Taber, *Tanbark, Alcohol, and Lumber*, Logging Railroad Era of Lumbering in Pennsylvania no. 10. (Williamsport, Pa.: Lycoming Printing Company, 1974), 1082–91; Gordon G. Whitney, *From Coastal Wilderness to Fruited Plain: A History of Environmental Change in Temperate North America from 1500 to the Present* (Cambridge: Cambridge University Press, 1996), 186–88.

26. Barbara McMartin, *Hides, Hemlocks and Adirondack History* (Utica, N.Y.: North Country Books, 1992), 5–6, 8–10.

27. Franklin B. Hough, *Report on Forestry*, vol. 3 (Washington, D.C.: GPO, 1882), 68–103.

28. Ibid., 104, 107–8, 110, 117–28.

29. Nelson C. Brown, *Forest Products: Their Manufacture and Use* (New York: John Wiley and Sons, 1919), 65–66.

30. Rob Neufeld, "Portrait of the Past: Hans Rees Tannery in Asheville," *Asheville Citizen-Times*, October 23, 2014, accessed July 8, 2019, https://www.citizen-times.com/story/life/2014/10/23/portrait-past-hans-rees-tannery-asheville/17791999.

31. U.S. Bureau of the Census, *Forest Products of the United States, 1909*, vol. 3 (Washington, D.C.: GPO, 1911), 3–6. In 1909, the top five consumers of chestnut extract were Pennsylvania (69,527,427 pounds), North Carolina (18,072,965 pounds), West Virginia (14,745,548 pounds), Virginia (12,156,061 pounds), and New York (11,951,965 pounds).

32. H. K. Benson, *By-Products of the Lumber Industry*, Department of Commerce Special Agent Series no. 110 (Washington, D.C.: GPO, 1916), 33–34.

33. Brown, *Forest Products*, 36, 73–74; Benson, *By-Products of the Lumber Industry*, 34.

34. U.S. Bureau of the Census, *Forest Products of the United States, 1909*, 4. See also Benson, *By-Products of the Lumber Industry*, 33–34.

35. Benson, *By-Products of the Lumber Industry*, 32–33.

36. Gifford Pinchot, *Forest Products of the United States, 1906*, USDA Forest Service Bulletin 77 (Washington, D.C.: GPO, 1908), 77; Ezra S. Grover, "Hide and Skin Markets," *Shoe and Leather Reporter* 108 (October 17, 1912): 58; U.S. Congress, "Tanning Extracts," in *Tariff Hearings before the Committee on Ways and Means of the House of Representatives*, 60th Congress, 2nd Session (Washington, D.C.: GPO, 1909), 7916–17.

37. Robert W. Griffith, "Chestnut Wood in the Tanning Industry," *Journal of Forestry* 22, no 5 (May 1924): 543. See also F. H. Small, "The Manufacture of Leather Belting," *Journal of the American Society of Mechanical Engineers* 37, no. 2 (December 1915): 679–82; Charles A. Oberfell, "Manufacture and Use of Chestnut Extract," *Hide and Leather* 64 (October 21, 1922): 26–28.

38. U.S. Census Bureau, *Fourteenth Census of the United States*, vol. 5, *General Report and Analytic Tables* (Washington, D.C.: GPO, 1923), 486–87. In 1919, 1,509,944,000 pounds of chestnut cordwood was consumed for extract or related dyestuffs in the United States (486). See also Griffith, "Chestnut Wood in the Tanning Industry" (544).

39. U.S. Census Bureau, *Fourteenth Census of the United States*, vol. 5, 487.

40. Samuel B. Detwiler, "The Farmer and the Chestnut Blight," in *Proceedings of the Farmers' Annual Normal Institute and Spring Meeting*, Commonwealth of Pennsylvania Bulletin no. 229 Commonwealth of Pennsylvania (Harrisburg, Pa.: C. E. Aughinbaugh, 1912), 74; Samuel B. Detwiler, "The Chestnut Blight and Its Remedy," in *Pennsylvania Arbor Day Manual, 1913*, ed. Pennsylvania Department of Public Instruction (Harrisburg, Pa.: C. E. Aughinbaugh, 1913), 172.

41. Joseph Shrawder, "Report of the Chemist," in Chestnut Tree Blight Commission, *Report of the Pennsylvania Chestnut Tree Blight Commission, July 1 to December 31, 1912* (Harrisburg, Pa.: C. E. Aughinbaugh, 1913), 49–50. See also Melville T. Cooke and Guy W. Wilson, *The Influence of the Tannin Content of the Host Plant on Endothia Parasitica and Related Species*, New Jersey Agricultural Experiment Stations Bulletin no. 291 (New Brunswick, N.J.: New Jersey Agricultural Experiment Stations, 1914), 3–47.

42. Shrawder, "Report of the Chemist," 50.

43. Mark A. Carleton, "The Fight to Save the Chestnut Trees: Final Report of the General Manager," in Chestnut Tree Blight Commission, *Final Report of the Pennsylvania Chestnut Tree Blight Commission, January 1 to December 31, 1913* (Harrisburg, Pa.: William Stanley Ray, 1914), 27; [Ralph T. Olcott], "Pennsylvania Chestnut Trees to Be Sold to Save Timber Left by Blight: Gifford Pinchot, Forester, Explains the Action," *American Nut Journal* 12, no. 6 (June 1920): 91.

44. On the volume of cordwood used in American tanneries in 1919, see U.S. Census Bureau, *Fourteenth Census of the United States*, vol. 5, 487. On the debarking of chestnut for processing, see Samuel J. Record, "Utilizing Blight

Killed Chestnut," *Lumber World Review* 23, no. 23 (December 10, 1912): 24–25; W. D. Sterrett, "Marketing Woodlot Products in Tennessee," in *The Resources of Tennessee*, vol. 7 (Nashville: Tennessee State Geologic Survey, 1918), 135.

45. Marion Extract Company broadside, "Chestnut Wood Requirements," photocopy in possession of author. The Marion Extract Company was based in Marion, Virginia, but was incorporated in Delaware to minimize tax payments and avoid legal liabilities.

46. Marion Extract Company to B. Y. Dickey of Sugar Valley, Georgia, January 24, 1922, photocopy of letter in possession of author.

47. U.S. Bureau of the Census, "Tanning Materials, Natural Dyestuffs, Mordants and Assistants, and Sizes," in *Biennial Census of Manufactures: 1925* (Washington, D.C.: GPO, 1928), 838.

48. U.S. Bureau of the Census, "Leather Tanned, Curried, and Finished," in *Biennial Census of Manufactures: 1923* (Washington, D.C.: GPO, 1926), 534, table 6.

49. Howard Long, *Kingsport: A Romance of Industry* (Kingsport, Tenn.: Sevier, 1928), 110. The Kingtan Extract Company made "sole leather; welting leather; bark extract; liquid chestnut extract; powdered chestnut extract; decolorized chestnut extract; special blends extracts" (107).

50. The first postblight mention of the term "chestnut belt" as a reference to the Appalachian area containing the largest concentration of healthy and merchantable trees is in *The Conference Called by the Governor of Pennsylvania to Consider Ways and Means for Preventing the Spread of the Chestnut Tree Bark Disease, Harrisburg, Pennsylvania, February 20 and 21, 1912*, Pennsylvania Chestnut Tree Blight Commission (Harrisburg, Pa.: C. E. Aughinbaugh, 1912), 24.

51. Reuben B. Robertson, "Oral History Interview," Asheville, North Carolina, February 15, 1959, transcript, Forest History Society, Durham, North Carolina, 4. Evidence Thompson planned to use chestnut prior to moving to Canton is found in the trade journal *Concrete*. A 1907 article discusses the various concrete structures at the site, all designed to "utilize a new process for extracting tannic acid from chestnut wood." Walter C. Boynton, "Twelve Acres of Reinforced Concrete Buildings," *Concrete* 7, no. 4 (April 1907): 26.

52. The patent was issued to "Oma Carr, of Buena Vista, Virginia." Oma Carr, Manufacturing paper-pulp, U.S. Patent *762,139, filed April 22, 1903, and issued June 7, 1904*. According to Robertson, the sale of chestnut extract made the Champion Fibre Company successful. "There was many years in which the sale of tanning materials paid for the wood," recalled Robertson, "so that was Champion's most profitable enterprise." Robertson, "Oral History Interview," 3.

53. In 1911, the Champion Fibre Company operated one of "the largest combined sulphite and soda mills in the world." Its ninety thousand acres of timber holdings were reported to contain 160,535,000 board feet of chestnut. *Commercial and Financial Chronicle* 93, no. 2407 (August 12, 1911): 410. See also

Henry H. Gibson, "North Carolina Timber Situation," *Hardwood Record* 34, no. 3 (May 25, 1912): 38–39; Elaine K. Lanning, "Robertson, Reuben Buck," in *Dictionary of North Carolina Biography*, vol. 5, *P–S*, ed. William S. Powell (Chapel Hill: University of North Carolina Press, 1994), 231–32.

54. The numbers, from *The 1916 Pictorial Story of Haywood County: Reprint of a Special Industrial and Resort Edition of the Carolina Mountaineer*, are quoted on "Travel Western North Carolina: Taking the Train: Canton," Hunter Library, Special Collections, Western North Carolina University, accessed December 3, 2018, http://www.wcu.edu/library/DigitalCollections/TravelWNC/1910s/1910canton.html.

55. Gibson, "North Carolina Timber Situation," 39; John S. Holmes, *Forest Conditions in Western North Carolina*, North Carolina Geological and Economic Survey Bulletin no. 23 (Raleigh, Va.: Edwards and Broughton, 1911), 66; W. J. Damtoft, "Activities of a Large Pulp Company and a Forest Engineer's Relation Thereto," in *Ames Forester*, vol. 11 (Ames: Iowa State College Forestry Club, 1923), 128.

56. Holmes, *Forest Conditions in Western North Carolina*, 65; Richard A. Bartlett, *Troubled Waters: Champion International and the Pigeon River Controversy* (Knoxville: University of Tennessee Press, 1995), 302; Paul Franklin, "The Rise and Fall of Champion Chestnut Extract," *Journal of the American Chestnut Foundation* 26, no. 3 (May/June 2012): 14–15. On extract exports to Europe, see U.S. Tariff Commission, "Tanning Materials and Natural Dyes," in *Tariff Information Surveys on the Articles in Paragraphs 30 and 31 of the Tariff Act of 1913* (Washington, D.C.: GPO, 1921), 13, 18, 20.

57. A. H. Lockwood, "War Creates Export Demand," *Shoe and Leather Reporter* 116, no. 1 (October 1914): 14. See also A. H. Lockwood, "The War and Quebracho Prices," *Shoe and Leather Reporter* 116, no. 3 (October 15, 1914): 14–15; H. Frederick Lesh, "Sole Leather Situation Analyzed," *Shoe and Leather Facts* 36, no. 2 (February 1916): 47–48; Robert W. Griffith, "Review of Tanning Extract Industry," *Hide and Leather* 63 (May 13, 1922): 17.

58. Franklin, "Rise and Fall of Champion Chestnut Extract," 15.

59. Champion Paper and Fibre Company, *The Story of Chestnut Extract* (Canton, N.C.: Champion Paper and Fibre Company, 1937), n.p.

60. *Jackson County Journal*, March 3, 1933, 1. By the early 1930s, the Armour Leather Company annually produced thirty million pounds of extract (ibid.).

61. George Henry Smathers, "Memoirs of George Henry Smathers," ed. Allen T. Roudebush (unpublished manuscript, August 1990), University of North Carolina Asheville, D. H. Ramsey Library, Special Collections, 113.

62. Margaret R. Wolfe, *Kingsport, Tennessee: A Planned American City* (Lexington: University Press of Kentucky, 1987), 64; Anna Ahola, "Comparing the Strategic Evolution of Georgia-Pacific, Mead, and Weyerhaeuser," in *The*

Evolution of Competitive Strategies in Global Forest Industries, ed. Juha-Antti Lamberg et al. (Dordrecht, Netherlands: Springer, 2006), 75–76. See also Philip L. Buttrick, "Chestnut as a Source of Pulp for Paper," *Paper: A Weekly Technical Journal for Paper and Pulp Mills* 17, no. 13 (December 8, 1915): 12–13, 32.

63. Ahola, "Comparing the Strategic Evolution," 76.

64. Ibid.

65. Harold S. Betts, *Chestnut*, USDA, U.S. Forest Service pamphlet, American Woods series (Washington, D.C.: GPO, 1945), 3. The figures reflect "long cords," each yielding nearly 700 pounds of extract (2).

66. In 1909, the State of Tennessee passed legislation to file a lawsuit against the Champion Fibre Company for depositing chemicals and waste in the Pigeon River. The resolution claimed the company had "blackened its waters, killed its fish, rendered it unfit and unhealthful to be drunk by man or beast, . . . [and] destroyed its purity and usefulness for every purpose to which it has heretofore been put by the thousands of citizens of this State along its course." House Joint Resolution no. 40, in *Acts of the State of Tennessee Passed by the Fifty-Sixth General Assembly, 1909* (Nashville, Tenn.: McQuiddy Printing Company, 1909), 2207–8.

67. Ruth I. Voris and Ethel L. Best, *Women in Tennessee Industries: A Study of Hours, Wages, and Working Conditions*, U.S. Department of Labor Bulletin of the Women's Bureau no. 56, (Washington, D.C.: GPO, 1927), 43–44, 211; Tom Lee, *The Tennessee-Virginia Tri-cities: Urbanization in Appalachia, 1900–1950* (Knoxville: University of Tennessee Press, 2005), 50–51, 57, 93; Bartlett, *Troubled Waters*, 44.

68. The Appalachian forest commons is discussed at length in Jefferson C. Boyer, "Reinventing the Appalachian Commons," in *The Global Idea of "the Commons,"* ed. Donald N. Nonini (New York: Berghahn Books, 2008), 92–114; Mary Hufford, "Reclaiming the Commons: Narratives of Progress, Preservation, and Ginseng," in *Culture, Environment, and Conservation in the Appalachian South*, ed. Benita J. Howell (Urbana: University of Illinois Press, 2002), 100–20; Kathyrn Newfont, *Blue Ridge Commons: Environmental Activism and Forest History in Western North Carolina* (Athens: University of Georgia Press, 2012), 15–48, 73–74, 94–124, 271–78.

69. *Daily News* (Frederick, Md.), October 24, 1911, 6.

70. C. S. Knapp, "The Chestnut Blight," in *Transactions of the Warren Academy of Sciences, 1912–1913*, vol. 2, part 2 (Warren, Pa.: Warren Academy of Sciences, 1915), 66.

71. Ralph H. Lutts, "Like Manna from God: The American Chestnut Trade in Southwestern Virginia," *Environmental History* 9, no. 3 (July 2004): 506–7, tables 1, 2.

72. Ibid., 507.

73. *Western North Carolina Democrat* (Hendersonville, N.C.), October 16, 1913, 8.

74. *Daily News* (Frederick, Md.), October 24, 1911, 6.

75. Samuel D. Perry, *Grandpa Wouldn't Lie: A Boyhood Memoir* (Bloomington, Ind.: AuthorHouse, 2013), 117.

76. Ibid., 118

77. *Watauga Democrat* (Boone, N.C.), October 2, 1913, 2.

78. Ibid. Dozens of newspaper and magazine articles chronicle chestnut roasts and parties during the first two decades of the twentieth century. See, for example, James Elverson, "How the Girls Went Chestnutting," *Los Angeles Herald*, August 11, 1907, 4, 6; Aunt Janet [pseud.], "What to Do on Halloween," *Woman's Home Companion* 35, no. 10 (October 1908): 43, 73; *Richmond (Va.) Times-Dispatch*, August 20, 1911, 5; *Harrisburg (Pa.) Telegraph*, October 19, 1914, 4.

79. *New York Times*, November 26, 1912, 19; *Lebanon (Pa.) Daily News*, October 14, 1914, 6; *Syracuse (N.Y.) Herald*, October 15, 1914, 20; *Greenfield (Ma.) Gazette and Courier*, November 13, 1915, 2; *Kingston (N.Y.) Daily Freeman*, October 26, 1916, 7; *New Castle (Pa.) News*, October 22, 1917, 8; *Bloomington (Ind.) Evening World*, December 5, 1917, 2.

80. "The Chestnut Bark Disease," *California State Commission of Horticulture Monthly Bulletin* 5, no. 12 (December 1916): 438–39.

81. John Craig, "The Chestnut—*Castanea Vesca*, Linn.," *American Nut Journal* 6, no. 5 (May 1917): 5.

82. See, for example, James W. Withers, "Wholesale Market Report, New York," *American Gardening* 21, no. 307 (November 10, 1900), 749; "Blight Hard on Chestnuts," *Cannelton (Ind.) Inquirer*, November 8, 1919, 6; *Western Canner and Packer* 11 (November 1919): 62; *Syracuse (N.Y.) Post*, October 10, 1926, 24.

83. *Western Canner and Packer* 11 (November 1919): 62.

84. *Syracuse (N.Y) Herald*, October 14, 1920, 9; *Kingston (N.Y) Daily Freeman*, October 25, 1920, 5. In 1922, southern chestnuts were $18.00 to $22.00 per bushel in New York produce markets. *Syracuse (N.Y) Herald*, October 1, 1922, 35.

85. *Charlotte Observer* quoted in *Statesville (N.C.) Landmark*, October 29, 1923, 3.

86. Ibid.

87. Ibid. "In former years," the article continued, "it was not until Christmas was near that chestnuts could be found in local stores."

88. George F. Gravatt, "The Chestnut Blight in the Southern Appalachians," *American Forestry* 26, no. 322 (October 1920): 606–7; Earl H. Frothingham, *Timber Growing and Logging Practice in the Southern Appalachian Region*, USDA Technical Bulletin no. 250 (Washington, D.C.: GPO, 1931), 49–54; Dow V. Baxter and Lake S. Gill, *Deterioration of Chestnut in the Southern Appalachians*, USDA Technical Bulletin no. 257 (Washington, D.C.: GPO, 1931), 2–4.

89. Haven Metcalf, "The Chestnut Bark Disease," *Journal of Economic Entomology* 5, no. 2 (April 1912): 228–30. In 1923, George Gravatt found chestnut blight in Henderson and Polk Counties, North Carolina, and Greenville County, South Carolina. Based on the size and intensity of the infections, he calculated

the blight had reached both areas in 1912. Gravatt, "The Chestnut Blight in North Carolina," in *Chestnut and the Chestnut Blight in North Carolina*, North Carolina Geological and Economic Survey Economic Paper no. 56 (Raleigh: North Carolina Geological and Economic Survey, 1925), 16.

90. Gravatt, "Chestnut Blight in North Carolina," 16, 24; George F. Gravatt and Rush P. Marshall, *Chestnut Blight in the Southern Appalachians*, U.S. Department of Agriculture Circular no. 370 (Washington, D.C.: GPO, 1926), 5, 6.

91. According to Craig Lorimer, chestnut mortality in Poplar Cove occurred between 1928 and 1938. Lorimer, "Age Structure and Disturbance History," 1180.

92. USDA, U.S. Forest Service, "Acquisition Examination Report," Gennett Lumber Company, Tracts 309f/g, August 10, 1935, n.p. The document is located at the U.S. Forest Service Supervisor's Office, Asheville, North Carolina, Lands Shop Building. Many thanks to U.S. Forest Service archaeologist Rodney Snedeker for locating the file.

93. Dozens of newspapers announced the dedication of the Joyce Kilmer Memorial Forest in 1936. The *Hattiesburg American* claimed it contained "one of the finest stands of virgin timber to be found in any of the 154 national forests throughout the United States." *Hattiesburg (Miss.) American*, July 2, 1936, 9. See also Shelley Smith Mastran and Nan Lowerre, *Mountaineers and Rangers: A History of Federal Forest Management in the Southern Appalachians, 1900–81*, USDA, U.S. Forest Service Publication FS-380 (Washington, D.C.: U.S. Forest Service, 1983), 38–39; Andrew Gennett, *Sound Wormy: Memoir of Andrew Gennett, Lumberman*, ed. Nicole Hayler (Athens: University of Georgia Press, 2007), 194, 199–200, 212–15.

94. Joyce Kilmer, *Trees and Other Poems* (New York: George H. Doran, 1914), 18.

95. The 30 percent figure is from a 1909 commercial timber cruise. When Craig Lorimer surveyed Poplar Cove in the mid-1970s, he found chestnut comprised 25 percent of the standing timber. However, in the 1940s, forest ecologist E. Lucy Braun found that one section of the watershed, an area she designated as the "North Slope chestnut community," contained 82 percent chestnut. E. Lucy Braun, *Deciduous Forests of Eastern North America* (Philadelphia: Blakiston, 1950), 219.

CHAPTER 11. A National Calamity

1. See, for example, C. S. Knapp, "The Chestnut Blight," in *Transactions of the Warren Academy of Sciences, 1912–1913*, vol. 2, part 2 (Warren, Pa.: Warren Academy of Sciences, 1915), 65–66.

2. Frederick J. Haskin, Topeka State Journal Information Bureau feature, *Topeka (Kans.) State Journal*, October 22, 1917, 4.

3. Ibid.

4. *Bridgeport (Conn.) Telegram*, December 12, 1921, 9.

5. Ibid. "As everyone knows," continues the article, "the American chestnut is becoming a thing of the past. . . . The trees are practically all dead, and up through Connecticut it is difficult to find a single living tree" (9).

6. Quoted in the *Mount Sterling (Ky.) Advocate*, January 20, 1920, 5. Besley also noted chestnut blight had been identified in Garrett County, Maryland, along the West Virginia border.

7. *Dubois (Pa.) Courier*, October 11, 1927, 1. "Chestnuts have failed to make their appearance to any extent in the Dubois markets this season," noted the report. "October, the month that has been heralded in poetry, prose, and song as the nutting time of the year . . . has yet showed no signs of producing the desired crop" (ibid.).

8. *Warren (Pa.) Tribune*, September 10, 1927, 13. In Lawrence County, Pennsylvania, chestnuts were so scarce the *New Castle News* reported the blight had "played havoc with the one-time yield of nuts. . . . A ride through the county at the present time shows scores of chestnut trees with bare limbs and gnarled trunks as a result of the blight." *New Castle (Pa.) News*, October 18, 1927, 8.

9. *Daily News* (Frederick, Md.), November 22, 1926, 2. The following year, the *Frederick Post* announced the American chestnut was indeed "a thing of the past" *Frederick (Md.) Post*, October 1, 1927, 1.

10. *Daily News* (Frederick, Md.), November 22, 1926, 2.

11. Ralph H. Lutts, "Like Manna from God: The American Chestnut Trade in Southwestern Virginia," in *Environmental History and the American South: A Reader*, ed. Paul S. Sutter and Christopher J. Manganiello (Athens: University of Georgia Press, 2009), 267–68.

12. Eliot Wigginton, ed., *Foxfire 9* (New York: Anchor Books, 1986), 197, 200; Anne D. Amerson, *The Best of "I Remember Dahlonega": Memories of Lumpkin County, Georgia* (Charleston, S.C.: History Press, 2006), 45, 73; Sara M. Gregg, *Managing the Mountains: Land Use Planning, the New Deal, and the Creation of a Federal Landscape in Appalachia* (New Haven, Conn.: Yale University Press, 2010), 16–17, 30–34; Will Sarvis, *The Jefferson National Forest: An Appalachian Environmental History* (Knoxville: University of Tennessee Press, 2011), 80–81.

13. Noel Moore in the chapter "Memories of the American Chestnut," in *Foxfire 6*, ed. Eliot Wigginton (New York: Doubleday/Anchor, 1980), 402.

14. Amerson, *The Best of "I Remember Dahlonega,"* 45.

15. The 30 percent figure is taken from Philip L. Buttrick's survey of North Carolina chestnut stands in 1912–13. According to Buttrick, Jackson County timberlands contained 36 percent chestnut and Swain County, 26 percent. Historian John R. Finger claimed chestnut comprised 60 percent of the forest in parts of the Cherokee reservation, however. See Philip L. Buttrick, "Chestnut in North Carolina," in *Chestnut and the Chestnut Blight in North Carolina*, North

Carolina Geologic and Economic Survey Economic Paper no. 56 (Raleigh: North Carolina Geologic and Economic Survey, 1925), 9; John R. Finger, *Cherokee Americans: The Eastern Band of Cherokees in the Twentieth Century* (Lincoln: University of Nebraska Press, 1991), 77.

16. Finger, *Cherokee Americans*, 77; Margaret Lynn Brown, *The Wild East: a Biography of the Great Smoky Mountains* (Gainesville: University Press of Florida, 2001), 25–26; Kathyrn Newfont, *Blue Ridge Commons: Environmental Activism and Forest History in Western North Carolina* (Athens: University of Georgia Press, 2012), 41.

17. James Mooney, *Myths of the Cherokee: Extract from the* Nineteenth Annual Report of the Bureau of *American* Ethnology (Washington, D.C.: GPO, 1902), 179, 516; Mary Ulmer and Samuel E. Beck, eds., *Cherokee Cooklore: To Make My Bread* (Asheville, N.C.: Stephens Press, 1951), 45; Mary Ulmer Chiltoskey, "Cherokee Indian Foods," in *Gastronomy: The Anthropology of Food and Food Habits*, ed. Margaret L. Arnott (The Hague, Netherlands: Mouton, 1975), 239; Florence Cope Bush, *Dorie: Woman of the Mountains* (Knoxville: University of Tennessee Press, 1992), 12.

18. Raymond D. Fogelson and Paul Kutsche, "Cherokee Economic Cooperatives: The Gadugi," in *Symposium on Cherokee and Iroquois Culture*, Smithsonian Institution Bureau of American Ethnology Bulletin no. 180, ed. William N. Fenton and John Gulick (Washington, D.C.: GPO, 1961), 106.

19. On the importance of chestnut mast in the Appalachian subsistence economy, see Ronald L. Lewis, *Transforming the Appalachia Countryside: Railroads, Deforestation, and Social Change in West Virginia, 1880–1920 (Chapel Hill: University of North Carolina Press, 1998)*, 27–28; Ted L. Gragson, Paul V. Bolstad, and Meredith Welch-Devine, "Agricultural Transformation of Southern Appalachia," in *Agrarian Landscapes in Transition: Comparisons of Long-Term Ecological and Cultural Change*, ed. Charles L. Redman and David R. Foster (New York: Oxford University Press, 2008), 100; Jodi A. Barnes, "From Farms to Forests: The Material Life of an Appalachian Landscape" (PhD diss., American University, 2008), 38–39.

20. Maggie Axe Wachacha, interview by Lois Calonehuskie, *Journal of Cherokee Studies* 12 (1987): 46

21. Frederick Law Olmsted, *A Journey in the Back Country* (New York: Mason Brothers, 1863), 224.

22. Quoted in Stephen Nash, "The Blighted Chestnut," *National Parks* 62, nos. 7–8 (July/August 1988): 16.

23. Charles R. Burnham, "The Restoration of the American Chestnut," *American Scientist* 76, no. 5 (September–October 1988): 478–79; James M. Hill, "Wildlife Value of *Castanea dentata* Past and Present: The Historical Decline of the Chestnut and Its Future Use in Restoration of Natural Areas," in *Proceedings*

of the International Chestnut Conference, ed. Mark L. Double and William L. McDonald (Morgantown: West Virginia University Press, 1994), 186–93; Jason Van Driesche and Roy Van Driesche, *Nature Out of Place: Biological Invasions in the Global Age* (Washington, D.C.: Island Press, 2000), 123–24.

24. Bethany N. Baxter, "An Oral History of the American Chestnut in Southern Appalachia" (master's thesis, University of Tennessee at Chattanooga, 2009), 46.

25. Ibid., 45–49, 81.

26. Raymond D. Fogelson, "The Conjuror in Eastern Cherokee Society" (master's thesis, University of Pennsylvania, 1958), 48–49. See also John Gulick, *Cherokees at the Crossroads* (Chapel Hill: Institute for Research in Social Science, University of North Carolina, 1960), 20–21; Finger, *Cherokee Americans*, 77.

27. Tribble quoted in Nyoka Hawkins, "Building Community through Grassroots Democracy," *Local Voices* 10 (February–March 1993): 5. The Tribble quote has been cited by numerous authors, but seldom gets proper attribution. Susan Freinkel used the phrase in *American Chestnut* and even titled one of her chapters "A Whole World Dying." Susan Freinkel, *American Chestnut: The Life, Death, and Rebirth of a Perfect Tree* (Berkeley: University of California Press, 2007), 71–92. See also Donald Davis and Margaret Brown, "I Thought the Whole World Was Going to Die," *Now and Then* 10, no. 1 (Spring 1995): 30–31; Ralph H. Lutts, "Like Manna from God: The American Chestnut Trade in Southwestern Virginia," *Environmental History* 9, no. 3 (July 2004): 515; Newfont, *Blue Ridge Commons*, 39.

28. Verna Mae Sloan, interview by author, Hindman, Kentucky, June 12, 1998.

29. Dean Cornett and Nina Cornett, *American Chestnut: Appalachian Apocalypse* (Cornett Media, 2011), accessed April 1, 2019, http://www.cornettmedia.com/documentaries.html.

30. L. J. Peet and R. V. Reynolds, "Types of Land Utilization," in Bureau of Agricultural Economics, Bureau of Home Economics, and U.S. Forest Service, *Economic and Social Problems and Conditions of the Southern Appalachians*, USDA Miscellaneous Publication no. 205 (Washington, D.C.: GPO, 1935), 35.

31. *Evening Star* (Washington, D.C.), August 30, 1929, 3. See also Katrina M. Powell, *The Anguish of Displacement: The Politics of Literacy in the Letters of Mountain Families in Shenandoah National Park* (Charlottesville: University of Virginia Press, 2007), 26, 44.

32. See Powell, *Anguish of Displacement*, 26, 44.

33. Leland R. Cooper and Mary Lee Cooper, *The Pond Mountain Chronicle: Self-Portrait of a Southern Appalachian Community* (Jefferson, N.C.: McFarland, 1998), 29.

34. See, for example, Lutts, "Like Manna from God," 516–17; Sarvis, *Jefferson National Forest*, 82, 155–56; Susan Freinkel, *American Chestnut: The Life, Death, and*

Rebirth of a Perfect Tree (Berkeley: University of California Press, 2007), 83–85; Newfont, *Blue Ridge Commons*, 42–43, 91.

35. In 1911, two-thirds of Graham County, North Carolina, was owned in tracts of a thousand acres or more—"principally by lumber interests"—and one timber company alone held title to one-third of the entire county. John S. Holmes, *Forest Conditions in Western North Carolina*, North Carolina Geological and Economic Survey Bulletin no. 23 (Raleigh, Va.: Edwards and Broughton, 1911), 36.

36. Peet and Reynolds, "Types of Land Utilization," 24–31.

37. Dow V. Baxter and Lake S. Gill, *Deterioration of Chestnut in the Southern Appalachians*, USDA Technical Bulletin no. 257 (Washington, D.C.: GPO, 1931); *Indiana (Pa.) Evening Gazette*, November 4, 1930, 4; *Kingsport (Tenn.) Times*, December 13, 1934, 4; *Anniston (Ala.) Star*, April 12, 1936, 4; *Statesville (N.C.) Record*, December 4, 1936, 2.

38. John S. Holmes, "Foreword," in *Chestnut and the Chestnut Blight*, 6.

39. Clarence F. Korstian and Paul W. Stickel, "The Natural Replacement of Blight-Killed Chestnut in the Hardwood Forests of the Northeast," *Journal of Agricultural Research* 34, no. 7 (April 1927): 637.

40. Ibid., 644–45. A study in Somerset County, New Jersey, found chestnut oak, pignut hickory, and northern red oak the dominant replacement species, particularly on ridgetops and adjoining slopes Edward C. M. Richards, "Reforesting Cut-Over Chestnut Lands," *Forestry Quarterly* 12, no. 2 (June 1914): 204–10. See also Norma F. Good, "A Study of Natural Replacement of Chestnut in Six Stands in the Highlands of New Jersey," *Bulletin of the Torrey Botanical Club* 95, no. 3 (May–June 1968): 240–53.

41. Korstian and Stickel, " Natural Replacement of Blight-Killed Chestnut," 647.

42. Ibid., 632, 636.

43. Brenda R. Myers, Jeffrey L. Walck, and Kurt E. Blum, "Vegetation Change in a Former Chestnut Stand on the Cumberland Plateau of Tennessee during an 80-Year Period (1921–2000)," *Castanea* 69, no. 2 (June 2004): 81–91.

44. Donald Caplenor, "The Vegetation of the Gorges of the Fall Creek Falls State Park in Tennessee," *Journal of the Tennessee Academy of Science* 40, no. 1 (1965): 27–39.

45. Myers, Walck, and Blum, "Vegetation Change in a Former Chestnut Stand," 87.

46. Catherine Keever, "Present Composition of Some Stands of the Former Oak-Chestnut Forest in the Southern Blue Ridge Mountains," *Ecology* 34, no. 1 (January 1953): 44–54. According to Keever, "If northern red oak, chestnut oak, hickory and white oak continue to be the most abundant overstory trees in the forests of the southern Blue Ridge mountains, it should be called 'oak-hickory'" (53).

47. J. Frank McCormick and Robert B. Platt, "Recovery of an Appalachian Forest following the Chestnut Blight, or Catherine Keever—You Were Right!," *American Midland Naturalist* 104, no. 2 (October 1980): 264–73. A study examining more than one hundred former chestnut stands in southwest Virginia found other trees dominant in the overstory, including oaks and maples. Steven L. Stephenson, Harold S. Adams, and Michael L. Lipford, "The Present Distribution of Chestnut in the Upland Forest Communities of Virginia," *Bulletin of the Torrey Botanical Club* 118, no. 1 (January–March 1991): 24–32.

48. Good, "Study of Natural Replacement of Chestnut," 247; Craig G. Lorimer, "Age Structure and Disturbance History of a Southern Appalachian Virgin Forest," *Ecology* 61, no. 5 (October 1980): 1177–80.

49. Frank W. Woods, "Natural Replacement of Chestnut by Other Species in the Great Smoky Mountains" (PhD diss., Knoxville: University of Tennessee, 1957), 37.

50. Ibid., 26.

51. Ibid., 38. Woods believed his findings were supported by studies done elsewhere in the United States, thus disagreeing with Keever. According to Woods, the southern Appalachian forest was more likely to become an "oak association-complex" of tree species, which might include some hickory species, but in fewer numbers proposed by Keever (39). See also Richard T. Busing, "A Half Century of Change in a Great Smoky Mountains Cove Forest," *Bulletin of the Torrey Botanical Club* 116, no. 3 (July–September 1989): 283–88.

52. E. Lucy Braun, "An Ecological Transect of Black Mountain, Kentucky," *Ecological Monographs* 10, no. 2 (April 1940): 224.

53. Charles C. Rhoades, "The Influence of American Chestnut (*Castanea dentata*) on Nitrogen Availability, Organic Matter and Chemistry of Silty and Sandy Loam Soils," *Pedobiologia* 50, no. 6 (2007): 557.

54. Stephen M. Wagener, Sandra E. Slemmer, and Breamond Ostrander, "Decomposition of American Chestnut Leaves," *Journal of the American Chestnut Foundation* 21, no. 2 (Fall 2007): 44.

55. Leonard A. Smock and Christina M. MacGregor, "Impact of the American Chestnut Blight on Aquatic Shredding Macroinvertebrates," *Journal of the North American Benthological Society* 7, no. 3 (September 1988): 212.

56. Dana R. Warren et al., "Forest-Stream Interactions in Eastern Old-Growth Forests," in *Ecology and Recovery of Eastern Old-Growth Forests*, ed. Andrew M. Barton and William S. Keeton (Washington, D.C.: Island Press, 2018), 166–67.

57. Ibid., 169.

58. Craig W. Hedman, David H. Van Lear, and Wayne T. Swank, "In-Stream Large Woody Debris Loading and Riparian Forest Seral Stage Associations in the Southern Appalachian Mountains," *Canadian Journal of Forest Research* 26, no. 7 (July 1996): 1218–27.

59. J. Bruce Wallace et al., "Large Woody Debris in a Headwater Stream: Long-Term Legacies of Forest Disturbance," *International Review of Hydrobiology* 86, nos. 4–5 (July 2001): 501–13.

60. The debilitating substances in European chestnut leaves are quercetin, rutin, and apigenin, which prevent seed germination and growth. See Adriana Basile et al., "Antibacterial and Allelopathic Activity of Extract from *Castanea sativa* Leaves," *Fitoterapia* 71, supp. 1 (August 2000): S110–S16.

61. Good, "Study of Natural Replacement of Chestnut," 247–48; David B. Vandermast, David H. Van Lear, and Barton D. Clinton, "American Chestnut as an Allelopath in the Southern Appalachians," *Forest Ecology and Management* 165, nos. 1–3 (July 15, 2002): 173–81.

62. Kim D. Coder, "Potential Allelopathy in Different Tree Species," University of Georgia, Daniel B. Warnell School of Forest Resources Extension Publication FOR99–003, available via https://www.walterreeves.com/landscaping/uga-tree-publications (accessed February 8, 2021).

63. Vandermast, Van Lear, and Clinton, "American Chestnut as an Allelopath," 176.

64. Catherine Keever, "Present Composition of Some Stands," 51. Frank W. Woods found sixteen different ferns or fern allies and twenty-four herbaceous species in chestnut stands. Frank W. Woods and Royal E. Shanks, "Natural Replacement of Chestnut by Other Species in the Great Smoky Mountains National Park," *Ecology* 40, no. 3 (July 1959): 354–55.

65. Braun, "Ecological Transect of Black Mountain," 202, 205, table 2.

66. Wagener, Slemmer, and Ostrander, "Decomposition of American Chestnut Leaves," 44. See also Rhoades, "Influence of American Chestnut," 557–58.

67. Jonathan M. Palmer, *Morphological and Molecular Characterization of Mycorrhizal Fungi Associated with a Disjunct Stand of American Chestnut (Castanea dentata) in Wisconsin* (master's thesis, University of Wisconsin–La Crosse, 2006), 32–37; Bryn T. M. Dentinger, "Systematics and Evolution of Porcini and Clavarioid Mushrooms" (PhD diss., University of Minnesota, 2007), 32; Jonathan M. Palmer, Daniel L. Lindner, and Thomas J. Volk, "Ectomycorrhizal Characterization of an American Chestnut (*Castanea dentata*)-Dominated Community in Western Wisconsin," *Mycorrhiza* 19, no. 1 (December 2008): 27–36.

68. William A. Murrill, "Illustrations of Fungi–VI," *Mycologia* 2, no. 2 (March 1910): 46. In 1916, Murrill saw sulphur shelf (chicken of the woods) fungi growing on chestnut logs in the Blue Ridge Mountains of Virginia. See Murrill, "Some Fungi Collected in Virginia," *Mycologia* 9, no. 1 (January 1917): 34–35.

69. Murrill, "Illustrations of Fungi–VI," 45. See also S. T. Rorer, "Foods of the Woods," *Ladies' Home Journal*, August 1898, 22.

70. Seth J. Diamond et al., "Hard Mast Production before and after the

Chestnut Blight," *Southern Journal of Applied Forestry* 24, no. 4 (November 2000): 198, table 3, and 199, table 4.

71. Katie L. Burke, "Chestnuts and Wildlife," *Journal of the American Chestnut Foundation* 27, no. 2 (March/April 2013): 9. See also Keith E. Gilland, Carolyn H. Keiffer, and Brian C. McCarthy, "Seed Production of Mature Forest-Grown American Chestnut (*Castanea dentata* (Marsh.) Borkh)," *Journal of the Torrey Botanical Society* 139, no. 3 (July–September 2012): 283–89.

72. Harmony J. Dalgleish and Robert K. Swihart, "American Chestnut Past and Future: Implications of Restoration for Resource Pulses and Consumer Populations of Eastern U.S. Forests," *Restoration Ecology* 20, no. 4 (July 2012): 490–97.

73. C. L. Newcombe, "An Ecological Study of the Allegheny Cliff Rat (*Neotoma Pennsylvanica* Stone)," *Journal of Mammalogy* 11, no. 2 (May 1930): 204–11. Allegheny woodrat nests were constructed from the bast fibers of chestnuts and their middens contained "acorns, chestnut burs, chestnuts, leaves, stems and fungi" (207). See also Kathleen LoGiudice, "Multiple Causes of the Allegheny Woodrat Decline: A Historical-Ecological Examination," in *The Allegheny Woodrat: Ecology, Conservation, and Management of a Declining Species*, ed. John D. Peles and Janet Wright (New York: Springer, 2008), 23–41; Rita M. Blythe, "Ecological Interactions Affecting American Chestnut Restoration and Allegheny Woodrat Conservation in Indiana" (master's thesis, Purdue University, 2014).

74. Walter Cole, interview by Charles Grossman, 1965. Transcript in Oral History collection, Great Smoky Mountains National Park Archives, Sugarlands Visitor Center, Gatlinburg, Tennessee, n.p.

75. Effler quoted in Vic Weals, *Last Train to Elkmont: A Look Back at Life on Little River in the Great Smoky Mountains* (Knoxville, Tenn.: Olden Press, 1991), 128. The largest trees in the Great Smoky Mountains, recalled Weals, "would drop ten or more bushels of chestnuts" (129).

76. Cady quoted ibid., 129.

77. Maynard Ledbetter, interview by Bill Landry, 1989. Transcript in Oral History collection, Great Smoky Mountains National Park Archives, Sugarlands Visitor Center, Gatlinburg, Tennessee, n.p.

78. See Dalgleish and Swihart, "American Chestnut Past and Future," 495–96.

79. Hill, "Wildlife Value of *Castanea dentata*," 190.

80. Jim Lee, "Chestnut: Tree with a Past," *North Carolina Wildlife* 21, no. 11 (November 1957): 10. See also Wilma Dykeman, "Our Trees Are Wonderful Gift," *Knoxville (Tenn.) News-Sentinel*, January 3, 1963, 18.

81. Paul A. Opler, "Insects of the American Chestnut: Possible Importance and Conservation Concern," in *Proceedings of the American Chestnut Symposium, College of Agriculture and Forestry, West Virginia University, Morgantown, West*

Virginia, January 4–5, 1978, ed. William L. McDonald et al. (Morgantown: West Virginia University Books, 1978), 84–85.

82. Wagner quotation from the Connecticut Chapter of the American chestnut homepage, accessed June 6, 2019, https://www.acf.org/ct/news-and-updates/chestnut-and-invertebrate-extinctions. For more on the impact of chestnut blight on insect populations, see David L. Wagner and Roy G. Van Driesche, "Threats Posed to Rare or Endangered Insects by Invasions of Nonnative Species," *Annual Review of Entomology* 55 (2010): 547–68.

83. Opler, "Insects of the American Chestnut," 85.

84. Douglas W. Tallamy, *Bringing Nature Home: How Native Plants Sustain Wildlife in Our Gardens* (Portland, Ore.: Timber Press, 2007), 126.

CHAPTER 12. Genes for Blight Resistance

1. Russell Clapper, "Chestnut Breeding, Techniques and Results: I. Breeding Material and Pollination Techniques," *Journal of Heredity* 45, no. 3 (May/June 1954): 106–8; Russell Clapper, "Chestnut Breeding, Techniques and Results: II. Inheritance of Characters, Breeding for Vigor, and Mutations," *Journal of Heredity* 45, no. 4 (July 1954): 201–2; U.S. Department of the Interior, *Crab Orchard National Wildlife Refuge Annual Report, January–December 1964* (Crab Orchard, Ill.: U.S. Fish and Wildlife Service, 1964), 24.

2. Douglas J. Buege, *If a Tree Falls: Rediscovering the Great American Chestnut* (self-pub., XLibris, 2008), 40; Oana L. Spitzer, "A Chestnut Grows in Illinois," *St. Louis Post-Dispatch*, September 8, 1968, 227; Russell B. Clapper, "A Promising New Forest-Type Chestnut Tree," *Journal of the American Chestnut Foundation* 21, no. 1 (Spring 2007): 12–14.

3. R. Kent Beattie and Jesse D. Diller, "Fifty Years of Chestnut Blight in America," *Journal of Forestry* 52, no. 5 (May 1954), 328. Beattie and Diller also mention the potential of the Nanking chestnut, a Chinese cultivar introduced in 1924. By 1954, forty-five thousand Nanking seedlings had been planted across the former range of the American chestnut. Jesse D. Diller, *A Potential Timber-Type Chinese Chestnut*, Northeastern Forest Experiment Station Research Note no. 37 (Darby, Pa.: USDA, U.S. Forest Service, 1954), 2.

4. U.S. Forest Service press release, reprinted in American Chestnut Foundation, "Diary of the Clapper Tree," *Journal of the American Chestnut Foundation* 11, no. 1 (Summer 1997): 19. For additional commentary on the Clapper hybrid, see Russell B. Clapper, "Breeding New Chestnuts for Southern Forests," *Forest Farmer* 9, no. 11 (1950): 8; Clapper, "Chestnut Breeding, Techniques and Results: II," 205–7; Jesse D. Diller and Russell B. Clapper, "A Progress Report on Attempts to Bring Back the Chestnut Tree in the Eastern United States, 1954–1964," *Journal of Forestry* 63, no. 3 (March 1965): 186–88.

5. Hans Nienstaedt and Arthur H. Graves, *Blight Resistant Chestnuts: Culture and Care*, Connecticut Agricultural Experiment Station Circular no. 192 (New Haven, Conn.: Connecticut Agricultural Experiment Station, 1955), 7–8; Richard A. Jaynes and Arthur H. Graves, *Connecticut Hybrid Chestnuts and Their Culture*, Connecticut Agricultural Experiment Station Bulletin no. 657 (New Haven, Conn.: Connecticut Agricultural Experiment Station, 1963), 8–9; Susan Freinkel, *American Chestnut: The Life, Death, and Rebirth of a Perfect Tree* (Berkeley: University of California Press, 2007), 101–3.

6. Jesse D. Diller, Russell B. Clapper, and Richard A. Jaynes, *Cooperative Test Plots Produce Some Promising Chinese and Hybrid Chestnut Trees*, U.S. Forest Service Research Note NE-25 (Darby, Pa.: USDA, U.S. Forest Service, Northeastern Forest Experiment Station, 1964), 5. See also Jill Jonnes, *Urban Forests: A Natural History of Trees and People in the American Cityscape* (New York: Viking, 2016), 122.

7. U.S. Department of the Interior, *Crab Orchard National Wildlife Refuge Annual Report*, 24.

8. Buege, *If A Tree Falls*, 41. See also Jesse D. Diller and Russell B. Clapper, "Asiatic and Hybrid Chestnut Trees in the Eastern United States," *Journal of Forestry* 67, no. 5 (May 1969): 328–31; Beattie and Diller, "Fifty Years of Chestnut Blight," 328.

9. Buege, *If A Tree Falls*, 41. See also Nienstaedt and Graves, *Blight Resistant Chestnuts*, 7–8; Jaynes and Graves, *Connecticut Hybrid Chestnuts*, 8–9; Sandra L. Anagnostakis, "The Pathogens and Pests of Chestnuts," in *Advances in Botanical Research: Incorporating Advances in Plant Pathology*, vol. 21, ed. John H. Andrews and Inez C. Tommerup (New York: Academic Press, 1995), 141.

10. Evelyn Fox Keller, *A Feeling for the Organism: The Life and Work of Barbara McClintock* (New York: W. H. Freeman, 1983), 48–49; Nathaniel C. Comfort, *The Tangled Field: Barbara McClintock's Search for the Patterns of Genetic Control* (Cambridge, Mass.: Harvard University Press, 2003), 54–55, 58, 68, 226.

11. Ronald L. Phillips et al., "Charles R. Burnham: Long-Time Contributor to Maize Breeding," in *Maize Genetics and Breeding in the 20th Century*, ed. Peter A. Peterson and Angelo Bianchi (River Edge, N.J.: World Scientific, 1999), 115–18

12. Charles R. Burnham, *Discussions in Cytogenetics* (Minneapolis, Minn.: Burgess, 1962). See also Phillips et al., "Charles R. Burnham," 116; Paul Sisco, "Chestnut Cytogenetics: Faridi, Burnham, and McClintock," *Compass* 11 (June 2008): 13 (*Compass* is published by the Science Delivery Group of the Southern Research Station, U.S. Forest Service, Asheville, North Carolina).

13. Charles R. Burnham, Phillip A. Rutter, and David W. French, "Breeding Blight-Resistant Chestnuts," in *Plant Breeding Reviews*, vol. 4, ed. Jules Janick (New York: John Wiley & Sons, 1986), 347–97; Peter Friederici, *Nature's*

Restoration: People and Places on the Front Lines of Conservation (Washington, D.C.: Island Press, 2006), 64–69.

14. Charles Burnham, "Blight-Resistant American Chestnut: There's Hope," *Plant Disease* 65, no. 6 (June 1981): 459–60.

15. Ibid., 460.

16. Ibid. See also Charles Burnham, "Historical Overview of Chestnut Breeding in the United States," *Journal of the American Chestnut Foundation* 2, no. 1 (December 1987): 9–11. Personal correspondence, Charles Burnham to Barbara McClintock, June 19, 1983, Barbara McClintock Papers, "Correspondence," U.S. National Library of Medicine, Bethesda, Maryland.

17. Mary A. Hosier, Charles R. Burnham, and Paul E. Read, "Breeding Strategy for Blight-Resistant American Chestnut," *Journal of the American Chestnut Foundation* 1, no. 1 (July 1985): 6–7. See also Charles R. Burnham, "Backcross Breeding: I. Blight-Resistant American Chestnuts," *Journal of the American Chestnut Foundation* 2, no. 1 (December 1987): 12–16.

18. Phillip Rutter, "The President's Message," *Journal of the American Chestnut Foundation* 1, no. 1 (July 1985): 13–14.

19. Phillip Rutter, "A National Chestnut Research Center," *Journal of the American Chestnut Foundation* 2, no. 1 (December 1987): 21.

20. Ibid.

21. Fred V. Hebard, "The Backcross Breeding Program of the American Chestnut Foundation," in *Restoration of American Chestnut to Forest Lands: Proceedings of a Conference and Workshop Held at the North Carolina Arboretum, Asheville, N.C., U.S.A., May 4–6, 2004*, ed. Kim C. Steiner and John E. Carlson (Washington, D.C.: U.S. Department of the Interior, National Park Service, 2006), 61–77; Friederici, *Nature's Restoration*, 68; Buege, *If a Tree Falls*, 58.

22. Fred V. Hebard and Peter B. Kaufman, "Chestnut Callus-Cultures: Tannin Content and Colonization by *Endothia parasitica*," in *Proceedings of the American Chestnut Symposium, College of Agriculture and Forestry, West Virginia University, Morgantown, West Virginia, January 4–5, 1978*, ed. William McDonald et al. (Morgantown: West Virginia University College of Agriculture and Forestry, 1978), 63–70; Gary J. Griffin et al., "Survival of American Chestnut Trees: Evaluation of Blight Resistance and Virulence in *Endothia parasitica*," *Phytopathology* 73, no. 7 (1983): 1084–92.

23. Hebard, "Backcross Breeding Program," 2–3; Freinkel, *American Chestnut*, 185; Buege, *If a Tree Falls*, 55–57.

24. Fred V. Hebard, Paul H. Sisco, and Peter A. Woods, "Meadowview Notes, 1999–2000," *Journal of the American Chestnut Foundation* 14, no. 1 (Summer 2000): 7–17; Fred V. Hebard, "Meadowview Notes, 2000–2001," *Journal of the American Chestnut Foundation* 15, no. 1 (Summer/Fall 2001): 14.

25. Hebard, Sisco, and Wood, "Meadowview Notes, 1999–2000," 9.

26. Ibid., 10–11. See also Fred V. Hebard, "Research Objectives of the American Chestnut Foundation, 2004–2014," *Journal of the American Chestnut Foundation* 18, no. 2 (Fall 2004): 16–17.

27. Shawn A. Mehlenbacher, Ronald L. Phillips, and J. P. van Buijtenen, "The TACF Breeding Program, Condensed," *Journal of the American Chestnut Foundation* 14, no. 1 (Summer 2000): 24.

28. An excellent discussion of the hybrid problem in chestnut restoration is found in Friederici, *Nature's Restoration*, 71–74. The concept of genetic purity informed the management policies of the National Park Service, which, between 2001 and 2011, restricted the release of TACF hybrids on Department of Interior lands. See John Dennis, "National Park Service Management Policy Guidance for Restoration of American Chestnut to National Park System Units," in Steiner and Carlson, *Restoration of American Chestnut to Forest Lands*, 3–11.

29. See, for example, Hugh Irwin, "The Road to American Chestnut Restoration," *Journal of the American Chestnut Foundation* 16, no. 2 (Spring 2003): 6–13; Friederici, *Nature's Restoration*, 71–72; Freinkel, *American Chestnut*, 201, 216.

30. Charles S. Elton, *The Ecology of Invasions by Animals and Plants* (London: Methuen, 1958). See also David M. Richardson, ed., *Fifty Years of Invasion Ecology: The Legacy of Charles Elton* (Oxford: Wiley-Blackwell, 2011), xi–xii, xiv.

31. Hebard, Sisco, and Wood, "Meadowview Notes, 1999–2000," 11.

32. Ibid. See also Yan Shi and Fred V. Hebard, "Male Sterility in the Progeny Derived from Hybridizations between *Castanea dentata* and *C. Mollissima*," *Journal of the American Chestnut Foundation* 11, no. 1 (Summer 1997): 38–47; Richard Jaynes, "Interspecific Crosses in the Genus Castanea," *Silvae Genetica* 13 (1964): 146–54; Arif Soylu, "Heredity of Male Sterility in Some Chestnut Cultivars (*Castanea sativa* Mill.)," *Acta Horticulturae* 317 (1992): 181–83; Michael L. Arnold, *Divergence with Genetic Exchange* (Oxford: Oxford University Press, 2016), 88–99, 123–29, 138, 173.

33. Shi and Hebard, "Male Sterility in the Progeny," 41.

34. John B. S. Haldane, "Sex Ratio and Unisexual Sterility in Hybrid Animals," *Journal of Genetics* 12, no. 2 (1922): 101–9.

35. See Amanda N. Brothers and Lynda F. Delph, "Haldane's Rule is Extended to Plants with Sex Chromosomes," *Evolution* 64, no. 12 (December 2010): 3643–48; Menno Schilthuizen, Maartje C. W. G. Giesbers, and Leo W. Beukeboom, "Haldane's Rule in the 21st Century," *Heredity* 107, no. 2 (August 2011): 95–102; Jeffrey P. Demuth, Rebecca J. Flanagan, and Lynda F. Delph, "Genetic Architecture of Isolation between Two Species of *Silene* with Sex Chromosomes and Haldane's Rule," *Evolution* 68, no. 2 (February 2014): 332–42.

36. Zoë Hoyle, "A Chromosomal Conundrum," *Compass* 11 (June 2008): 11–12.

37. Ibid., 12. See also Zoë Hoyle, "Solutions from the Double Helix," *Compass* 11 (June 2008): 6–9; Douglass F. Jacobs, Harmony J. Dalgleish, and C. Dana

Nelson, "A Conceptual Framework for Restoration of Threatened Plants: The Effective Model of American Chestnut (*Castanea dentata*) Reintroduction," *New Phytologist* 197, no. 2 (January 2013): 378–93.

38. Although the first nuts establishing the BC_3F_3 line of trees were produced in 2006, the seedlings tested for blight resistance were not available until 2008. Paul H. Sisco, "Outlook for Blight-Resistant American Chestnut Trees," in *National Proceedings: Forestry and Conservation Nursery Associations—2008*, ed. R. Kasten Dumroese and L. E. Riley (Fort Collins, Colo.: USDA, U.S. Forest Service, Rocky Mountain Research Station, 2009), 61–68.

39. Matthew Diskin, Kim C. Steiner, and Fred V. Hebard, "Recovery of American Chestnut Characteristics Following Hybridization and Backcross Breeding to Restore Blight-Ravaged *Castanea dentata*," *Forest Ecology and Management* 223, nos. 1–3 (March 2006): 439–47; Benjamin O. Knapp et al., "Leaf Physiology and Morphology of *Castanea dentata* (Marsh) Borkh., *Castanea mollissima* Blume, and Three Backcross Breeding Generations Planted in the Southern Appalachians, USA," *New Forests* 45, no. 2 (March 2014): 283–93.

40. Stacy L. Clark et al., "Lessons from the Field: The First Tests of Restoration American Chestnut (*Castanea dentata*) Seedlings Planted in the Southern Region," in *Proceedings of the 16th Biennial Southern Silvicultural Research Conference*, USDA Forest Service General Technical Report SRS-156, ed. John R. Butnor (Asheville, N.C.: U.S. Forest Service, Southern Research Station, 2012), 69–70; Stacy L. Clark, Scott E. Schlarbaum, and Fred V. Hebard, "The First Research Plantings of Third-Generation, Third-Backcross American Chestnut (*Castanea dentata*) in the Southeastern United States," in *Proceedings of the Fifth International Chestnut Symposium, International Society for Horticultural Science*, ed. Mark L. Double and William L. MacDonald, *Acta Horticulturae* 1019 (December 2014): 39–44.

41. Stacy L. Clark et al., "Reintroduction of American Chestnut in the National Forest System," *Journal of Forestry* 112, no. 5 (September 2014): 502–12.

42. Ibid., 509.

43. Gary J. Griffin, "Blight Control and Restoration of the American Chestnut," *Journal of Forestry* 98, no. 2 (February 2000): 22–27; Gary Griffin et al., "Integrated Use of Resistance, Hypovirulence, and Forest Management to Control Blight on American Chestnut," in *Restoration of American Chestnut to Forest Lands*, 97–108; Virginia Shepherd, "Pursuing an American Dream: Restoring the American Chestnut to Our Forests—and Our Wildlife," *Virginia Wildlife* 70, no. 2 (February 2009): 4–9.

44. Griffin et al., "Integrated Use of Resistance," 99–101.

45. "The American Chestnut Cooperators' Foundation," American Chestnut Cooperators' Foundation, last updated November 1, 2015, accessed November 4, 2018, http://www.accf-online.org/brochure.htm.

46. Entry for 2008 on the American Chestnut Cooperators' Foundation newsletters webpage, accessed November 5, 2018, http://www.accf-online.org/news.html.

47. "How to Get Seed Nuts: Participate," American Chestnut Cooperators' Foundation, accessed November 5, 2018, http://www.accf-online.org/seednuts.htm.

48. Entry for 2017 on the American Chestnut Cooperators' Foundation newsletters webpage, accessed June 8, 2019, http://www.accf-online.org/news.html.

49. Ibid.

50. Ibid.

51. Pascal P. Pirone, *Diseases and Pests of Ornamental Plants*, 5th ed. (New York: John Wiley & Sons, 1978), 182–83; Simone Prospero and Daniel Rigling, "Chestnut Blight," in *Infectious Forest Diseases*, ed. Paolo Gonthier and Giovanni Nicolotti (Oxfordshire, UK: CAB International, 2013), 318–33.

52. Friederici, *Nature's Restoration*, 75; Freinkel, *American Chestnut*, 191–92; Buege, *If a Tree Falls*, 86–90.

53. American Chestnut Cooperators' Foundation homepage, accessed November 12, 2018, http://www.accf-online.org (my emphasis). See also Sandra L. Anagnostakis, "Chestnut Breeding in the United States for Disease and Insect Resistance," *Plant Disease* 96, no. 10 (October 2012): 1392–1403.

54. Quotation taken from SUNY-ESF, American Chestnut Research and Restoration Project's webpage, accessed November 20, 2018, http://www.esf.edu/chestnut. See also the online press release "Powell Named Forest Biotechnologist of the Year," December 10, 2013, Office of Communications, State University of New York, College of Environmental Science and Forestry, accessed November 20, 2019, http://www.esf.edu/communications/view.asp?newsID=2469.

55. Andrew E. Newhouse et al., "Transgenic American Chestnuts Show Enhanced Blight Resistance and Transmit the Trait to T1 Progeny," *Plant Science* 228 (November 2014): 88.

56. SUNY College of Environmental Science and Forestry, "Blight-Resistant American Chestnut Trees Take Root," *Science Daily*, November 6, 2014, accessed November 19, 2019, http://www.sciencedaily.com/releases/2014/11/141106082032.htm; Newhouse et al., "Transgenic American Chestnuts," 88–97.

57. William A. Powell, Andrew E. Newhouse, and Vernon Coffey, "Developing Blight-Tolerant American Chestnut Trees," *Cold Harbor Perspectives in Biology* 11, no. 7 (July 2019): a034587, https://doi.org/10.1101/cshperspect.a034587; Steve Featherstone, "A CNY Scientist's Work Could Change the World. But He Might Not Live Long Enough to See it Happen," NewYorkUpState.com, January 23, 2023, accessed November 12, 2024, https://www.newyorkupstate.com/outdoors/2023/06/a-cny-scientists-work-might-change-the-world-but-he-might-not-live-long-enough-to-see-it-happen.html.

58. Paul Sisco, personal communication, October 2, 2014. Sisco's remarks were

circulated in an email to TACF members with the subject heading, "Why I oppose releasing fully-fertile genetically-engineered chestnut trees into our forests."

59. Martha L. Crouch, letter to the editor, *Syracuse Post-Standard*, November 14, 2014, accessed March 23, 2018, http://www.syracuse.com/opinion/index.ssf/2014/11/suny_esf_plan_to_release_genetically_engineered_chestnut_tree_is_too_hasty_your.html.

60. Sarah Fecht, "Genetically Modified Chestnuts Roasting on an Open Fire," *Popular Science*, December 24, 2014, accessed January 13, 2018, https://www.popsci.com/transgenic-chestnuts-roasting-open-fire.

61. William A. Powell, SUNY-ESF *Annual Report*, June 1, 2014, to May 31, 2015, accessed April 3, 2018, http://www.esf.edu/efb/annualreports/1415/Powell1415.pdf.

62. William A. Powell et al., "Petition for Determination of Nonregulated Status for Blight-Tolerant Darling 58 American Chestnut (*Castanea dentata*)," State University of New York College of Environmental Science and Forestry (Syracuse, N.Y.: American Chestnut Research and Restoration Project, 2020), 3–5, 148–57; State University of New York College of Environmental Science and Forestry, "Petition for Determination of Nonregulated Status for Blight-Tolerant Darling 58 American Chestnut (*Castanea dentata*)," *Federal Register* 85, no. 161 (August 19, 2020): 51008.

63. Global Justice Ecology Project, "GE American Chestnuts: Save the Forest by Destroying It?," Global Justice Ecology Project blog, December 15, 2015, accessed April 4, 2018, http://globaljusticeecology.org/ge-chestnuts-ho-ho-no-the-nightmare-before-christmas-future-2.

64. FOIA document, Center for Food Safety vertical files, San Francisco, California. The APHIS correspondence is related to SUNY-ESF field trial no. 05–116–03n. In response to the violation, APHIS told SUNY-ESF researchers to "use aluminum bags and reevaluate their procedures to secure the site from squirrels." Christine Stella, personal communication, January 26, 2016.

65. Ibid.

66. Andrew Kimbrell, personal communication, April 25, 2020. Similar arguments were made by Anne Petermann at a February 9, 2016, press conference. "Forest ecosystems," noted Petermann, "are extremely complex, having evolved over millions of years and including not only trees and plants, but fungi, soil microorganisms, insects, birds, wildlife and other biodiversity. . . . The impact of engineered traits being released into or contaminating these forests are [*sic*] impossible to predict, but potentially extremely serious." "GE Trees in 2016 TelePress Conference Remarks by Anne Petermann," part of Global Justice Ecology Project press package, Genetically Engineered Trees in 2016 Telepress Conference, February 9, 2016, http://globaljusticeecology.org/pressroom/ge-trees-in-2016-press-package.

67. Andrew Kimbrell, personal communication, April 25, 2020.

68. Anne Petermann, "Overwhelming Opposition to USDA Proposal to Legalize Genetically Engineered Eucalyptus Trees," Global Forest Coalition, August 10, 2017, accessed June 12, 2019, https://globalforestcoalition.org/oposicion-masiva-la-propuesta-del-usda-de-legalizar-arboles-de-eucalipto-geneticamente-modificados-gm; Center for Food Safety, "Overwhelming Opposition to USDA Proposal to Legalize Genetically Engineered Eucalyptus Trees," Center for Food Safety, July 6, 2017, accessed June 12, 2019, https://www.centerforfoodsafety.org/issues/310/ge-trees/press-releases/5008/overwhelming-opposition-to-usda-proposal-to-legalize-genetically-engineered-eucalyptus-trees.

69. Rachel Smolker and Anne Petermann, *Biotechnology for Forest Health? The Test Case of the Genetically Engineered American Chestnut* (Buffalo, N.Y.: Campaign to Stop GE Trees, 2019).

70. Ibid., 17. See also Anne Petermann, "GE Chestnuts: Ho Ho No: The Nightmare Before Christmas—Future," Indigenous Environmental Network online webpage, December 15, 2015, accessed March 22, 2018, http://www.ienearth.org/ge-chestnuts-ho-ho-no-the-nightmare-before-christmas-future.

71. Ibid., 10–11. See also Andrew E. Newhouse et al., "Transgenic American Chestnuts Do Not Inhibit Germination of Native Seeds or Colonization of Mycorrhizal Fungi," *Frontiers in Plant Science* 9 (July 19, 2018): 1046, https://doi.org/10.3389/fpls.2018.01046.

72. Suzuki quoted in *A Silent Forest: The Growing Threat of Genetically Engineered Trees*, dir. Ed Schehl (Three Americas, 2006), accessed June 14, 2019, https://vimeo.com/51481514.

73. Ibid.

74. Ibid.

75. Smolker and Petermann, *Biotechnology for Forest Health?*, 3. Three timber certification programs do not recognize GE trees: the Forest Stewardship Council, the Programme for the Endorsement of Forest Certification, and the Sustainable Forest Initiative.

76. American Chestnut Foundation, *2023 Annual Report* (Asheville, N.C.: The American Chestnut Foundation, 2024), 2–5, accessed November 14, 2024, https://tacf.org/about-us/financials/.

77. Agricultural Marketing Resource Center, "Chestnuts," an occasional online document published by the USDA and Iowa State University, last updated October 2018, https://www.agmrc.org/commodities-products/nuts/chestnuts.

78. Kathleen Lavey, "Chestnuts: A Michigan Tradition Makes a Comeback," *Detroit Free Press*, December 19, 2016, accessed June 14, 2019, https://www.freep.com/story/life/2016/12/19/chestnuts-michigan-tradition/95606316.

79. American Chestnut Foundation, *2023 Annual Report*, 33–23 accessed November 14, 2024, https://tacf.org/about-us/financials/.

80. Starr Anderson, "Saving the American Chestnut Tree," Virginia

Department of Conservation and Recreation, March 8, 2024, accessed November 9, 2024, https://www.dcr.virginia.gov/insights/saving-the-american-chestnut-tree.

81. Pennsylvania Chapter of the American Chestnut Foundation, "Quick Facts," accessed November 27, 2018, http://www.patacf.org/wp-content/uploads/2013/06/pub_quick_fact_sheet_3.pdf; Pennsylvania Chapter of the American Chestnut Foundation, "PA Chapter Staff," accessed June 14, 2018, https://patacf.org/staff.

82. Abigail Curtis, "Revival of Forest Giant: Mainers Work to Bring Back American Chestnut," *Bangor Daily News*, February 20, 2016, accessed April 2, 2018, http://bangordailynews.com/2016/02/20/homestead/revival-of-forest-giant-mainers-work-to-bring-back-american-chestnut; Maine Chapter of the American Chestnut Foundation, "Chapter History," accessed June 12, 2019, https://www.acf.org/me/about-us/history.

83. Nick McCrea, "Tallest Chestnut Tree in North American Found in Maine," *Bangor Daily News*, November 28, 2015, accessed April 1, 2019, http://bangordailynews.com/2015/11/28/news/state/tallest-chestnut-tree-in-north-america-found-in-maine. See also Meredith Goad, "After Decades of Blight, Mainers Could Help Save the American Chestnut Tree," *Portland Press Herald*, August 27, 2017, accessed June 14, 2019, https://www.pressherald.com/2017/08/27/saving-an-american-classic.

84. Georgia Chapter of the American Chestnut Foundation, "Partnerships," accessed November 27, 2019, https://acf.org/ga/about/partnerships; Georgia Chapter of the American Chestnut Foundation, "Research and Breeding," accessed June 15, 2019, https://acf.org/ga/research-breeding.

85. Appalachian Regional Reforestation Initiative, "About ARRI," U.S. Office of Surface Mining Reclamation and Enforcement, U.S. Department of the Interior, accessed December 3, 2018, http://arri.osmre.gov/About/AboutARRI.shtm.

86. Appalachian Regional Reforestation Initiative, "ARRI's First Two Years," ARRI *Newsletter* 1 (December 2005): 1.

87. Ibid., 1–2. See also Jim Burger et al., *The Forestry Reclamation Approach*, Forest Reclamation Advisory no. 2, Appalachian Regional Reforestation Initiative, U.S. Office of Surface Mining (Pittsburgh, Pa.: U.S. Office of Surface Mining, 2005), 1–4.

88. Leslie Middleton, "American Chestnuts Rise Where Other Trees Failed to Make a Stand," *Bay Journal* (Seven Valleys, Pa.), October 28, 2015, 2. An online version was published by the Chesapeake Media Service, accessed December 2, 2018, https://www.bayjournal.com/news/wildlife_habitat/american-chestnuts-rise-where-other-trees-failed-to-make-a-stand/article_4a38a21c-efe5-5c70-85a7-4cbbc0fd97d0.html.

89. Patrick N. Angel et al., "Reforesting Unused Surface Mined Lands by Replanting with Native Trees," in *National Proceedings: Forest and Conservation*

Nursery Associations—2011, ed. D. L. Haase, J. R. Pinto, and L. E. Riley (Fort Collins, Colo.: U.S. Forest Service, Rocky Mountain Research Station, 2012), 12.

90. American Chestnut Foundation, *2015 Annual Report* (Asheville, N.C.: American Chestnut Foundation, 2015), 7–8; American Chestnut Foundation, *2014 Annual Report* (Asheville, N.C.: American Chestnut Foundation, 2014), 18–19; Appalachian Regional Reforestation Initiative, "Restoring the American Chestnut on Mined Lands in Appalachia," U.S. Office of Surface Mining Reclamation and Enforcement, U.S. Department of the Interior, accessed December 3, 2018, http://arri.osmre.gov/AC/OS08.shtm.

91. Middleton, "American Chestnuts Rise Where Other Trees Failed," 2.

92. Survival rates for backcross hybrids on mine lands range from a low of 23 percent to a high of 73 percent. Figures also vary according to planting methods. See Michael French et al., *Re-establishing American Chestnut on Mined Lands in the Appalachian Coalfields*, Forest Reclamation Advisory no. 12, Appalachian Regional Reforestation Initiative, U.S. Office of Surface Mining (Pittsburgh, Pa.: U.S. Office of Surface Mining, 2015), 1–6; Jeff G. Skousen et al., "Plantation Performance of Chestnut Hybrids and Progenitors on Reclaimed Appalachian Surface Mines," *New Forests* 49, no. 5 (September 2018): 604.

93. Skousen et al., "Plantation Performance of Chestnut Hybrids," 604.

94. Ibid. See also Melissa A. Thomas-Van Gundy et al., "Mortality, Early Growth, and Blight Occurrence in Hybrid, Chinese, and American Chestnut Seedlings in West Virginia," in *Proceedings: 20th Central Hardwood Forest Conference, Columbia, Mo., March 28–April 1, 2016*, General Technical Report NRS-P-167, ed. John M. Kabrick et al. (Newton Square, Pa.: U.S. Forest Service, Northern Research Station, 2017), 222–39.

95. See, for example, Diskin, Steiner, and Hebard, "Recovery of American Chestnut Characteristics," 439–47; G. Geoff Wang et al., *The Silvics of Castanea dentata (Marsh.) Borkh., American Chestnut, Fagaceae (Beech Family)*, USDA, Forest Service, Southern Research Station, General Technical Report SRS-173 (Asheville, N.C.: U.S. Forest Service, Southern Research Station, 2013), 13; Martin Cipollini et al., "Evaluation of Phenotypic Traits and Blight-Resistance in an American Chestnut Backcross Orchard in Georgia," *Global Ecology and Conservation* 10 (April 2017): 1–8.

96. ARRI does not find the release of advanced hybrid trees on private lands problematic, even though the level of disease resistance in TACF's hybrid chestnuts "will not be known for several years." French et al., *Re-Establishing American Chestnut on Mined Lands*, 5. TACF views the plantings as progeny tests, but does not explain how the trees will be removed if the experiments prove unsuccessful. American Chestnut Foundation, *2014 Annual Report*, 10. The science and rationale for establishing chapter-based seed orchards is found in Sarah Fitzsimmons et al., "Regionally Adapted Seed Orchards within TACF's State

Chapters," *Journal of the American Chestnut Foundation* 28, no. 1 (January/February 2014): 15–19.

97. Such claims were made by those favoring the transgenic approach to chestnut restoration, including William Powell of SUNY-ESF. Powell argued GE trees will contain more American chestnut genes than the TACF hybrids and will more closely resemble the native tree. See William Powell, "New Genetically Engineered Chestnut Will Help Restore the Decimated, Iconic Tree," *Conversation*, January 19, 2016, accessed January 26, 2019, https://theconversation.com/new-genetically-engineered-american-chestnut-will-help-restore-the-decimated-iconic-tree-52191.

98. Lisa Thomson, May 1, 2019, email message to TACF collaborators, partners, and chapter members.

99. Skousen et al., "Plantation Performance of Chestnut Hybrids," 608. See also Douglass F. Jacobs, "Toward Development of Silvical Strategies for Forest Restoration of American Chestnut (*Castanea dentata*) Using Blight-Resistant Hybrids," *Biological Conservation* 137, no. 4 (July 2007): 497–506.

100. Skousen et al., "Plantation Performance of Chestnut Hybrids," 608.

101. Barbara O'Connell et al., *The Forest Inventory and Analysis Database: Database Description and User Guide Version 6.0.2 for Phase 2* (Washington, D.C.: USDA, U.S. Forest Service, 2015), accessed April 4, 2018, available at http://www.fia.fs.fed.us/library/database-documentation; Harmony Dalgleish et al., "Consequences of Shifts in Abundance and Distribution of American Chestnut for Restoration of a Foundation Forest Tree," *Forests* 7, no. 4 (2016): 1.

102. Sara Fitzsimmons, "Out-Crossing of TACF Restoration Chestnut 1.0 Trees," American Chestnut Foundation newsletter, January 31, 2015, accessed April 4, 2017, http://acf.org/newsletter1.31.15Outcrossing.php.

103. Ibid.

104. John Scrivani, "Forest Inventory and Analysis," *Journal of the American Chestnut Foundation* 25, no. 3 (May/June 2011): 18.

105. Kathleen Marmet and Sara Fitzsimmons. "The Appalachian Trail MEGA-Transect Chestnut Project: A Preliminary Pilot Project Report," *Journal of the American Chestnut Foundation* 22, no. 2 (Fall/Winter 2008): 10–18; American Chestnut Foundation, *2015 Annual Report*, 13, 122; Fitzsimmons, "Out-Crossing of TACF Restoration Chestnut 1.0 Trees."

106. Of the 32,500 trees documented growing on or near the Appalachian Trail in 2015, fewer than two hundred were trees thirteen or more inches in circumference. American Chestnut Foundation, *2015 Annual Report*, 13.

107. Ulysse Guillaume Liénard, *Cataloque des essences forestières composant l'Arboretum géographique de Tervuren* (Brussels: Administration de la Donation Royale, 1958), 38.

108. Kevin Knevels, email communication to author, July 2, 2019.

109. The law of unintended consequences is attributed to the sociologist Robert K. Merton. See Merton, "The Unanticipated Consequences of Purposive Social Action," *American Sociological Review* 1, no. 6 (December 1936): 894–904.

110. See the comments made by Lucille Griffin in the entry for 2008 on the American Chestnut Cooperators' Foundation newsletters webpage, accessed June 8, 2019, http://www.accf-online.org/news.html.

111. The Ozark Chinquapin Association embraces a similar approach, breeding only those trees showing a natural resistance to the blight. See Robert Langellier, "A Legendary Ozark Chestnut Tree, Thought Extinct, Is Rediscovered," *National Geographic*, June 24, 2019, accessed September 24, 2020, https://www.nationalgeographic.com/environment/2019/06/saving-chestnut-trees-ozarks.

112. American chestnut enthusiasts should avoid using "racialized" language when discussing species preservation. Condoning the protection of "uncontaminated" American chestnut germplasm, while promoting the elimination of "foreign" or "alien" genetic material, can also be mistaken for nativist politics. In the past, language used to describe nonnative species has been used to describe foreign immigrants, and vice versa. See Peter Coates, *American Perceptions of Immigrant and Invasive Species: Strangers on the Land* (Berkeley: University of California Press, 2006); Harriet Ritvo, "Going Forth and Multiplying: Animal Acclimatization and Invasion," *Environmental History* 17, no. 2 (April 2012): 404–14.

113. Douglass F. Jacobs, "Silvicultural and Logistical Considerations Associated with the Pending Reintroduction of American Chestnut," in *Proceedings of the 14th Biennial Southern Silvicultural Research Conference*, USDA Forest Service General Technical Report SRS-121, ed. John A. Stanturf (Asheville, N.C.: U.S. Forest Service, Southern Research Station, 2010), 207–9; Helen Thompson, "Plant Science: The Chestnut Resurrection," *Nature* 490, no. 7418 (October 3, 2012): 22–23.

114. Edmund Russell, *Evolutionary History: Uniting History and Biology to Understand Life on Earth* (Cambridge: Cambridge University Press, 2011), xv–xviii, 1–5.

115. Ibid., xv–xviii, 3–4, 66, 78–82, 103–5, 108.

116. G. Seetharaman, "These Two Issues Could Put the Brakes on the Bt Cotton Story," *Economic Times*, January 21, 2018, accessed July 10, 2019, https://economictimes.indiatimes.com/news/economy/agriculture/the-brakes-are-applied-on-the-bt-cotton-story/articleshow/62583116.cms. See also Mario Soberón, Yulin Gao, and Alejandra Bravo, eds., *Bt Resistance: Characterization and Strategies for* GM *Crops Producing Bacillus thuringiensis Toxins* (Oxfordshire, UK: CAB International, 2015).

CONCLUSION. The Giving Tree

1. Jason Van Driesche and Roy Van Driesche, *Nature Out of Place: Biological Invasions in the Global Age* (Washington, D.C.: Island Press, 2000), 123; Rachel J. Collins et al., "American Chestnut: Re-examining the Historical Attributes of a Lost Tree," *Journal of Forestry* 116, no. 1 (January 2018): 68–75.

2. Albert Roth Photograph Collection, University of Tennessee Libraries, Knoxville, Tennessee, Special Collections, Great Smoky Mountains Regional Project, image no. roth0149.

3. Alonzo B. Brooks, "Castanea dentata," *Castanea: The Journal of the Southern Appalachian Botanical Club* 2, no. 5 (May 1937): 61–67. The Southern Appalachian Botanical Club changed its name to the Southern Appalachian Botanical Society in 1992.

4. Ibid., 65. According to the accompanying caption, the tree measured "eight feet, eight inches in diameter at 4.5 feet above the ground."

5. Ovid Butler, "Big Trees," *American Forests* 48, no. 11 (November 1942): 484.

6. Ibid.

7. Brooks, "Castanea dentata," 65.

8. Robert Van Pelt, *Champion Trees of Washington State* (Seattle: University of Washington Press, 2003), 41; John Dodge, "Tumwater Chestnut Trees Attract National Attention," *Olympian* (Olympia, Washington), April 13, 2004, A1; John Dodge, "Tumwater Legacy Trees Receive Their Due," *Olympian* (Olympia, Wash.), February 11, 2015, accessed March 17, 2018, http://www.theolympian.com/2015/02/11/3572594/tumwater-legacy-trees-receive.html.

9. Harvey K. Kines, *An Illustrated History of the State of Washington, Containing Biographical Mention of Its Pioneers and Prominent Citizens* (Chicago: Lewis, 1893), 367–68; John E. Ayer, "George Bush: The Voyageur," *Washington Historical Quarterly* 7, no 1 (January 1916): 40–45.

10. Roger W. Pease, *Growing Chestnuts from Seed*, West Virginia University Agricultural Experiment Station Circular no. 90 (Morgantown: West Virginia University Agricultural Experiment Station, 1954), 3–9.

11. Van Pelt, *Champion Trees of Washington State*, 41. American Forests provides a description of the Tumwater tree at its online Champion Tree National Register, accessed June 30, 2019, https://www.americanforests.org/big-trees/american-chestnut-castanea-dentata.

12. Craig D. Tiedemann and Edward R. Hasselkus, "The American Chestnut in Wisconsin," *Transactions of the Wisconsin Academy of Sciences, Arts, and Letters* 63 (1975), 87.

13. *La Crosse (Wis.) Tribune*, November 23, 1952, 9; R. Bruce Allison, *Every Root an Anchor: Wisconsin's Famous and Historic Trees* (Madison: Wisconsin Historical Society Press, 2005), 72–73.

14. Tiedemann and Hasselkus, "American Chestnut in Wisconsin," 96; Harry D. Tiemann, "The Lamented Chestnut: Possibility of its Resuscitation in Wisconsin," *Southern Lumberman* 179, no. 2249 (December 15, 1949): 239–43.

15. Tiedemann and Hasselkus, "American Chestnut in Wisconsin," 96

16. William P. Corsa, comp., *Nut Culture in the United States, Embracing Native and Introduced Species* (Washington, D.C.: GPO, 1896), 77.

17. Jane Cummings-Carlson et al., "West Salem: A Research Update," *Journal of the American Chestnut Foundation* 12, no. 1 (Winter/Spring 1998): 24–26; Lee Bergquist, "Chestnut Trees May Get a Dose of Good Health," *Journal Sentinel* (Milwaukee, Wis), December 31, 2011, accessed June 3, 2018, http://www.jsonline.com/news/wisconsin/chestnut-trees-may-get-a-dose-of-good-health-pg3i8e4-136488753.html; Brad Bryan, "Scientists Treat Rare Chestnut Trees for Blight," *LaCrosse (Wis.) Tribune*, July 12, 2015, accessed June 30, 2019, https://lacrossetribune.com/news/local/scientists-treat-rare-chestnut-trees-for-blight/article_b62583a4-425c-5671-8879-92132760567f.html.

18. Susan Freinkel, *American Chestnut: The Life, Death, and Rebirth of a Perfect Tree* (Berkeley: University of California Press, 2007), 111. See also Michael G. Milgroom and Paolo Cortesi, "Biological Control of Chestnut Blight with Hypovirulence: A Critical Analysis," *Annual Review of Phytopathology* 42 (2004): 311–38; Mark Double et al., "Recapping Twenty Years of Biological Control Efforts in a Stand of American Chestnut in Western Wisconsin," *Journal of the American Chestnut Foundation* 27, no. 4 (July/August 2013): 19–23.

19. G. H. Choi et al., "Hypovirulence of Chestnut Blight Fungus Conferred by an Infectious Viral CDNA," *Science* 257, no. 5071 (August 1992): 800–3; Christine D. Smart and Dennis W. Fulbright, "Molecular Biology of Fungal Diseases," in *Molecular Biology of the Biological Control of Pests and Diseases of Plants*, ed. Muthukumaran Gunasekaran and Darrell J. Weber (Boca Raton, Fla.: CRC Press, 1996), 60–62; Martin Weichert and André Fleissner, "Anastomosis and Heterokaryon Formation," in *Genetic Transformation Systems in Fungi*, vol. 2, ed. Marco A. van den Berg and Karunakaran Maruthachalam (New York: Springer, 2015), 3–22.

20. Choi et al., "Hypovirulence of Chestnut Blight," 197–98; Michael G. Milgroom and Bradley I. Hillman, "The Ecology and Evolution of Fungal Viruses," in *Studies in Viral Ecology*, vol. 1, *Microbial and Botanical Host Systems*, ed. Christon J. Hurst (Hoboken, N.J.: John Wiley and Sons, 2011), 241–43; Simone Prospero and Daniel Rigling, "Chestnut Blight," in *Infectious Forest Diseases*, ed. Paolo Gonthier and Giovanni Nicolotti (Oxfordshire, UK: CAB International, 2013), 326–31.

21. Charles A. Davis, "Suggestions for Securing Quick Returns from Forest Plantations," in *Report of the Michigan Forestry Commission for the Years 1903–4*, Michigan Forestry Commission (Lansing, Mich.: Wynkoop Hallenbeck

Crawford, 1905), 99; Dennis Fulbright et al., "Chestnut Blight and Recovering American Chestnut Trees in Michigan," *Canadian Journal of Botany* 61, no. 12 (December 1983): 3164–71.

22. Lawrence G. Brewer, "Ecology of Survival and Recovery from Blight in American Chestnut Trees (*Castanea dentata* (Marsh.) Borkh.) in Michigan," *Bulletin of the Torrey Botanical Club* 122, no. 1 (January/March 1995): 43. See also Anita L. Davelos and Andrew M. Jarosz, "Demography of American Chestnut Populations: Effects of a Pathogen and a Hyperparasite," *Journal of Ecology* 92, no. 4 (August 2004): 675–85.

23. Walter Sullivan, "Respite from Chestnut Blight Seen," *Dispatch* (Lexington, N.C.), April 25, 1977, 8; Lawrence G. Brewer, "The Distribution of Surviving American Chestnuts in Michigan," in *Proceedings of the* USDA *Forest Service American Chestnut Cooperators' Meeting, Morgantown, West Virginia, January 5–7, 1982, ed. H. Clay Smith and William L. MacDonald (Morgantown: West Virginia University Books, Office of Publications, 1982),* 97; Richard A. Jaynes and John E. Elliston, "Hypovirulent Isolates of *Endothia parasitica* Associated with Large American Chestnut Trees," *Plant Disease* 66, no. 9 (1982): 769–72.

24. John Flesher, "Northern Michigan Growers among Those Who Seek Comeback of the Mighty Chestnut Tree," *Ludington (Mich.) Daily News*, September 30, 1995; 4; Brewer, "Ecology of Survival and Recovery from Blight," 54.

25. See, for example, Pascal P. Pirone, *Diseases and Pests of Ornamental Plants*, 5th ed. (New York: John Wiley & Sons, 1978), 182–83; Bob Wallace, "Research Report: The Dunstan Hybrid Chestnuts," *Agroforestry Review* 3, no. 2 (Spring 1981): 5–8.

26. Paul H. Russell, "We Still Have Chestnut," *Southern Lumberman* 179, no. 2249 (December 15, 1949), 237.

27. *Washington (D.C.) Evening Star*, November 19, 1950, 34. By 1954, 80 percent of all vegetable tannins used in the United States were imported. R. Kent Beattie and Jesse D. Diller, "Fifty Years of Chestnut Blight in America," *Journal of Forestry* 52, no. 5 (May 1954): 328.

28. Laurence C. Walker and Brian P. Oswald, *The Southern Forest: Geography, Ecology, and Silviculture* (Boca Raton, Fla.: CRC Press, 2000), 109–10; Laurence C. Walker, *The North American Forests: Geography, Ecology, and Silviculture* (Boca Raton, Fla.: CRC Press, 1999), 68.

29. William A. Shires, "Only Skeletons of Trees Remain," repr. in *High Point (N.C.) Enterprise*, July 18, 1964, 7.

30. Kathryn Newfont, *Blue Ridge Commons: Environmental Activism and Forest History in Western North Carolina* (Athens: University of Georgia Press, 2012), 42. In 1993, the U.S. government even recommended crafting wormy chestnut into items for resale. See Margaret G. Thomas and David R. Schumann, *Income Opportunities in Special Forest Products: Self-Help Suggestions for Rural*

Entrepreneurs, USDA, U.S. Forest Service, Agriculture Information Bulletin no. 666 (Washington, D.C: GPO, 1993), 49.

31. See, for example, K. D. Woods Company, "Reclaimed Chestnut—Remilled," online catalog, accessed June 2021, http://kdwoodscompany.com/flooring/antique-reclaimed-chestnut-remilled.html; Wormy Chestnut Lumber Company, online catalog, accessed June 1, 2019, http://www.wormychestnutlumber.com.

32. Shel Silverstein, *The Giving Tree* (New York: Harper & Row, 1964).

33. Eoin O'Carroll, "GMO Could Bring Back the American Chestnut, But Should It?," *Christian Science Monitor*, April 24, 2019, accessed June 20, 2019, https://www.csmonitor.com/Science/2019/0424/GMO-could-bring-back-the-American-chestnut.-But-should-it. See also Charles C. Mann, "Let's Farm Chestnuts Again," *Wall Street Journal*, May 9, 2019, accessed June 20, 2019, https://www.wsj.com/articles/lets-farm-chestnuts-again-11557411615; Meghna Chakrabarti, host, "How GMOs Might Save the American Chestnut Tree," National Public Radio, *On Point*, aired April 29, 2019, accessed June 21, 2019, https://www.wbur.org/npr/718391784/gmos-genetics-ethics-chestnut-tree.

34. Stanton B. Gelvin, "*Agrobacterium*-Mediated Plant Transformation: The Biology behind the 'Gene-Jockeying' Tool," *Microbiology and Molecular Biology Reviews* 67, no. 1 (March 2003): 16–37; David A. Somers and Irina Makarevitch, "Transgene Integration in Plants: Poking or Patching Holes in Promiscuous Genomes?," *Current Opinion in Biotechnology* 15, no. 2 (April 2004): 126–31; Allison K. Wilson, Jonathan R. Latham, and Ricarda A. Steinbrecher, "Transformation-Induced Mutations in Transgenic Plants: Analysis and Biosafety Implications," *Biotechnology and Genetic Engineering Reviews* 23 (2006): 209–38, https://doi.org/10.1080/02648725.2006.10648085;

35. Bo Zhang et al., "A Threshold Level of Oxalate Oxidase Transgene Expression Reduces *Cryphonectria parasitica*–Induced Necrosis in a Transgenic American Chestnut (*Castanea dentata*) Leaf Bioassay," *Transgenic Research* 22, no. 5 (October 2013): 973–82.

36. National Academies of Sciences, Engineering, and Medicine, *Gene Drives on the Horizon: Advancing Science, Navigating Uncertainty, and Aligning Research with Public Values* (Washington, D.C.: National Academies Press, 2016), 31–38, 67, 109, 118. Although gene drive technology is not being used in chestnut restoration, the end result—permanently modified GE trees—is the same.

37. TACF members assume the progeny of the BC_3F_3 hybrids (after breeding with wild populations) will be genetically more American, but will retain the Chinese trait of blight resistance. However, not all offspring will possess high levels of resistance, which means, over time, that evolution will favor those trees with Chinese chestnut genes. See Lisa M. Worthen, Keith E. Woeste, and Charles H. Michler, "Breeding American Chestnuts for Blight Resistance," in

Plant Breeding Reviews, vol. 33, ed. Jules Janick (Hoboken, N.J.: John Wiley and Sons, 2010), 324–31.

38. World Rainforest Movement, "Indigenous Peoples Unite to Stop Genetically Engineered Trees," October 13, 2014, accessed March 30, 2018, http://wrm.org.uy/meetings-and-events/indigenous-peoples-unite-to-stop-genetically-engineered-trees. Montelongo's statement partly refers to the assumption that GE chestnuts will be clonally propagated in order to speed up the restoration process, an opinion that has some basis in fact.

39. Eastern Band of the Cherokee Indians (EBCI), Council Resolution No. 31, October 14, 2015, Cherokee, North Carolina.

40. Julia Rosen, "Should We Resurrect the American Chestnut Tree with Genetic Engineering?," *Los Angeles Times*, June 25, 2019, accessed July 31, 2019, https://www.latimes.com/science/sciencenow/la-sci-genetically-modified-trees-american-chestnut-20190625-htmlstory.html.

41. G. Geoff Wang et al., *The Silvics of Castanea dentata (Marsh.) Borkh., American Chestnut, Fagaceae (Beech Family)*, USDA Department of Agriculture, Forest Service, Southern Research Station, General Technical Report SRS-173 (Asheville, N.C.: Southern Research Station, 2013), 12.

42. A study measuring the effects of fire on seedlings found significant resprouting after a single burn. However, debris was removed from around the seedlings prior to the burn, which was conducted in early spring before the leaves appeared. See Ethan P. Belair, Mike R. Saunders, and Stacy L. Clark, "Effects of Simulated Prescribed Fire on American chestnut and Northern Red Oak Regeneration," in *Proceedings, 19th Central Hardwood Conference, Carbondale, Illinois, March 10–12, 2014*, USDA, U.S. Forest Service, Northern Research Station, General Technical Report NRS-P-142, ed. John Groninger et al. (Newtown Square, Pa.: Northern Research Station, 2014), 133–36.

43. Willy Blackmore, "General Mills Will Label GMOs on Products Nationwide," *Alternet*, March 20, 2016, accessed June 19, 2018, https://www.alternet.org/2016/03/general-mills-will-label-gmos-products-nationwide; Chris Prentice, "USDA Outlines First-Ever Rule for GMO Labeling, Sees Implementation in 2020," *Reuters*, December 20, 2018, accessed April 27, 2020, https://www.reuters.com/article/us-usa-gmo-labeling/usda-outlines-first-ever-rule-for-gmo-labeling-sees-implementation-in-2020-idUSKCN1OJ2TF.

44. United Nations General Assembly, "Principle 15," in *Report of the United Nations Conference on Environment and Development (Rio de Janeiro, 3–14 une 1992: Annex I*, 6, available at https://www.un.org/en/development/desa/population/migration/generalassembly/docs/globalcompact/A_CONF.151_26_Vol.I_Declaration.pdf.

index

Page numbers in italics represent references to images.